IRELAND

and the

IRISH QUESTION

A Collection of Writings by
Karl Marx and Frederick Engels

INTERNATIONAL PUBLISHERS

New York

Prepared by L. I. Golman and V. E. Kunina, with the
assistance of M. A. Zhelnova.

TRANSLATORS: from the German, Angela Clifford, D. Danemanis,
S. Ryazanskaya, and V. Schneierson; from the French, K. Cook;
and from the Italian, B. Bean.

Edited by R. Dixon.

Copyright © *by* INTERNATIONAL PUBLISHERS CO., INC. 1972
All Rights Reserved
First Edition, 1972

Library of Congress Catalog Card Number: 73-188754
ISBN: (Cloth) 0-7178-0341-4; (paperback) 0-7178-0342-2
Printed in the United States of America

CONTENTS

KARL MARX AND FREDERICK ENGELS

EXCERPTS FROM LETTERS ON IRELAND
WRITTEN BETWEEN 1867 AND 1868

KARL MARX AND FREDERICK ENGELS
EXCERPTS FROM LETTERS ON IRELAND WRITTEN
BETWEEN 1869 AND 1872

FOREWORD

There are many reasons why the publication in one volume of the writings of Marx and Engels on Ireland is to be welcomed. It is timely since the myth that the Irish national struggle was over has been exploded by events in "Northern Ireland", the area, since the partition of 1921, still held within the Union. The present tangled position is totally incomprehensible to those who lack knowledge of the historical events which brought it about. That in their day Marx and Engels faced and solved problems which are essentially those that still lie before us today will be apparent to any reader of this book who has freed himself even modestly from that limitation. Consequently it provides numerous guide-lines which, *mutatis mutandis*, have high relevance today.

It should perform two valuable functions. First it should give the Marxists, of England and the world, a fresh interest in the Irish question, when they see how seriously the founders of Marxism regarded it, and the sheer volume of the work they devoted to it. And it should help restore vision in what has tended to become one of the blindest spots of the English labour movement. England has centuries of imperialism (in the broad sense) behind her. Consequently most of her best radicals have tended, in their struggle against the chauvinism surrounding them on all sides, to identify all national struggles with reaction. Perhaps since Irish people resemble them so much, in contrast to peoples further afield, they find it hard to believe that Irish nationalism has not the same content as that of their own ruling class, which they have rejected. Wisdom begins in frank, total and unconditional recognition of the right of the Irish nation to determine its own international destiny, and it is *from that point* that the identity of interest between the working class of England and that of Ireland begins to operate in practice.

Second, it should arouse the interest of the Irish people in Marxism, which the present ruling classes have tried to represent as something alien to them. Not that the masses have altogether believed it. There were many old craftsmen in country parts, even in the dark days of the forties and fifties of this century, who knew well that Marx defended Ireland in *Capital*. The book has been available in public libraries. But a nation seeking national freedom thirsts after politics, not economics. Many would know, few would read. The collection made by the late Ralph Fox contained some of the letters and exercised some influence in the thirties, but it has been long out of print, except for a few copies zealously guarded in Dublin.

At the turn of the century the great Irish Marxist James Connolly, a working man who taught himself German, spread some knowledge of the Letters to Kugelmann. Connolly's life work was indeed the revival and application of Marx's teaching on the primacy of the national independence struggle within Ireland. He had become aware of this teaching while a member of the Social Democratic Federation in Edinburgh, at a time when Engels was still alive and influencing the theoretical development of the world proletarian movement.

The extension of the International Workingmen's Association to Ireland is referred to in Dr. Golman's admirably compact but comprehensive introduction. It must not be lightly assumed, however, that it arose from a foreign importation. There were a number of left radical and socialist-oriented groups in Ireland during the times of the Chartists. The attempts of O'Connell's followers to extinguish Chartism in Drogheda received attention in the *Northern Star*. It is surely no accident that Feargus O'Connor, who became the leader of Chartism in England during one of its most virile periods, was the nephew of Arthur O'Connor, the United Irishman, and that he chose for the name of his paper the name given by the United Irishmen to their own paper in Belfast.

The deepest and most abiding tradition in Ireland is that of Republicanism, best expressed in Marx's day by the Fenians. It originated in the Irish response to the French revolution. As Wolfe Tone wrote: "In a little time the French Revolution became a test of every man's political creed, and the nation was fairly divided into two great parties, the Aristocrats and the Democrats. . . . It is needless, I believe, to say I was a Democrat from the beginning."

In his centenary book *The Internationale* Mr. R. Palme Dutt remarks that "The international communist movement developed in direct line of descent from the left wing of the democratic revolution and the first beginnings of the working-class movement". He quotes Marx in 1848: "The Jacobin of 1793 has become the Communist of today." Because of their common origins in the political movements of the oppressed classes of Europe in the early nineteenth century, Republicanism, pragmatic rather than scientific, but consistently revolutionary, showed a constant receptivity to socialist ideas. Thus it is said that Stephens, the Young Ireland revolutionary forced to flee to Paris after the failure of the rising of 1848, had contact with revolutionary socialist groups, before returning to Ireland (after a stay in America) to found Fenianism. The great Fenian John Devoy, while scarcely describable as a Marxist, worked closely with the First International. Why then did the United Irishman of 1798 not become the Communist of 1848? The full answer must await a study that the publication of this book must surely stimulate, namely, the analysis of ideological developments in the Irish national movement during the nineteenth century. But a decisive factor must surely have been the diaspora following the "famine" and the scattering of Irish revolutionaries over the face of the earth. The emigrant ship was English imperialism's strongest safeguard alike against revolution and against revolutionary ideas.

The radical wing of Republicanism was constantly attracted towards the revolutionary working class. Thus Clarke, Pearse and MacDermott were drawn into an alliance with Connolly in 1916. Similar forces came together in 1921-22. The bourgeoisie, backed by a confused and fearful Labour leadership, was prepared to accept partition, which Lloyd George had forced on Ireland at the point of the gun. The

Republican party, Sinn Fein, split. Among those who opposed the monstrous settlement were leaders such as Liam Mellows, strongly influenced by Connolly's teachings, and Marxists within the Republican movement, such as Peadar O'Donnell. Outside the Republican ranks the only party to oppose the Treaty was the young Communist Party of Ireland, led by James Connolly's son, and a number of its members took part in the fighting that ensued. In the twenties and thirties the Republican newspaper *An Phoblacht* regularly carried articles by internationally known Marxists. It may therefore be said, and the Irish reader of this book can judge for himself from his own experience, that so little is Marxism alien to the Irish tradition that reactionary ruling classes, actual or prospective, have always sought special means for insulating the people from its influence.

There was perhaps instinctive recognition of this fact in Engels's envious cry when O'Connell was parading Ireland with two hundred thousand followers about him. "Give *me*" (surely this emphasis is implied) "two hundred thousand Irishmen and I could overthrow the entire British monarchy." Engels's favourite resort was the home of the Burns family, who had Fenian connections. After the death of his constant companion Mary Burns in 1863, her sister Lizzie became his second wife, and accompanied him on his visit to Ireland in 1869. The fascination which the country held for him was expressed in his description of the Irish climate: "The weather, like the inhabitants, has a more acute character, it moves in sharper, more sudden contrasts; the sky is like an Irish woman's face: here also rain and sunshine succeed each other suddenly and unexpectedly and there is none of the grey English boredom." There is a research subject for some young historian in the details of Engels's connections with Ireland and the Fenians of Lancashire.

Indeed, the publication of this collection prepared by Dr. Golman and Dr. Valeria Kunina might well spur much research in Irish and English political and economic history. That Marx postulated a special variant of the universal law of capitalist accumulation in Ireland has been very little appreciated even in Ireland, where it still appears to be generally accepted that such economic categories as prices of capital goods followed the same pattern in Ireland as in

England. It would be interesting to expand Marx's analysis
of the eighteen sixties, in *Capital*, to cover a longer period
and to collate the data throughout the whole field of Irish
economic life. It would also not be impossible to find evidence
that the special mode of capital accumulation discerned by
Marx over a hundred years ago is by no means defunct today.

Marx and Engels never had the opportunity to arrange and
systematise all their ideas on Ireland and Irish history in a
state suitable for publication. As Dr. Golman points out,
despite the obvious existence of a completely ordered outlook,
it has to be "gleaned from handwritten notes and fragments".
But there is no subject on which Marx's and Engels's views
are not provocative of further thought. And this is going to
be the great value of this compilation. The problems of
Ireland today, internally, and in her relations with England,
Europe and the world, are complex and thorny. Viewed as
dogma Marx's or Engels's writings will not help to solve
them. But their writings were not intended as dogma. Viewed
as examples of the analytical methods of a scientific genius,
revealing the way problems were approached and how
thought out and solved, the contents of this book make a most
important contribution to the equipment of the Labour and
Republican movements in Ireland, and to their progressive
counterparts in Britain.

C. Desmond Greaves
Liverpool, September 1970

INTRODUCTION

This is the first endeavour to make a comprehensive collection of works of Karl Marx and Frederick Engels on Ireland and the Irish question—letters and articles, whole or in part, excerpts from bigger works, transcripts and outlines of speeches and reports, a number of synopses and rough sketches. The supplement contains articles by Marx's daughter Jenny and documents of the international labour movement.

This collection is based on texts from the latest editions of the works of Marx and Engels and of documents of the International Working Men's Association—the second Russian edition of the *Collected Works* of Marx and Engels, the *Marx-Engels Archives*, Vol. X (in Russian), and the five-volume edition, *The General Council of the First International. Minutes*, issued recently by the Institute of Marxism-Leninism of the C.C. of the C.P.S.U., in Moscow.* Included in the collection are a number of previously unknown works, letters and documents casting additional light on the views of Marx and Engels on the Irish question and affording the student a more complete knowledge of their heritage.

Marx and Engels wrote more about Ireland than has been collected between these covers. Their *précis* and notes on Irish history and economy are given in part only, since substantial sections of these preliminary summaries are still in process of decyphering. However, this book comprises their most pertinent statements on the subject at hand and presents a comprehensive picture of their views.

---

* *Documents of the First International*, 5 vols., Progress Publishers, Moscow.—*Ed.*

Ireland claimed the attention of Marx and Engels from the 1840s onward. They followed the developments there, responded to them in the press, spoke on Ireland at meetings of the General Council of the First International, and discussed the subject in letters to each other and to others. Both made a close study of Ireland's economic and political situation, and of her history and social relations. Marx used his findings in his main work, *Capital*, while Engels, who paid several visits to Ireland, wrote an account of his travels of 1856 and 1869 in some of his letters to Marx. Engels intended to write a history of Ireland, gathered a store of material for it, wrote the opening chapters, but did not live to complete his project.

Marx's and Engels's interest in Ireland ranged far afield. As students of capitalist society, and as economists, sociologists and historians, they seized the opportunity to examine the operation of capitalist laws of development in an agrarian and poorly developed land. The example of Ireland demonstrated the influence exerted by what was then the classic capitalist country, England, from which they derived their main data for analysing the laws of capitalist production. Peasant countries and the peasant masses were of continuous concern to the two theorists of the socialist revolution, who arrived scientifically at the conception of a workers' alliance with the working peasants and of the peasants' involvement in the socialist reconstruction of society under the guidance of the proletariat.

However, Ireland was not simply an agrarian country. Marx and Engels saw her as a land subject to colonial domination by a more powerful neighbour. Engels described Ireland as England's first colony (see p. 83), whose conquest dated back to the latter half of the twelfth century. Social oppression became interlaced with national oppression, since landlords descended from the conquerors were the main exploiters of the Irish peasants. The plunder of Ireland was one of the sources of England's industrial development, contributing to the rapid growth of her capitalist economy. Along with Marx and Engels, Lenin described the process as follows:

"Britain owes her 'brilliant' economic development and the 'prosperity' of her industry and commerce largely to her treatment of the Irish peasantry, which recalls the misdeeds of the Russian serf-owner Saltychikha.

"While Britain 'flourished', Ireland moved towards extinction and remained an undeveloped, semi-barbarous, purely agrarian country, a land of poverty-stricken tenant farmers." (*Collected Works*, Vol. 20, p. 148.)

Ireland shared the lot of other colonial countries ruthlessly exploited as agrarian appendages and suppliers of raw material. Marx referred to her in *Capital* as "an agricultural district of England" supplying the latter with "corn, wool, cattle, industrial and military recruits" (see p. 105). Ireland provided Marx and Engels, opponents of all social and national oppression, with abundant material for their indictment of colonialism and for drawing up the workers' programme of struggle against it.

For them, Ireland was of special interest as the seat of continual resistance to oppression. Centuries of Irish resistance convinced them of the boundless vitality of the national liberation movement of even such a relatively small people, and of the futility of even the most systematic and cruel measures to suppress it. "After the most savage suppression, after every attempt to exterminate them," Engels wrote in one of his fragments, "the Irish, following a short respite, stood stronger than ever before" (see p. 211). Marx and Engels were deeply impressed by the inexhaustible revolutionary energy of the Irish. "Give me two hundred thousand Irishmen," young Engels said, "and I could overthrow the entire British monarchy" (see p. 33). In Ireland Marx and Engels traced the origins of the anti-colonial forces in an oppressed country, the gradual development of the national liberation movement, its specific features and trends.

The writings here collected show how thoroughly Marx and Engels studied Irish history. They confuted many a biased notion traceable to the chauvinist prejudices of English bourgeois historians, economists and geographers, and brought down to earth the romanticism of Irish nationalist historians. Bias, often disguised as objectivism, and distortions of history to suit the class interest of privileged social groups, infuriated Marx and Engels. In one of his sketches, Engels wrote: "The bourgeoisie turns everything into a commodity, hence also the writing of history. It is part of its being, of its condition for existence, to falsify all goods: it falsified the writing of history. And the best-paid historiography is that which is best falsified for the purposes of the bourgeoisie" (see p. 211).

The treatment by Marx and Engels of some of the key problems of Irish history is a credit to their scholarship. They created an essentially new conception of Irish history based on the analytical method of historical materialism. Regrettably, neither published a complete study, elucidating the results of their investigation. We are compelled to glean their conclusions from handwritten notes and fragments, some of which first saw light only recently, and from their references to Ireland in articles and letters. Taken as a whole, however, these provide the basis of a scientific interpretation of the history of Ireland, defining its main periods and explaining at least the most important from antiquity to modern times. They likewise demonstrated the close interconnection of Irish and English history, the links between events on either side of St. George's Channel. True, many archaeological and historical discoveries have been made since Marx's and Engels's time, some facts clarified and many new sources published. But their general observations and judgments— and not only those of method, but many of purely historical significance—are still entirely valid, since the new material has corroborated them, adding to their value.

Permanently valid, for example, is Engels's assessment, in his *History of Ireland* (1870), of the main sources of knowledge about the early periods of Irish history, and equally so is his characterisation of the social and political system of the Celts in Ireland, their customs and culture. He exposed the chauvinist content of the argument that the Irish were culturally backward and incapable of running a state of their own. He demonstrated the significant contribution to Europe's culture made in the early Middle Ages by Irish Christian missionaries and scholars, among whom he makes special mention of the philosopher Johannes Scotus Erigena, whose doctrine, he wrote, "was very bold for the time", and led "close to Pantheism" (see p. 202). Irish history before the Anglo-Norman conquest, too, Engels stressed, abounded in heroic resistance to foreign invaders—the Vikings, culminating in the Irish victory over the Norsemen at Clontarf in 1014.

Engels's study of the survivals of clan relationships among the Irish Celts enabled him to anticipate in 1870 some of the conclusions about primitive societies drawn by Lewis Henry Morgan, the distinguished American ethnologist, whose book *Ancient Society* did not appear until 1877. Engels used the results of his investigations of Irish history in his book, *The Origin of the Family, Private Property and the State*, published in 1884. In his appraisal of the degree of social development of Ireland in early feudal times, he recorded traits typical of that period (political disunity, feudal decentralisation, etc.), which made it relatively easy for the Anglo-Norman conquerors, representatives of a more developed feudal society, to invade Ireland in the latter half of the twelfth century.

In Marx's notes for a report on Ireland dated December 16, 1867, in Engels's fragments and notes for his *History of Ireland*, in Engels's letter to Marx's daughter Jenny dated February 24, 1881, and in other works in this collection, we find an account of the main stages of the colonial subjection of Ireland, showing how, over the centuries, from the day the English base later known as the Pale was established in the south-eastern part of the island until the subjugation of the entire island in the 16th-17th centuries, Ireland became a colony of English landlords and capitalists, the "bulwark", as Marx put it, "of English landlordism" (see p. 161). Destructive wars were fought to conquer Ireland, national risings were brutally suppressed, the clan system forcibly exterminated, and land confiscated and appropriated by aliens. The tragic consequences are shown in bold relief. "The more I study the subject, the clearer it is to me," Engels wrote to Marx on January 19, 1870, "that Ireland has been stunted in her development by the English invasion and thrown centuries back. And this ever since the 12th century" (see p. 286).

The decisive phase in the conquest of Ireland, Marx and Engels stressed, came at the time of English absolutism and the 17th-century bourgeois revolution, when the capitalist system gradually expanded and the feudal barriers to its growth were torn down. Ireland then fell prey to the rising bourgeoisie and the gentry who were turning bourgeois. Marx and Engels demonstrated that every change in England in that stormy age—the Tudor and early Stuart era, the English revolution and Cromwell's protectorate, the Stuart Restoration, and the "Glorious Revolution" of 1688—was for Ireland an intensification of suppression, a new phase in which her colonial ravishers used the most bloody methods for the purposes of "primitive accumulation", forcible redivision of estates among the invaders and consolidation and expansion of English landlordism. "The Irish people were completely crushed by Elizabeth, James I, Oliver Cromwell and William of Orange, their landholdings robbed and given to English invaders, the Irish people outlawed in their own land and transformed into a nation of outcasts," Engels wrote (see p. 270).

Marx and Engels studied the forms of English colonial rule in the subsequent periods. For them the penal laws issued at the end of the seventeenth century on the pretext of combating Catholic plots, and enforced almost throughout the eighteenth century, were a tool for the final expropriation and enslavement of the Irish people, robbing them of political and civil rights, rooting out their national culture, customs and traditions. Marx and Engels demonstrated the colonialist nature of the Act of Union of 1801, and stressed that it was a sequel to the suppression of the Irish rising of 1798, the military occupation and the pressure brutally brought to bear on the Irish Parliament. The Union robbed the Irish of the gains made during the national revival in the latter decades of the eighteenth century, when the English Government was compelled, under pressure of the American and French revolutions, to grant important concessions, to repeal most of the penal laws and to recognise Irish parliamentary autonomy. The Union abolished the Irish Parliament and ushered in a new phase in Britain's colonial rule. The protective tariffs passed by the Irish Parliament were lifted as a result, and Ireland's budding industries were crippled. Farming became practically the only activity to which the local population could apply itself. "The people had now before them," Marx wrote, "the choice between the occupation of land, *at any rent,* or *starvation*" (see p. 132).

The Union established a system of plunder of the Irish peasants by landlords and middlemen (Marx called it a "system of rack-renting") which combined the worst features of capitalist exploitation with appropriation, by semi-feudal methods, of the surplus (and all too often the necessary) product. This is shown by Marx in *Capital* (Vol. III) and by Engels in his letter to N. F. Danielson of June 10, 1890. The English ruling classes, Marx wrote in his article, "The Indian Question—Irish Tenant Right", created in Ireland "those abominable 'conditions of society' which enable a small *caste* of rapacious lordlings to dictate to the Irish people the terms on which they shall be allowed to hold the land and to live upon it" (see p. 61).

This system of exploitation of the small tenant reduced the Irish population to appalling poverty, described in Engels's *The Condition of the Working-Class in England* and in other works of the founders of Marxism. Recurrent crop failures resulted in periodic famines. That of 1845-47 surpassed, however, anything that had been previously experienced. The almost total failure of the potato crop was rendered immeasurably more disastrous by the continued export of grain that was the basis of the landlords' rent. No event has so impressed itself on the memory of the Irish people. "The Irish population," Marx wrote, "decreased by two millions, some of whom starved, while others fled across the Atlantic" (see p. 95).

In the mid-19th century the Irish were struck by a new disaster, in part precipitated by the famine. Landowners began to refuse to rent out the small strips of land customarily sown to grain or potatoes. Instead, they took up large-scale grazing, to the accompaniment of wholesale evictions.

Marx and Engels saw this process as a fresh source of acute social and national contradictions. They demonstrated the causes, nature and consequences of the agrarian upheaval, showing that it was pursued in the interest of the big landlords, big tenant farmers and the English bourgeoisie who, after the repeal of the corn laws in 1846 and the fall in bread prices, wanted cheaper animal products. For the bulk of the Irish nation, however, the agrarian change meant loss of livelihood, even of hearth and home. Marx described it as "quiet business-like extinction" (see p. 123), marking the beginning of a continuous decline in the population of Ireland mainly by forced emigration. As Marx's daughter

Eleanor put it, Ireland became the Niobe of the nations, a country that was losing her children.

"In 1855-66," Marx wrote, "1,032,694 Irishmen were displaced by 996,877 head of cattle." The most urgent need was to end the forcible eviction of peasants and to stop the landlords, backed by the English authorities, from robbing the Irish farmers of their livelihood. This, Marx and Engels stressed, must be the first objective for the Irish national liberation movement.

They focussed their attention on the Irish fight for emancipation, tracing it back to the stout resistance of the clans to the invasion of the Anglo-Norman feudal knights in the Middle Ages, which was repeated at the time of Britain's 16th- and 17th-century colonial expansion. Thus, in 1641-52 and 1689-91, Ireland was the scene of general insurrections against English rule. The national resurgence at the end of the eighteenth century was another important landmark, culminating in the 1798 rising. Marx and Engels wrote in glowing terms of the Irish revolutionaries of that period, the founders of the patriotic Society of United Irishmen, Wolfe Tone, Edward Fitzgerald and others.

The works of Marx and Engels provide something of a chronicle of the Irish national liberation movement. Its main features are closely examined in the light of their scientific outlook, and assessment given to the revolutionary deeds of the Irish peasantry (Whiteboys, Ribbonism, etc.); the movement for repeal of the Anglo-Irish Union; the Irish revolutionary organisations—Young Ireland and the Irish Confederation—which attempted insurrection in 1848; the petit-bourgeois Fenians who repeated the attempt in 1867; the Home Rule movement; the agitation of the Land League, etc.

Their accoutn of the Fenian movement is particularly thorough. Marx maintained that it "took root only in the mass of the people, the lower orders" (see p. 126), expressing the farmers' protest against evictions and the nation's craving for national independence and social emancipation. Marx and Engels alike praised the bourgeois Home Rule Confederation and its leader, Charles Stewart Parnell (whose work is examined in many of Marx's and Engels's letters of the late seventies and early eighties and by Engels after

Marx's death), who, at the height of the movement, sought contact with the Irish masses. They praised Parnell for his part in founding the Land League. It was popular support, indeed, that helped the champions of Home Rule to gain their impressive political strength and to profit by the contradictions between the English ruling parties.

Marx and Engels admired the freedom-loving traditions of the Irish. Moreover, they believed that the Irish people's struggle had a fruitful influence on the English public mind. "The Irish are teaching our leisurely John Bull to get a move on," Engels wrote to one of his correspondents (see p. 333). And in "The English Elections", an article written in 1874, he described the fighting Irish and the demonstrations of the English workers as the "motive forces of English political development" (see p. 311).

At the same time, they were far from idealising the national movement in Ireland, or anywhere else, for they were always keenly aware of its weak sides at different stages of its growth and of its heterogeneous class composition. In a letter to Eduard Bernstein dated June 26, 1882, Engels observed that the Irish movement consisted of two trends: the radical agrarian that erupted into spontaneous peasant actions and was represented by democrats and revolutionaries, on the one hand, and the *"liberal-national* opposition of the *urban bourgeoisie"* (see p. 334), on the other. The sympathies of Marx and Engels lay naturally with the radical wing which was oriented towards the revolutionary liberation of Ireland and expressed the social demands of the people. The policy of the Irish liberals (their narrow national programme, their fear of setting loose the revolutionary energy of the people, their appeals for moderation, their love of conciliation and of deals with the English ruling classes) evoked severe criticism, which was, indeed, often extremely sharp, as in the case of Daniel O'Connell.

But the class narrowness of the radicals, whose illusions and errors Marx and Engels criticised, did not escape them either. They disapproved of the Fenian leaders' adventurist plotting, their inability to pick the right tactical means suited to a given situation, their national narrowness, typical of so many other Irish leaders, and their refusal to understand the importance of contacts with the English democratic, especially proletarian, movement. Engels wrote scathingly that "to these gentry the whole labour movement is pure heresy and the Irish peasant must not on any account be allowed to know that the socialist workers are his sole allies in Europe" (see p. 283).

The rising Irish proletariat must, of course, play a prominent role in the country's national liberation movement. Marx and Engels welcomed the early signs of its awakening as a class and, among other things, its participation in the Land and Labour League founded in the autumn of 1869 on the initiative of the General Council of the International. They welcomed the establishment of Irish sections of the International Working Men's Association in Ireland, Great Britain and the United States, and gave rebuff to the attempts of the reformists from the British Federal Council to gain control over these sections. They cultivated the friendship of Joseph Patrick MacDonnell, Corresponding Secretary of the General Council for Ireland. Their contacts with him continued after he settled in the United States, where he was active among the Irish immigrant workers.

In the late eighties and early nineties Engels was keenly interested in the fact that Irish workers were joining the Union of Gasworkers and General Labourers, one of the new trade unions with branches in Ireland. He censured the leaders of the Social-Democratic Federation and the Independent Labour Party for their sectarian disregard of the national interests of the Irish workers.

Marx and Engels were untiring in their exposure of English police brutality against Irish revolutionaries. Their public statements, together with the articles of Jenny, Marx's elder daughter, and those of other leaders of the International denouncing English reprisals against Irish freedom fighters, inhuman treatment in English jails and the humiliating conditions imposed by the English authorities as the price for amnesty, etc., were, in effect, a scathing indictment of English rule. There were rigid police control and coercion laws, on the one hand, and petty half-hearted reforms, on the other, designed to distract attention from the crying need for more fundamental changes. These were coupled with small concessions to the upper crust of Irish society—the Irish bourgeoisie, landowners and top Catholic clergy—designed to split and weaken the national camp. Gross brutality was accompanied by demagogy and half-measures that did little to alleviate the lot of the Irish people. Thus Marx and Engels assessed the agrarian and other English-sponsored reforms in Ireland, particularly the Land Acts of 1870 and 1881. "But something had to be done to pull the wool over the eyes of the public. It was essential to appear to be doing something for Ireland," Marx wrote (see pp. 167-68).

The only solution consistent with the basic interests of the Irish people and the principles of true democracy is contained in documents written by Marx and Engels. Their programme for Ireland's national liberation and social revival was based on recognition of the right to self-determination of the Irish and other oppressed peoples. It demanded repeal of the forcibly imposed union of Great Britain and Ireland, and independence for Ireland including the right of secession. As Marx and Engels saw it, those should have been the slogans of the English labour movement. They did not rule out a future voluntary and free federation of Ireland and England in the event of propitious radical social and political changes in the latter country. But they insisted that the choice of the appropriate form of relationship between the two countries should rest with the oppressed nation. The duty of the English workers, they pointed out, was to back the right of the Irish to this choice and to support the Irish fight for independence, working for the elimination in Ireland of all forms of national oppression and coercion. Lenin described this as "a splendid example of the attitude the proletariat of the oppressor nations should adopt towards national movements, an example which has lost none of its immense *practical* importance." (*Collected Works*, Vol. 20, p. 442.)

Marx's recommendations are still valid. He pointed out that the struggle for Ireland's national independence should fuse with that for the revolutionary remaking of the agrarian system (he spoke of it as "agrarian revolution", see p. 148), stressing that political independence in itself, necessary though it is, cannot guarantee that the former colony is relieved of every element of dependence on the former metropolitan country. To be genuinely independent it must achieve economic independence. Marx advised the Irish to introduce protective tariffs against the destructive competition of English industry (see Marx's letter to Engels, November 30, 1867, p. 148).

The liberation of colonial and dependent countries in recent years, and the subsequent history of Ireland, bear out these views of Marx and Engels. Their orientation towards revolutionary methods is proved correct by the experience of the colonial and dependent peoples who have shaken off their oppressors. And so is the idea that the struggle for independence must be combined with internal social reconstruction in the interest of the masses, sweeping away the social and economic consequences of colonial rule and building up an independent economy.

In their writings on Ireland, Marx and Engels demonstrated the essential connection between the national liberation movement and the workers' struggle for the socialist reconstruction of the world. They called on the English workers and the fighters for Ireland's independence to join hands. Engels, for example, advocated close bonds between the Chartists and the Irish liberation movement as early as in the 1840s (see his articles, "The Coercion Bill for Ireland and the Chartists" and "Feargus O'Connor and the Irish People"). At the time of the First International, Marx told the English workers time and time again that *"for them the national emancipation of Ireland* is no question of abstract justice or humanitarian sentiment, but *the first condition of their own social emancipation"* (see p. 294). Marx and Engels emphasised the ruinous effects on the workers of the chauvinist ideology and national strife that the capitalists were eager to cultivate among them. They demonstrated—graphically in the case of Anglo-Irish relations—that colonial oppression was a brake on the progressive development of the oppressor nation as well, because it strengthened the hand of the ruling exploiting class. Looking back, they showed, among other things, that the plunder of Irish land and the implantation of the new English aristocracy in Ireland under Oliver Cromwell paved the way for the Stuart Restoration. "By engaging in the conquest of Ireland, Cromwell threw the English Republic out the window," Marx wrote (see p. 128). "Any nation that oppresses another forges its own chains", was how he worded one of his key postulates that show the importance for the workers to adopt an internationalist position in national colonial matters (see Marx's "Confidential Communication", p. 163).

For Marx and Engels the national liberation movement was an ally of the working-class struggle against the system of exploitation. They saw the interaction of the liberative processes in the colonies and the metropolitan countries. The working class, as they saw it, was the decisive factor in the emancipation of the human race from all exploitation, but it must ally itself with the peasants in the fight against feudal and capitalist oppression and with the national liberation movement in freeing the oppressed peoples.

Ireland's fight for independence was to have a revolutionising effect on the English workers, rousing them to action. Whereas in the forties and fifties Marx had held that Ireland would gain her freedom through the victory of the English working class, in the sixties he considered it more probable that Irish victory would spark off the English workers' fight for socialism (see above-mentioned letter by Marx to Engels, December 10, 1869). Once Ireland was lost, Marx wrote to Paul and Laura Lafargue on March 5, 1870, "the class war in England, till now somnolent and chronic, will assume acute forms" (see p. 290).

Lenin explained the reasons for the change in Marx's viewpoint as follows: "At first Marx thought that Ireland would not be liberated by the national movement of the oppressed nation, but by the working-class movement of the oppressor nation....

"However, it so happened that the English working class fell under the influence of the liberals for a fairly long time, became an appendage to the liberals, and by adopting a liberal-labour policy left itself leaderless. The bourgeois liberation movement in Ireland grew stronger and assumed revolutionary forms. Marx reconsidered his view and corrected it. 'What a misfortune it is for a nation to have subjugated another.' The English working class will never be free until Ireland is freed from the English yoke. Reaction in England is strengthened and fostered by the enslavement of Ireland (just as reaction in Russia is fostered by her enslavement of a number of nations!)." (*Collected Works,* Vol. 20, p. 440.)

While allowing for the inevitable variety in the forms and course of the revolutionary processes by which the oppressed peoples would liberate themselves—for these depend on varying situations—Marx and Engels were convinced that the internationalist unity of the working class and the fighters for national independence was essential in all cases. They championed the idea of fraternal union between the Irish working people and the English workers, a union between them and all the workers of the world as a condition for a free, progressive future for England, as well as Ireland.

*   *   *

The works included in this collection are arranged chronologically. At the end of each are indicated the date of its first publication, the language of the original from which the English translation was made and, wherever the text was originally English, the edition here reproduced. The collection is supplied with notes, and with name and subject indexes, though some minor characters mentioned in the "Chronology of Ireland" are not listed.

L. I. Golman
Moscow

Karl Marx
and
Frederick Engels

IRELAND
AND THE IRISH QUESTION

IV

At present the whole town is talking about nothing but O'Connell and the Irish Repeal (i.e., the abrogation of the Anglo-Irish Union).[2] O'Connell, the wily old barrister, who under the Whig government sat placidly in the House of Commons and helped to pass "liberal" measures, which were then rejected by the House of Lords—O'Connell has suddenly departed from London leaving the parliamentary debates and has once again taken up his old demand for repeal of the Union. No one had given any more thought to it when Old Dan* appeared in Dublin and again began to stir up the stale obsolete rubbish. It is not surprising that the old fermenting junk now produces strange bubbles. The wily old fox gets around from town to town always surrounded by two hundred thousand men, a bodyguard such as no king can boast of. How much could be achieved if a sensible man possessed O'Connell's popularity, or if O'Connell had a little more sense and a little less egoism and vanity! Two hundred thousand men, and what kind of men! Men who have nothing to lose, two-thirds of them not having a shirt to their backs, they are real proletarians and sansculottes, and moreover Irishmen—wild, headstrong, fanatical Gaels. If one has not seen the Irish, one does not know them. Give me two hundred thousand Irishmen and I could overthrow the entire British monarchy. The Irishman is a light-hearted, cheerful, potato-eating child of na-

---

* Daniel O'Connell.—*Ed.*

ture. Straight from the moorland, where he grew up under a leaky roof fed on weak tea and short commons, he is suddenly flung into our civilisation. Hunger impels him to go to England, and in the mechanistic, frigid and egoistic bustle of the English industrial town his passions are awakened. What does this raw youth—whose early years were spent playing on the moorland or begging in the street—know of thriftiness? He squanders what he earns, and then goes hungry till the next pay-day or till he finds new work. He is quite used to being hungry. So he returns, looks for members of his family in the street, where they are scattered begging for alms and from time to time gather around the tea-kettle which mother carries around with her. But he has learned a great deal in England. He has attended public meetings and visited workers' clubs, he knows what the Repeal is and what Sir Robert Peel signifies; he is bound to have had many scuffles with the police and can tell a tale of the callousness and infamy of the "peelers" (policemen). He has also heard a lot about O'Connell. Now he returns to his old hut and the potato patch. The potatoes are ready for harvesting, he digs them up and is thus provided for the winter. At that moment the principal tenant[3] arrives on the scene and demands the rent. Saints in Heaven, where is the money to come from? Since the tenant is responsible to the landowner for the rent, he has his things attached. The Irishman resists and is imprisoned. In the end he is set free, but not long afterwards the principal tenant, or someone else who took part in the attachment, is found dead in a ditch.

Such events are commonplace in the life of the Irish proletarian. His half-wild upbringing and the wholly civilised surroundings in which he finds himself later, engender an internal conflict, a continuous irritation, a rage which constantly smoulders within him, making him capable of anything. He is, moreover, burdened by the weight of five hundred years of oppression with all its consequences. It is therefore not surprising that whenever an opportunity presents itself he hits out blindly and furiously like any half-wild creature, that he is consumed with a desire for revenge, a spirit of destruction, and the object against which this is turned is quite immaterial provided he can hit out and destroy. To this has to be added the passionate national hatred of the Gael for the Saxon, and the Roman Catholic fanaticism, which is fanned by the clergy, against the Protestant-Episcopalian arrogance. Anything can be accomplished with such elements, and they are under O'Connell's control, and what huge numbers are moved by him! The day before yesterday, 150,000 men at Cork; yesterday, 200,000 men in Nenagh; today, 400,000 men in Kilkenny, and so it is everywhere. A triumphal procession lasting a fortnight, such as no Roman emperor ever experienced. If O'Connell really wanted to further the welfare of the people, if he were really concerned with the elimination of misery—and not with his miserable, petty middle-class[4] objectives which are at the bottom of all the shouting and the agitation for the Repeal—I should like to know what demand advanced by O'Connell representing the power that is at present at his disposal could be refused by Sir Robert Peel. But what does O'Connell do with all his power and with his millions of militant and desperate Irishmen? He is unable to attain even the wretched Repeal of the Union. Of course, solely because he does not really mean to achieve it, since he uses the impoverished, oppressed Irish people to embarrass the Tory ministers and to help his middle-class friends to get back into office. Sir Robert Peel knows this very well and that is why 25,000 soldiers are quite sufficient to keep Ireland in check. If O'Connell were really the man of the people, if he had sufficient courage *and were not himself frightened of the people*, i.e., if he were not a two-faced Whig but an upright, consistent democrat, the last English soldier would have left Ireland long since and there would no longer be any idle Protestant pastor in purely Catholic areas or any Norman baron in an Irish castle. But there's the rub. If the people were set free even for a moment, Daniel O'Connell and his moneyed aristocrats would soon find themselves in the wilderness, where O'Connell himself would like to drive the Tories. This is the reason for O'Connell's close association with the Catholic clergy; that is why he exhorts the Irish to be on their guard against the dangerous socialists; that is why he rejects the assistance offered by the Chartists,[5] although for form's sake he speaks occasionally of democracy—just as Louis Philippe used to speak of republican institutions—consequently, the only thing he will achieve is the political education of the Irish people, and this is ultimately for no one more dangerous than for himself.

Published in the magazine *Der Schweizerische Republikaner* No. 39, June 27, 1843

Translated from the German

Frederick Engels

THE CONDITION OF THE WORKING-CLASS IN ENGLAND[6]

From the Chapter THE GREAT TOWNS

Let us investigate some of the slums in their order. *London* comes first,* and in London the famous rookery of *St. Giles* which is now, at last, about to be penetrated by a couple of broad streets. St. Giles is in the midst of the most populous part of the town, surrounded by broad, splendid avenues in which the gay world of London idles about, in the immediate neighbourhood of Oxford Street, Regent Street, of Trafalgar Square and the Strand. It is a disorderly collection of tall, three or four-storied houses, with narrow, crooked, filthy streets, in which there is quite as much life as in the great thoroughfares of the town, except that, here, people of the working-class only are to be seen. A vegetable market is held in the street, baskets with vegetables and fruits, naturally all bad and hardly fit to use, obstruct the sidewalk still further, and from these, as well as from the fish-dealers' stalls, arises a horrible smell. The houses are occupied from cellar to garret, filthy within and without, and their appearance is such that no human being could possibly wish to live in them. But all this is nothing in comparison with the dwellings in the narrow courts and alleys between the streets, entered by covered passages between the houses, in which the filth and tottering ruin surpass all description. Scarcely a whole window-pane can be

---

* The description given below had already been written when I came across an article in the *Illuminated Magazine* (October 1844) dealing with the working-class districts in London which coincides—in many places almost literally and everywhere in general tenor—with what I had said. The article is entitled "The Dwellings of the Poor, from the note-book of an M.D."

found, the walls are crumbling, door-posts and window-frames loose and broken, doors of old boards nailed together, or altogether wanting in this thieves' quarter, where no doors are needed, there being nothing to steal. Heaps of garbage and ashes lie in all directions, and the foul liquids emptied before the doors gather in stinking pools. Here live the poorest of the poor, the worst paid workers with thieves and the victims of prostitution indiscriminately huddled together, the majority Irish, or of Irish extraction, and those who have not yet sunk in the whirlpool of moral ruin which surrounds them, sinking daily deeper, losing daily more and more of their power to resist the demoralising influence of want, filth, and evil surroundings.

...But the most horrible spot (if I should describe all the separate spots in detail I should never come to the end) lies on the Manchester side, immediately south-west of Oxford Road, and is known as Little Ireland. In a rather deep hole, in a curve of the Medlock and surrounded on all four sides by tall factories and high embankments, covered with buildings, stand two groups of about two hundred cottages, built chiefly back to back, in which live about four thousand human beings, most of them Irish. The cottages are old, dirty, and of the smallest sort, the streets uneven, fallen into ruts and in part without drains or pavement; masses of refuse, offal and sickening filth lie among standing pools in all directions; the atmosphere is poisoned by the effluvia from these, and laden and darkened by the smoke of a dozen tall factory chimneys. A horde of ragged women and children swarm about here, as filthy as the swine that thrive upon the garbage heaps and in the puddles. In short, the whole rookery furnishes such a hateful and repulsive spectacle as can hardly be equalled in the worst court on the Irk. The race that lives in these ruinous cottages, behind broken windows, mended with oilskin, sprung doors, and rotten door-posts, or in dark, wet cellars, in measureless filth and stench, in this atmosphere penned in as if with a purpose, this race must really have reached the lowest stage of humanity. This is the impression and the line of thought which the exterior of this district forces upon the beholder. But what must one think when he hears that in each of these pens, containing at most two rooms, a garret and perhaps a cellar, on the average twenty human beings live; that in the whole region, for each one hundred and twenty persons, one usually inaccessible privy is provided; and that in spite of all the preachings of the physicians, in spite of the excitement into which the cholera epidemic plunged the sanitary police by reason of the condition of Little Ireland, in spite of everything, in this year of grace 1844, it is in almost the same state as in 1831! Dr. Kay asserts* that not only the cellars but the first floors of all the houses in this district are damp; that a number of cellars once filled up with earth have now been emptied and are occupied once more by Irish people; that in one cellar the water constantly wells up through a hole stopped with clay, the cellar lying below the river level, so that its occupant, a hand-loom weaver, had to bale out the water from his dwelling every morning and pour it into the street!

From the Chapter THE AGRICULTURAL PROLETARIAT

If England illustrates the results of the system of farming on a large scale and Wales on a small one, *Ireland* exhibits the consequences of overdividing the soil. The great mass of the population of Ireland consists of small tenants who occupy a sorry hut without partitions, and a potato patch just large enough to supply them most scantily with potatoes through the winter. In consequence of the great competition which prevails among these small tenants, the rent has reached an unheard-of height, double, treble, and quadruple that paid in England. For every agricultural labourer seeks to become a tenant-farmer, and though the division of land has gone so far, there still remain numbers of labourers in competition for plots. Although in Great Britain 32,000,000 acres of land are cultivated, and in Ireland but 14,000,000; although Great Britain produces agricultural products to the value of £150,000,000, and Ireland of but £36,000,000, there are in Ireland 75,000 agricultural proletarians *more* than in the neighbouring island.** How

---

* Dr. Kay, *loc. cit.*[7]
** Report of the Poor Law Commission on Ireland [Parliamentary Session of 1837].

great the competition for land in Ireland must be is evident
from this extraordinary disproportion, especially when one
reflects that the labourers in Great Britain are living in the
utmost distress. The consequence of this competition is that
it is impossible for the tenants to live much better than the
labourers, by reason of the high rents paid. The Irish people
is thus held in crushing poverty, from which it cannot free
itself under our present social conditions. These people live
in the most wretched clay huts, scarcely good enough for
cattle-pens, have scant food all winter long, or, as the report
above quoted expresses it, they have potatoes half enough
thirty weeks in the year, and the rest of the year nothing.
When the time comes in the spring at which this provision
reaches its end, or can no longer be used because of its
sprouting, wife and children go forth to beg and tramp
the country with their kettle in their hands. Meanwhile the
husband, after planting potatoes for the next year, goes in
search of work either in Ireland or England, and returns at
the potato harvest to his family. This is the condition in
which nine-tenths of the Irish country folks live. They are
poor as church mice, wear the most wretched rags, and
stand upon the lowest plane of intelligence possible in a half-
civilised country. According to the report quoted, there are,
in a population of $8^1/_2$ millions, 585,000 heads of families
in a state of total destitution; and according to other authorities, cited by Sheriff Alison,* there are in Ireland 2,300,000
persons who could not live without public or private assistance—or 27 per cent of the whole population paupers!

The cause of this poverty lies in the existing social conditions, especially in competition here found in the form of
the subdivision of the soil. Much effort has been spent in
finding other causes. It has been asserted that the relation
of the tenant to the landlord who lets his estate in large
lots to tenants, who again have their sub-tenants, and sub-
sub-tenants, in turn, so that often ten middlemen come
between the landlord and the actual cultivator—it has been
asserted that the shameful law which gives the landlord the
right of expropriating the cultivator who may have paid

---

* Archibald Alison, *The Principles of Population, and their Connection with Human Happiness*, Vol. II, London, 1840.—*Ed.*

his rent duly, if the first tenant fails to pay the landlord, that this law is to blame for all this poverty. But all this determines only the *form* in which the poverty manifests itself. Make the small tenant a landowner himself and what follows? The majority could not live upon their holdings even if they had no rent to pay, and any slight improvement which might take place would be lost again in a few years in consequence of the rapid increase of population. The children would then live to grow up under the improved conditions who now die in consequence of poverty in early childhood. From another side comes the assertion that the shameless oppression inflicted by the English is the cause of the trouble. It is the cause of the somewhat *earlier* appearance of this poverty, but not of the poverty itself. Or the blame is laid on the Protestant Church forced upon a Catholic nation; but divide among the Irish what the Church takes from them, and it does not reach six shillings a head. Besides, tithes are a tax upon *landed property*, not upon the tenant, though he may nominally pay them; now, since the Commutation Bill of 1838,[8] the landlord pays the tithes directly and reckons so much higher rent, so that the tenant is none the better off. And in the same way a hundred other causes of this poverty are brought forward, all proving as little as these. This poverty is the result of our social conditions; apart from these, causes may be found for the manner in which it manifests itself, but not for the fact of its existence. That poverty manifests itself in Ireland thus and not otherwise, is owing to the character of the people, and to their historical development. The Irish are a people related in their whole character to the Latin nations, to the French, and especially to the Italians. The bad features of their character we have already had depicted by Carlyle.[9] Let us now hear an Irishman, who at least comes nearer to the truth than Carlyle, with his prejudice in favour of the Teutonic character:

"They are restless, yet indolent, clever and indiscreet, stormy, impatient, and improvident; brave by instinct, generous without much reflection, quick to revenge and forgive insults, to make and to renounce friendships, gifted with genius prodigally, sparingly with judgement."*

* *The State of Ireland*, London, 1807; 2nd ed., 1821. Pamphlet.

With the Irish, feeling and passion predominate; reason must bow before them. Their sensuous, excitable nature prevents reflection and quiet, persevering activity from reaching development—such a nation is utterly unfit for manufacture as now conducted. Hence they held fast to agriculture, and remained upon the lowest plane even of that. With the small subdivisions of land, which were not here artificially created, as in France and on the Rhine, by the division of great estates,* but have existed from time immemorial, an improvement of the soil by the investment of capital was not to be thought of; and it would, according to Alison, require 120 million pounds sterling to bring the soil up to the not very high state of fertility already attained in England. The English immigration, which might have raised the standard of Irish civilisation, has contented itself with the most brutal plundering of the Irish people; and while the Irish, by their immigration into England, have furnished England a leaven which will produce its own results in the future, they have little for which to be thankful to the English immigration.

The attempts of the Irish to save themselves from their present ruin, on the one hand, take the form of crimes. These are the order of the day in the agricultural districts, and are nearly always directed against the most immediate enemies, the landlords' agents, or their obedient servants, the Protestant intruders, whose large farms are made up of the potato patches of hundreds of ejected families. Such crimes are especially frequent in the South and West. On the other hand, the Irish hope for relief by means of the agitation for the repeal of the Legislative Union with England.[10] From all the foregoing, it is clear that the uneducated Irish must see in the English their worst enemies; and their first hope of improvement in the conquest of national independence. But quite as clear is it, too, that Irish distress cannot be removed by any Act of Repeal. Such an Act would, however,

---

* Mistake. Small-scale agriculture had been the prevailing form of farming ever since the Middle Ages. Thus the small peasant farm existed even before the Revolution. The only thing the latter changed was its *ownership*; that it took away from the feudal lords and transferred, directly or indirectly, to the peasants. [*Added by Engels in the German edition of 1892.*]

at once lay bare the fact that the cause of Irish misery, which now seems to come from abroad, is really to be found at home. Meanwhile, it is an open question whether the accomplishment of repeal will be necessary to make this clear to the Irish. Hitherto, neither Chartism nor Socialism has had marked success in Ireland.

I close my observations upon Ireland at this point the more readily, as the Repeal Agitation of 1843 and O'Connell's trial[11] have been the means of making the Irish distress more and more known in Germany.

Published in the book: Friedrich Engels, *Die Lage der arbeitenden Klasse in England,* Leipzig, 1845

Printed according to the authorised English edition, London, 1892

Frederick Engels

From [THE COMMERCIAL CRISIS IN ENGLAND— THE CHARTIST MOVEMENT—IRELAND[12]]*

In the meantime starving Ireland is writhing in the most terrible convulsions. The workhouses are overflowing with beggars, the ruined property owners are refusing to pay the Poor Tax, and the hungry people gather in their thousands to ransack the barns and stables of the farmers and even of the Catholic priests, whom they still worshipped a short time ago.

It looks as though the Irish will not die of hunger as calmly next winter as they did last winter. Irish emigration to England is getting more alarming each day. It is estimated that an average of 50,000 Irish arrive each year; the number so far this year is already over 220,000. In September, 345 were arriving daily and in October this figure increased to 511. This means that the competition between the workers will become stronger, and it would not be at all surprising if the present crisis caused such an uproar that it compelled the government to grant reforms of a most important nature.

Published in the newspaper
La Réforme,
October 26, 1847

Translated from the French

---

* Headings set in square brackets have been provided by the Institute of Marxism-Leninism in Moscow.—*Ed.*

Frederick Engels

[THE COERCION BILL FOR IRELAND AND THE CHARTISTS]

The Irish Coercion Bill[13] came into force last Wednesday. The Lord Lieutenant was not slow in taking advantage of the despotic powers with which this new law invests him; the Act has been applied all over the counties of Limerick and Tipperary and to several baronies in the counties of Clare, Waterford, Cork, Roscommon, Leitrim, Cavan, Longford and *King's County*.[14]

It remains to be seen what the effect of these odious measures will be. In this connection we already have the opinion of the class in whose interests the measures were taken, namely, the Irish landowners. They announce to the world in their organs that the measures will have no effect whatsoever. And in order to achieve this a whole country is being placed in a state of siege! To achieve this nine-tenths of the Irish representatives have deserted their country!

This is a fact. The desertion has been a general one. During the discussion of the Bill the O'Connell family itself became divided: John and Maurice, two of the deceased *"Liberator's"** sons, remained faithful to their homeland, whereas their cousin,*** Morgan O'Connell, not only voted for the Bill, but also spoke in its support on several occasions. There were only eighteen members who voted outright for the rejection of the Bill, and only twenty supported the amendment put forward by Mr. Wakley, the Chartist member for a borough on the outskirts of London, who demanded that the Coercion Bill should also be accompanied by meas-

---

* Daniel O'Connell.—*Ed.*
** A mistake in *La Réforme*; it should read "brother".—*Ed.*

ures aimed at reducing the causes of the crimes which it was proposed to repress. And among these eighteen and twenty voters there were also four or five English Radicals and two Irishmen representing English boroughs, meaning that out of the hundred members which Ireland has in Parliament there were only a dozen who put up serious opposition to the Bill.

This was the first discussion on an important question affecting Ireland which had been held since the death of O'Connell. It was to decide who would take the place of the great agitator in leading Ireland. Up to the opening of Parliament Mr. John O'Connell had been tacitly acknowledged in Ireland as his father's successor. But it soon became evident after the debate had begun that he was not capable of leading the party and, what is more, that he had found a formidable rival in Feargus O'Connor. This democratic leader about whom Daniel O'Connell said, "We are happy to make the English Chartists a present of Mr. F. O'Connor", put himself at the head of the Irish party in a single bound. It was he who proposed the outright rejection of the Coercion Bill; it was he who succeeded in rallying all the opposition behind him; it was he who opposed each clause, who held up the voting whenever possible; it was he who in his speeches summed up all the arguments of the opposition against the Bill; and finally it was he who for the first time since 1835 reintroduced the motion for *Repeal of the Union*,[15] a motion which none of the Irish members would have put forward.

The Irish members accepted this leader with a bad grace. As simple Whigs in their heart of hearts they fundamentally detest the democratic energy of Mr. O'Connor. He will not allow them to go on using the campaign for repeal as a means for overthrowing the Tories in favour of the Whigs and to forget the very word "repeal" when the latter come to power. But the Irish members who support repeal cannot possibly do without a leader like O'Connor and, although they are trying to undermine his growing popularity in Ireland, they are obliged to submit to his leadership in Parliament.

When the parliamentary session is over O'Connor will probably go on a tour of Ireland to revive the agitation for repeal and to found an Irish Chartist Party. There can be no doubt that if O'Connor is successful in doing this he will be the leader of the Irish people in less than six months. By uniting the democratic leadership of the three kingdoms[16] in his hands, he will occupy a position which no agitator, not even O'Connell, has held before him.

We will leave it to our readers to judge the importance of this future alliance between the peoples of the two islands. British democracy will advance much more quickly when its ranks are swelled by two million brave and ardent Irish, and poverty-stricken Ireland will at last have taken an important step towards her liberation.

Published in the newspaper
La Réforme,
January 8, 1848

Translated from the French

Frederick Engels

FEARGUS O'CONNOR AND THE IRISH PEOPLE[17]

The first issue of the *Northern Star*[18] for 1848 contains an address to the Irish people by *Feargus O'Connor*, the well-known leader of the English Chartists who also represents them in the House of Commons. The whole address deserves to be read and carefully considered by every democrat, but our restricted space prevents us from reproducing it in full.

We would, however, be remiss in our duty if we were to pass it over in silence. The momentous consequences of this forceful appeal to the Irish will very soon be clearly evident. O'Connor—who is of Irish descent, a Protestant, and who has been for over ten years a leader and main pillar of the great labour movement in England—must henceforth be regarded as the virtual chief of the Irish Repealers[19] and advocates of reform. The part he played in opposing the latest of the ignominious Irish Coercion Bills has given him the first claim to this status, and his continuous agitation for the Irish cause has shown that Feargus O'Connor is just the man Ireland needs.

O'Connor is indeed seriously concerned about the well-being of the millions in Ireland, Repeal—the abolition of the Union, that is, the achievement of an independent Irish Parliament—is for *him* not an empty word, a pretext for obtaining posts for himself and his friends and for making profitable business transactions.

In his address he shows the Irish people that Daniel O'Connell, this political juggler, led them by the nose and deceived them for thirteen years by means of the word "Repeal".

He correctly elucidates the conduct of John O'Connell, who has taken possession of his father's political heritage and who like his father is prepared to sacrifice millions of credulous Irishmen for the sake of his personal ventures and interests. All O'Connell's orations at the Dublin Conciliation Hall[20] and all his hypocritical protestations and beautiful phrases will not obliterate the disrepute he has brought upon himself by his earlier actions and in particular now by the way he acted during the debates on the Irish Coercion Bill.

The Irish people must and will in the end grasp the real position, and then it will kick out the entire gang of so-called Repealers, who under cover of this cloak laugh up their sleeves and in their purses, and John O'Connell, the fanatical papist and political mountebank, will be kicked out first of all.

If this were all the address contained, we should not have especially referred to it. But it is of much wider importance. For Feargus O'Connor speaks in it not only as an Irishman but also, and primarily, as an English democrat and a Chartist.

With a lucidity which even the most obtuse mind cannot fail to notice, O'Connor shows that the Irish people must fight strenuously, and in close association with the English working classes and the Chartists, in order to win the six points of the People's Charter—annual parliaments, universal suffrage, vote by ballot, abolition of the property qualification for members of parliament, payment of M.P.s and the establishment of equal electoral districts. Only after these six points are won will the achievement of the Repeal have any advantages for Ireland.

Furthermore O'Connor pointed out that justice for Ireland had been demanded even earlier by the English workers in a petition which had received three and a half million signatures,[21] and that now the English Chartists again protested against the Irish Coercion Bill in numerous petitions. He finally stressed that the oppressed classes in both England and Ireland must fight together and conquer together or continue to languish under the same burden and live in the same misery and dependence on the privileged and ruling capitalist class.

Henceforth the mass of the Irish people will undoubtedly unite ever more closely with the English Chartists and will act in accordance with a common plan. This will bring the victory of the English democrats, and hence the liberation of Ireland, considerably nearer. That is the significance of O'Connor's address to the Irish people.

Published in *Deutsche-Brüsseler-Zeitung* No. 3, January 9, 1848

Translated from the German

Karl Marx

From [THE SPEECH ON THE POLISH QUESTION
FEBRUARY 22, 1848[22]]

The Cracow revolution has provided the whole of Europe with a fine example by identifying the national cause with that of democracy and the liberation of the oppressed class.

Repressed for a time by the bloody hands of hired assassins, this revolution has now emerged glorious and triumphant in Switzerland and Italy. The correctness of its principles is borne out by the situation in Ireland, where the party with narrow national interests was buried together with O'Connell, and where the new national party is first and foremost a reformist and democratic one.[23]

Published in the collection
Célébration, à Bruxelles, du deuxième anniversaire de la Révolution Polonaise du 22 Février 1846, Brussels, 1848

Translated from the French

Frederick Engels

From COLOGNE IS IN DANGER[24]

Cologne, June 10. Whitsun, the enchanting festival, is approaching, the fields wear green and the trees are in bloom, and wherever there are people who confuse the dative with the accusative preparations are being made to pour the holy spirit of reaction upon all countries in a *single* day.[25]

The moment has been cleverly chosen. In Naples lieutenants of the guards regiments and Swiss mercenaries have succeeded in drowning the recently established freedom in the blood of the people. In France an assembly composed of capitalists has gagged the Republic by means of Draconic laws and has appointed General Perrot, who at the Hôtel Guizot on February 23 gave the order to fire, commandant of Vincennes. Chartists and Repealers are thrown en masse into prison in England and Ireland, and dragoons are used to disperse unarmed meetings.[26] In Frankfurt we find that the triumvirate proposed by the late Bundestag and rejected by the Committee of Fifty[27] has now been installed by the National Assembly itself. The Right in Berlin has won battle after battle as a result of its numerical superiority and its drum beating, and the Prince of Prussia has pronounced the revolution null and void by taking possession of the "property of the entire nation".[28]

Published in *Neue Rheinische Zeitung* No. 11, June 11, 1848

Translated from the German

Karl Marx

From ELECTIONS—FINANCIAL CLOUDS— THE DUCHESS OF SUTHERLAND AND SLAVERY[29]

The process of *clearing estates* which, in Scotland, we have just now described, was carried out in England in the 16th, 17th and 18th centuries. Thomas Morus already complains of it in the beginning of the 16th century. It was performed in Scotland in the beginning of the 19th, and in Ireland it is now in full progress. The noble Viscount Palmerston, too, some years ago cleared of men his property in Ireland, exactly in the manner described above.

If of any property it ever was true that it was *robbery*, it is literally true of the property of the British aristocracy. Robbery of Church property, robbery of commons, fraudulous transformation, accompanied by murder, of feudal and patriarchal property into private property—these are the titles of British aristocrats to their possessions. And what services in this latter process were performed by a servile class of lawyers, you may see from an English lawyer of the last century, Dalrymple, who, in his *History of Feudal Property*,[30] very naïvely proves that every law or deed concerning property was interpreted by the lawyers, in England, when the middle class rose in wealth, in favour of the *middle class*—in Scotland, where the nobility enriched themselves, in favour of the *nobility*—in either case it was interpreted in a sense hostile to the *people*.

Published in *The New-York Daily Tribune* No. 3687, February 9, 1853, and in *The People's Paper* No. 45, March 12, 1853

Printed according to the text of *The New-York Daily Tribune* and verified with the text of *The People's Paper*

From FORCED EMIGRATION—KOSSUTH AND MAZZINI—THE REFUGEE QUESTION—ELECTION BRIBERY IN ENGLAND—Mr. COBDEN

From the accounts relating to trade and navigation for the years 1851 and 1852, published in Feb. last, we see that the total declared value of *exports* amounted to £68,531,601 in 1851, and to £71,429,548 in 1852; of the latter amount, £47,209,000 go to the export of cotton, wool, linen and silk manufactures. The quantity of *imports* for 1852 is below that for the year 1851. The proportion of imports entered for home consumption not having diminished, but rather increased, it follows that England has re-exported, instead of the usual quantity of colonial produce, a certain amount of gold and silver.

The Colonial Land Emigration Office gives the following return of the emigration from England, Scotland, and Ireland, to all parts of the world, from Jan. 1, 1847 to June 30, 1852:

Year	English	Scotch	Irish	Total
1847	34,685	8,616	214,969	258,270
1848	58,865	11,505	177,719	248,089
1849	73,613	17,127	208,758	299,498
1850	57,843	15,154	207,852	280,849
1851	69,557	18,646	247,763	335,966
1852 (till June) . .	40,767	11,562	143,375	195,704
Total	335,330	82,610	1,200,436	1,618,376

"Nine-tenths," remarks the Office, "of the emigrants from Liverpool are assumed to be Irish. About three-fourths of the emigrants from Scotland are Celts, either from the Highlands, or from Ireland through Glasgow."

Nearly four-fifths of the whole emigration are, accordingly, to be regarded as belonging to the Celtic population of Ireland and of the Highlands and islands of Scotland. The London *Economist* says of this emigration:

"It is consequent on the breaking down of the system of society founded on small holdings and potato cultivation"; and adds: "The departure of the redundant part of the population of Ireland and the Highlands of Scotland is an indispensable preliminary to every kind of improvement.... The revenue of Ireland has not suffered in any degree from the famine of 1846-47, or from the emigration that has since taken place. On the contrary, her *net revenue* amounted in 1851 to £4,281,999, being about £184,000 greater than in 1843."

Begin with pauperising the inhabitants of a country, and when there is no more profit to be ground out of them, when they have grown a burden to the revenue, drive them away, and sum up your Net Revenue! Such is the doctrine laid down by Ricardo in his celebrated work, *The Principles of Political Economy*.[31] The annual profits of a capitalist amounting to £2,000, what does it matter to him whether he employs 100 men or 1,000 men? "Is not," says Ricardo, "the real income of a nation similar?" The net real income of a nation, rents and profits, remaining the same, it is no subject of consideration whether it is derived from ten millions of people or from twelve millions. Sismondi, in his *Nouveaux principes d'économie politique*,[32] answers that, according to this view of the matter, the English nation would not be interested at all in the disappearance of the whole population, the King (at that time it was no Queen, but a King*) remaining alone in the midst of the island, supposing only that automatic machinery enabled him to procure the amount of *Net Revenue* now produced by a population of twenty millions. Indeed, that grammatical entity, "the national wealth", would in this case not be diminished.

In a former letter I have given an instance of the clearing of estates in the Highlands of Scotland. That emigra-

---

* George III.—*Ed.*

tion continues to be forced upon Ireland by the same process you may see from the following quotation from *The Galway Mercury*:

> "The people are fast passing away from the land in the West of Ireland. The landlords of Connaught are tacitly combined to weed out all the smaller occupiers, against whom a regular systematic war of extermination is being waged.... The most heart-rending cruelties are daily practised in this province, of which the public are not at all aware."

But it is not only the pauperised inhabitants of Green Erin* and of the Highlands of Scotland that are swept away by agricultural improvements, and by the "breaking down of the antiquated system of society". It is not only the able-bodied agricultural labourers from England, Wales, and Lower Scotland, whose passages are paid by the Emigration Commissioners. The wheel of "improvement" is now seizing another class, the most stationary class in England. A startling emigration movement has sprung up among the smaller English farmers, especially those holding heavy clay soils, who, with bad prospects for the coming harvest, and in want of sufficient capital to make the great improvements on their farms which would enable them to pay their old rents, have no other alternative but to cross the sea in search of a new country and of new lands. I am not speaking now of the emigration caused by the gold mania,[33] but only of the compulsory emigration produced by landlordism, concentration of farms, application of machinery to the soil, and introduction of the modern system of agriculture on a great scale.

In the ancient States, in Greece and Rome, compulsory emigration, assuming the shape of the periodical establishment of colonies, formed a regular link in the structure of society. The whole system of those States was founded on certain limits to the numbers of the population, which could not be surpassed without endangering the condition of antique civilisation itself. But why was it so? Because the application of science to material production was utterly unknown to them. To remain civilised they were forced to remain few. Otherwise they would have had to submit to

---

* Ireland.—*Ed.*

the bodily drudgery which transformed the free citizen into a slave. The want of productive power made citizenship dependent on a certain proportion in numbers not to be disturbed. Forced emigration was the only remedy.

It was the same pressure of population on the powers of production, that drove the barbarians from the high plains of Asia to invade the Old World. The same cause acted there, although under a different form. To remain barbarians they were forced to remain few. They were pastoral, hunting, war-waging tribes, whose manner of production required a large space for every individual, as is now the case with the Indian tribes in North-America. By augmenting in numbers they curtailed each other's field of production. Thus the surplus population was forced to undertake those great adventurous migratory movements which laid the foundation of the peoples of ancient and modern Europe.

But with modern compulsory emigration the case stands quite opposite. Here it is not the want of productive power which creates a surplus population; it is the increase of productive power which demands a diminution of population, and drives away the surplus by famine or emigration. It is not population that presses on productive power; it is productive power that presses on population.

Now I share neither in the opinion of Ricardo, who regards "Net-Revenue" as the Moloch to whom entire populations must be sacrificed, without even so much as complaint, nor in the opinion of Sismondi, who, in his hypochondriacal philanthropy, would forcibly retain the superannuated methods of agriculture and proscribe science from industry, as Plato expelled poets from his *Republic*.[34] Society is undergoing a silent revolution, which must be submitted to, and which takes no more notice of the human existences it breaks down than an earthquake regards the houses it subverts. The classes and the races, too weak to master the new conditions of life, must give way. But can there be anything more puerile, more short-sighted, than the views of those Economists who believe in all earnest that this woeful transitory state means nothing but adapting society to the acquisitive propensities of capitalists, both landlords and money-lords? In Great Britain the working of that process is most transparent. The application of modern science to production clears the land of its inhabitants, but it concentrates people in manufacturing towns.

"No manufacturing workmen," says *The Economist,* "have been assisted by the Emigration Commissioners, except a few Spitalfields and Paisley hand-loom weavers, and few or none have emigrated at their own expense."

The Economist knows very well that they could not emigrate at their own expense, and that the industrial middle class would not assist them in emigrating. Now, to what does this lead? The rural population, the most stationary and conservative element of modern society, disappears while the industrial proletariat, by the very working of modern production, finds itself gathered in mighty centres, around the great productive forces, whose history of creation has hitherto been the martyrology of the labourers. Who will prevent them from going a step further, and appropriating these forces, to which they have been appropriated before? Where will be the power of resisting them? Nowhere! Then, it will be of no use to appeal to the "rights of property". The modern changes in the art of production have, according to the Bourgeois Economists themselves, broken down the antiquated system of society and its modes of appropriation. They have *expropriated* the Scotch clansman, the Irish cottier and tenant, the English yeoman, the hand-loom weaver, numberless handicrafts, whole generations of factory children and women; they will expropriate, in due time, the landlord and the cotton-lord.

On the Continent heaven is fulminating, but in England the earth itself is trembling. England is the country where the real revulsion of modern society begins.

Published in *The New-York Daily Tribune* No. 3722, March 22, 1853, and in *The People's Paper* No. 50, April 16, 1853

Printed according to the text of *The New-York Daily Tribune* and verified with the text of *The People's Paper*

Karl Marx

THE INDIAN QUESTION—IRISH TENANT RIGHT

London, June 28, 1853

The debate on Lord Stanley's motion with respect to India commenced on the 23rd, continued on the 24th, and adjourned to the 27th inst., has not been brought to a close. When that shall at length have arrived, I intend to resume my observations on the Indian question.[35]

As the Coalition Ministry[36] depends on the support of the Irish party, and as all the other parties composing the House of Commons so nicely balance each other that the Irish may at any moment turn the scales which way they please, some concessions are at last about to be made to the Irish tenants. The "Leasing powers (Ireland) Bill", which passed the House of Commons on Friday last, contains a provision that for the improvements made on the soil and separable from the soil, the tenant shall have, at the termination of his lease, a compensation in money, the incoming tenant being at liberty to take them at the valuation, while with respect to improvements in the soil, compensation for them shall be arranged by contract between the landlord and the tenant.[37]

A tenant having incorporated his capital, in one form or another, in the land, and having thus effected an improvement of the soil, either directly by irrigation, drainage, manure, or indirectly by construction of buildings for agricultural purposes, in steps the landlord with demand for increased rent. If the tenant concede, he has to pay the interest for his own money to the landlord. If he resist, he will be very unceremoniously ejected, and supplanted by a new tenant, the latter being enabled to pay a higher rent by the very expenses incurred by his predecessors, until he also, in his turn, has become an improver of the land, and is replaced in the same way, or put on worse terms. In this easy way a class of absentee landlords has been enabled to pocket, not merely the labour, but also the capital, of whole generations, each generation of Irish peasants sinking a grade lower in the social scale, exactly in proportion to the exertions and sacrifices made for the raising of their condition and that of their families. If the tenant was industrious and enterprising, he became taxed in consequence of his very industry and enterprise. If, on the contrary, he grew inert and negligent, he was reproached with the "aboriginal faults of the Celtic race". He had, accordingly, no other alternative left but to become a pauper—to pauperise himself by industry, or to pauperise by negligence. In order to oppose this state of things, "Tenant Right" was proclaimed in Ireland— a right of the tenant, not in the soil but in the improvements of the soil effected at his cost and charges. Let us see in what manner *The Times*, in its Saturday's leader, attempts to break down this Irish "Tenant Right"[38]:

"There are two general systems of farm occupation. Either a tenant may take a lease of the land for a fixed number of years, or his holding may be terminable at any time upon certain notice. In the first of these events, it would be obviously his course to adjust and apportion his outlay so that all, or nearly all the benefit would find its way to him before the expiration of his term. In the second case it seems equally obvious that he should not run the risk of the investment without a proper assurance of return."

Where the landlords have to deal with a class of large capitalists who may, as they please, invest their stock in commerce, in manufactures or in farming, there can be no doubt but that these capitalist farmers, whether they take long leases or no time leases at all, know how to secure the "proper" return of their outlays. But with regard to Ireland the supposition is quite fictitious. On the one side you have there a small class of land monopolists, on the other, a very large class of tenants with very petty fortunes, which they have no chance to invest in different ways, no other field of production opening to them, except the soil. They are, therefore, forced to become tenants-at-will. Being once tenants-at-will, they naturally run the risk of losing their revenue, provided they do not invest their small capital. Investing it, in order to secure their revenue, they run the risk of losing their capital, also.

> "Perhaps," continues *The Times*, "it may be said, that in any case a tenantry could hardly expire without something being left upon the ground, in some shape or another, representing the tenant's own property, and that for this compensation should be forthcoming. There is some truth in the remark, but the demand thus created ought, under proper conditions of society, to be easily adjusted between landlord and tenant, as it might, at any rate, be provided for in the original contract. We say that the conditions of society should regulate these arrangements, because we believe that no Parliamentary enactment can be effectually substituted for such an agency."

Indeed, under "proper conditions of society", we should want no more Parliamentary interference with the Irish land-tenant, as we should not want, under "proper conditions of society", the interference of the soldier, of the policeman, and of the hangman. Legislature, magistracy and armed force, are all of them but the offspring of improper conditions of society, preventing those arrangements among men which would make useless the compulsory intervention of a third supreme power. Has, perhaps, *The Times* been converted into a social revolutionist? Does it want a *social* revolution, reorganising the "conditions of society", and the "arrangements" emanating from them, instead of "Parliamentary enactments"? England has subverted the conditions of Irish society. At first it confiscated the land, then it suppressed the industry[39] by "Parliamentary enactments", and lastly, it broke the active energy by armed force. And thus England created those abominable "conditions of society" which enable a small *caste* of rapacious lordlings to dictate to the Irish people the terms on which they shall be allowed to hold the land and to live upon it. Too weak yet for revolutionising those "social conditions", the people appeal to Parliament, demanding at least their mitigation and regulation. But "No," says *The Times*; if you don't live under proper conditions of society, Parliament can't mend that. And if the Irish people, on the advice of *The Times*, tried tomorrow to mend their conditions of society, *The Times* would be the first to appeal to bayonets, and to pour out sanguinary denunciations of the "aboriginal faults of the Celtic race", wanting the Anglo-Saxon taste for pacific progress and legal amelioration.

"If a landlord," says *The Times*, "deliberately injures one tenant, he will find it so much the harder to get another, and whereas his occupation consists in letting land, he will find his land all the more difficult to let."

The case stands rather differently in Ireland. The more a landlord injures one tenant, the easier he will find it to oppress another. The tenant who comes in, is the means of injuring the ejected one, and the ejected one is the means of keeping down the new occupant. That, in due course of time, the landlord, beside injuring the tenant, will injure himself and ruin himself, is not only a probability, but the very fact, in Ireland—a fact affording, however, a very precarious source of comfort to the ruined tenant.

"The relations between the landlord and tenant are those between two traders," says *The Times*.

This is precisely the *petitio principii* which pervades the whole leader of *The Times*. The needy Irish tenant belongs to the soil, while the soil belongs to the English lord. As well you might call the relation between the robber who presents his pistol, and the traveller who presents his purse, a relation between two traders.

"But," says *The Times*, "in point of fact, the relation between Irish landlords and tenants will soon be reformed by an agency more potent than that of legislation. The property of Ireland is fast passing into new hands, and, if the present rate of emigration continues, its cultivation must undergo the same transfer."

Here, at least, *The Times* has the truth. British Parliament does not interfere at a moment when the worked-out old system is terminating in the common ruin, both of the thrifty landlord and the needy tenant, the former being knocked down by the hammer of the *Encumbered Estates* Commission, and the latter expelled by compulsory emigration. This reminds us of the old Sultan of Morocco. Whenever there was a case pending between two parties, he knew of no more "potent agency" for settling their controversy, than by killing both parties.

"Nothing could tend," concludes *The Times* with regard to Tenant Right, "to greater confusion than such a *communistic distribution of ownership*. The only person with any right in the land, is the landlord."

The Times seems to have been the sleeping Epimenides of the past half century, and never to have heard of the hot controversy going on during all that time upon the claims of the landlord, not among social reformers and Communists, but among the very political economists of the British middle class. Ricardo, the creator of modern political economy in Great Britain, did not controvert the "right" of the landlords, as he was quite convinced that their claims were based upon fact, and not on right, and that political economy in general had nothing to do with questions of right; but he attacked the land-monopoly in a more unassuming, yet more scientific, and therefore more dangerous manner. He proved that private proprietorship in land, as distinguished from the respective claims of the labourer, and of the farmer, was a relation quite superfluous in, and incoherent with, the whole framework of modern production; that the economical expression of that relationship and the rent of land, might, with great advantage, be appropriated by the State; and finally that the interest of the landlord was opposed to the interest of all other classes of modern society. It would be tedious to enumerate all the conclusions drawn from these premises by the Ricardo School against the landed monopoly. For my end, it will suffice to quote three of the most recent economical authorities of Great Britain.

The London Economist, whose chief editor, Mr. J. Wilson, is not only a Free Trade oracle,[40] but a Whig one, too, and not only a Whig, but also an inevitable Treasury-appendage in every Whig or composite ministry, has contended in different articles that exactly speaking there can exist no title authorising any individual, or any number of individuals, to claim the exclusive proprietorship in the soil of a nation.

Mr. Newman, in his *Lectures on Political Economy*, London, 1851, professedly written for the purpose of refuting socialism, tells us:

"No man has, or can have, a natural right to *land,* except so long as he occupies it in person. His right is to the use, and to the use only. All other right is the creation of artificial law" (or Parliamentary enactments as *The Times* would call it).... "If, at any time, land becomes needed to *live upon,* the right of private possessors to withhold it comes to an end."

This is exactly the case in Ireland, and Mr. Newman expressly confirms the claims of the Irish tenantry, and in lectures held before the most select audiences of the British aristocracy.

In conclusion let me quote some passages from Mr. Herbert Spencer's work, *Social Statics,* London, 1851, also, purporting to be a complete refutation of communism, and acknowledged as the most elaborate development of the Free Trade doctrines of modern England.

"No one may use the earth in such a way as to prevent the rest from similarly using it. Equity, therefore, does not permit property in land, or the rest would live on the earth by sufferance only. The landless men might equitably be expelled from the earth altogether.... It can never be pretended, that the existing titles to such property are legitimate. Should anyone think so let him look in the Chronicles. The original deeds were written with the sword, rather than with the pen. Not lawyers but soldiers were the conveyancers: blows were the current coin given in payment; and for seals blood was used in preference to wax. Could valid claims be thus constituted? Hardly. And if not, what becomes of the pretensions of all subsequent holders of estates so obtained? Does sale or bequest generate a right where it did not previously exist?... If one act of transfer can give no title, can many?... At what rate per annum do invalid claims become valid?... The right of mankind at large to the earth's surface is still valid, all deeds, customs and laws notwithstanding. It is impossible to discover any mode in which land can become private property.... We daily deny landlordism by our legislation. Is a canal, a railway, or a turnpike road to be made? We do not scruple to seize just as many acres as may be requisite. We do not wait for consent.... The change required would simply be a change of landlords.... Instead of being in the possession of individuals, the country would be held by the great corporate body—society. Instead of leasing his acres from an isolated proprietor, the farmer would lease them from the nation. Instead of paying his rent to the agent of Sir John, or His Grace, he will pay to an agent, or deputy-agent of the community. Stewards would be public officials, instead of private ones, and tenantry the only land tenure.... Pushed to its ultimate consequences, a claim to exclusive possession of the soil involves landowning despotism."

Thus, from the very point of view of modern English political economists, it is not the usurping English landlord but the Irish tenants and labourers, who have the only right in the soil of their native country, and *The Times*, in opposing the demands of the Irish people, places itself into direct antagonism to British middle-class science.

Published in *The New-York Daily Tribune* No. 3816, July 11, 1853

Printed according to the text of the newspaper

Karl Marx

From FINANCIAL FAILURE OF GOVERNMENT— CABS—IRELAND—THE RUSSIAN QUESTION

Like the world in general, we are assured that Ireland in particular is becoming a paradise for the labourer, in consequence of famine and exodus. Why then, if wages really are so high in Ireland, is it that Irish labourers are flocking in such masses over to England to settle permanently on this side of the "pond",[41] while they formerly used to return after every harvest? If the social amelioration of the Irish people is making such progress, how is it that, on the other hand, insanity has made such terrific progress among them since 1847, and especially since 1851? Look at the following data from "the Sixth Report on the District Criminal and Private Lunatic Asylums in Ireland":

1851 — Sum total of admissions in Lunatic Asylums. (1,301 males and 1,283 females.)	2,584
1852 . . . . (1,276 males and 1,386 females.)	2,662
March, 1853 . . . . (1,447 males and 1,423 females.)	2,870

And this is the same country in which the celebrated Swift, the founder of the first Lunatic Asylum in Ireland,[42] doubted whether 90 madmen could be found.

Published in *The New-York Daily Tribune* No. 3844, August 12, 1853

Printed according to the text of the newspaper

Karl Marx

From THE WAR QUESTION—BRITISH POPULATION AND TRADE RETURNS—DOINGS OF PARLIAMENT

In its sitting of Aug. 9, the House of Lords had to decide on the fate of three Ireland Bills, carried through the Commons after ten months' deliberation, viz.: the *Landlord and Tenant Bill*, removing the laws concerning mortgages, which form at present an insuperable bar to the effective sale of the smaller estates not falling under the *Encumbered Estates Act*[43]; the *Leasing Powers Bill*, amending and consolidating more than sixty acts of Parliament which prohibit leases to be entered into for 21 years, regulating the tenant's compensation for improvements in all instances where contracts exist, and preventing the system of sub-letting; lastly, the *Tenant's Improvement Compensation Bill*, providing compensation for improvements effected by the tenant in the absence of any contract with the landlord, and containing a clause for the retrospective operation of this provision. The House of Lords could, of course, not object to parliamentary interference between landlord and tenant, as it has laden the statute book from the time of Edward IV to the present day, with acts of legislation on landlord and tenant, and as its very existence is founded on laws meddling with landed property, as for instance the *Law of Entail*. This time, the noble lords sitting as judges on their own cause, allowed themselves to run into a passion quite surprising in that hospital of invalids.

"Such a bill," exclaimed the Earl of Clanricarde, "as the Tenants' Compensation Bill, such a total violation and disregard of all contracts, was never before, he believed, submitted to Parliament, nor had he ever heard of any government having ventured to propose such a measure as was carried out in the retrospective clauses of the bill."

The Lords went as far as to threaten the Crown with the withdrawal of their feudal allegiance, and to hold out the prospect of a landlord rebellion in Ireland.

"The question," remarked the same nobleman, "touched nearly the *whole question of the loyalty* and confidence of the landed proprietors in Ireland in the Government of this country. If they saw landed property in Ireland treated in such a way, he would like to know what was to *secure their attachment to the Crown, and their obedience to its supremacy?*"

Gently, my lord, gently! What was to secure their obedience to the supremacy of the Crown? One magistrate and two constables. A landlord rebellion in Great Britain! Has there ever been uttered a more monstrous anachronism? But for a long time the poor Lords have only lived upon anachronisms. They naturally encourage themselves to resist the House of Commons and public opinion.

"Let not their lordships," said old Lord St. Leonards, "for the sake of preventing what was called a *collision* with the other House, or for the sake of popularity, or on account of a pressure from without, pass imperfect measures like these." "I do not belong to any party," exclaimed the Earl of Roden, "but I am highly interested in the welfare of Ireland."

That is to say, his lordship supposes Ireland to be highly interested in the welfare of the Earl of Roden. "This is no party question, but a Lords' question," was the unanimous shout of the House; and so it was. But between both parties, Whig Lords and Tory Lords, Coalition Lords[44] and Opposition Lords, there has existed from the beginning a secret understanding to throw the bills out, and the whole impassioned discussion was a mere farce, performed for the benefit of the newspaper reporters.

This will be evident when we remember that the bills which formed the subject of so hot a controversy were originated, not by the Coalition Cabinet, but by Mr. Napier, the Irish Attorney-General under the Derby Ministry, and that the Tories at the last elections in Ireland appealed to the testimony of these bills introduced by them. The only substantial change made by the House of Commons in the measures introduced by the Tory Government was the excluding of the growing crops from being distrained upon.

"The bills are not the same," exclaimed the Earl of Malmesbury, asking the Duke of Newcastle whether he did not believe him. "Certainly not," replied the Duke. "But whose assertion would you then believe?" "That of Mr. Napier," answered the Duke. "Now," said the Earl, "here is a letter from Mr. Napier, stating that the bills are not the same." "There," said the Duke, "is another letter of Mr. Napier, stating that they are."

If the Tories had remained in, the Coalition Lords would have opposed the Ireland Bills. The Coalition being in, on the Tories fell the task of opposing their own measures. The Coalition having inherited these bills from the Tories and having introduced the Irish party into their own cabinet, could, of course, not oppose the bills in the House of Commons; but they were sure of their being burked in the House of Lords. The Duke of Newcastle made a faint resistance but Lord Aberdeen declared himself contented with the bills passing formally through a second reading, and being really thrown out for the session. This accordingly was done. Lord Derby, the chief of the late ministry, and Lord Lansdowne, the nominal President of the present ministry, yet at the same time one of the largest proprietors of land in Ireland, managed, wisely, to be absent from indisposition.

Published in *The New-York Daily Tribune* No. 3854, August 24, 1853

Printed according to the text of the newspaper

From LORD PALMERSTON[45]

Let us now look at his exertions for Catholic Emancipation,[46] one of his great "claims" on the gratitude of the Irish people. I shall not dwell upon the circumstances, that, having declared himself for Catholic Emancipation, when a member of the Canning Ministry, he entered, nevertheless, the Wellington Ministry, avowedly hostile to that emancipation. Perhaps Lord Palmerston considered religious liberty as one of the Rights of Man, not to be intermeddled with by the Legislature. He may answer for himself,

"Although I wish the Catholic claims to be considered, I never will admit those claims to stand upon the ground of right.... If I thought the Catholics were asking for their right, I, for one, would not go into the committee." (*House of Commons,* March 1, 1813.)

And why is he opposed to their asking their right?

"Because the Legislature of a country has the right to impose such political disabilities upon any class of the community, as it may deem necessary for the safety and the welfare of the whole.... This belongs to the fundamental principles on which civilised government is founded." (*House of Commons,* March 1, 1813.)

There you have the most cynic confession ever made, that the mass of the people have no rights at all, but that they may be allowed that amount of immunities, the Legislature —or, in other words, the ruling class—may deem fit to grant them. Accordingly, Lord Palmerston declared in plain words, "Catholic Emancipation to be a measure of grace and favour". (*House of Commons,* Feb. 10, 1829.)

It was then entirely upon the ground of expediency that

he condescended to discontinue the Catholic disabilities. And what was lurking behind this expediency?

Being himself one of the great Irish proprietors, he wanted to entertain the delusion, that other remedies for Irish evils than Catholic Emancipation are impossible, that it would cure absenteeism, and prove a substitute for Poor Laws. (*House of Commons*, March 18, 1829.)

The great philanthropist, who afterwards cleared his Irish estates of their Irish natives, could not allow Irish misery to darken, even for a moment, with its inauspicious clouds, the bright sky of the landlords and moneylords.*

"It is true," he said, "that the peasantry of Ireland do not enjoy all the comforts which are enjoyed by all the peasantry of England" (only think of all the comforts enjoyed by a family at the rate of 7s. a week). Still, he continues, "still, however, the Irish peasant has his comforts. He is well supplied with fuel, and is seldom" (only four days out of six) "at a loss for food."

What a comfort! But this is not all the comfort he has— "he has a greater cheerfulness of mind than his English fellow-sufferer!" (*House of Commons*, May 7, 1829.)

As to the extortions of Irish landlords, he deals with them in as pleasant a way as with the comforts of the Irish peasantry.

"It is said that the Irish landlord insists on the highest possible rent that can be extorted. Why, Sir, I believe that is not a singular circumstance; certainly in England the landlord does the same thing." (*House of Commons*, March 7, 1829.)

Are we then to be surprised that the man, so deeply initiated in the mysteries of the "glories of the English constitution", and the "comforts of her free institutions", should aspire at spreading them all over the Continent?

Published in *The People's Paper* No. 77, October 22, 1853, and in *The New-York Daily Tribune* No. 3902, October 19, 1853

Printed according to the text of *The People's Paper* and verified with *The New-York Daily Tribune*

---

* In the version of this article which appeared in *The New-York Daily Tribune* of October 19, 1853, Marx worded the end of the sentence as follows: "The bright sky over the Parliament of landlords and moneylords."—*Ed.*

From [THE BLUE BOOKS—PARLIAMENTARY DEBATES OF FEBRUARY 6...—THE IRISH BRIGADE]

Mr. I. Butt, in yesterday's sitting of the Commons, gave notice

"that to-morrow he should move that there should be read by the clerk, at the table of the House, an article published in *The Times* of to-day, and the previous statements of *The Dublin Freeman's Journal,* imputing to the [*Irish*] members of the House a trafficking in places for money. He should also move for a Select Committee to inquire into the allegations of such trafficking as contained in these publications".

Why Mr. Butt is indignant only at the trafficking for money will be understood by those who remember that the legality of any other mode of trafficking was settled during last session. Since 1830 Downing-st. has been placed at the mercy of the Irish Brigade.[47] It is the Irish members who have created and kept in place the Ministers to their mind. In 1834 they drove from the Cabinet Sir J. Graham and Lord Stanley. In 1835 they compelled William IV to dismiss the Peel Ministry and to restore the Melbourne Administration. From the general election of 1837 down to that of 1841, while there was a British majority in the Lower House opposed to that Administration, the votes of the Irish Brigade were strong enough to turn the scale and keep it in office. It was the Irish Brigade again who installed the Coalition Cabinet. With all this power of Cabinet-making, the Brigade have never prevented any infamies against their own country nor any injustice to the English people. The period of their greatest power was at the time of O'Connell, from 1834-1841. To what account was it turned? The Irish agitation was never anything but a cry for the Whigs against the Tories, in order to extort places from the Whigs. Nobody who knows anything about the so-called Lichfield-House Contract,[48] will differ from this opinion—that contract by which O'Connell was to vote *for*, but licensed to spout *against*, the Whigs on condition that he should nominate his own Magistrates in Ireland. It is time for the Irish Brigade to put off their patriotic airs. It is time for the Irish people to put off their dumb hatred of the English and call their own representatives to an account for their wrongs.

Published in *The New-York Daily Tribune* No. 4008, February 21, 1854

Printed according to the text of the newspaper

Karl Marx

IRELAND'S REVENGE[49]

London, March 13. Ireland has revenged herself upon
England, *socially*—by bestowing an *Irish quarter* on every
English industrial, maritime or commercial town of any
size, and *politically*—by furnishing the English Parliament
with an "Irish Brigade". In 1833, Daniel O'Connell decried
the Whigs as "base, bloody and brutal". In 1835, he became
the most efficient tool of the Whigs; although the English
majority was opposed to the Melbourne Administration, it
remained in office from April 1835 to August 1841 because
of the support it received from O'Connell and his Irish
Brigade. What intervened between O'Connell of 1833 and
O'Connell of 1835? An agreement, known as the *Lichfield-
House Contract*, according to which the Whig Cabinet
granted O'Connell government "patronage" in Ireland, and
O'Connell promised the Whig Cabinet the votes of the Irish
Brigade in Parliament. "King Dan's" Repeal agitation[50]
began immediately the Whigs were overthrown, but as soon
as the Tories were defeated "King Dan" sank again to the
level of a common advocate. The influence of the Irish
Brigade by no means came to an end with O'Connell's death.
On the contrary, it became evident that this influence did
not depend on the talent of one person, but was a result of
the general state of affairs. The Tories and Whigs, the big
traditional parties in the English Parliament, were more or
less equally balanced. It is thus not surprising that the new,
numerically small factions, the Manchester School[51] and the
Irish Brigade, which took their seats in the reformed parliament, should play a decisive role and be able to turn the
scale. Hence the importance of the "Irish quarter" in the
English Parliament. After O'Connell left the scene it was

no longer possible to stir the Irish masses with the "Repeal" slogan. The "Catholic" problem, too, could be used only occasionally. Since the Catholic Emancipation[52] it could no longer serve as a permanent propaganda theme. Thus the Irish politicians were compelled to do what O'Connell had always avoided and refused to do, that is, to explore the real cause of the Irish malady and to make the relations of landed property and their reform the election slogan, in other words a slogan that would help them to get into the House of Commons. But having taken their seats in the House, they used the rights of the tenants, etc.—just as formerly the Repeal—as a means to conclude a new Lichfield-House Contract.

The Irish Brigade had overthrown the Derby ministry and had obtained a seat, even though a minor one, in the coalition government. How did it use its position? It helped the coalition to burke measures designed to reform landed ownership in Ireland. The Tories themselves, having taken the patriotism of the Irish Brigade for granted, had decided to propose these measures in order to gain the support of the Irish M.P.s Palmerston, who is an Irishman by birth and knows his "Irish quarter", has renewed the Lichfield-House Contract of 1835 on an all-embracing basis. He has appointed Keogh, the chief of the Brigade, Attorney-General of Ireland, Fitzgerald, also a liberal Catholic M.P. for Ireland, has been made Solicitor-General, and a third member of the Brigade has become legal counsel to the Lord Lieutenant of Ireland, so that the juridical general staff of the Irish government is now composed entirely of Catholics and Irishmen. Monsell, the Clerk of Ordnance in the coalition government, has been reappointed by Palmerston after some hesitation, although—as Muntz, deputy for Birmingham and an arms manufacturer, rightly observed— Monsell cannot distinguish a musket from a needle-gun. Palmerston has advised the lieutenants of the counties always to give preference to the protégés of Irish priests close to the Irish Brigade when nominating colonels and other high-ranking officers in the Irish militia. That Palmerston's policy is already exerting an influence is evident from the fact that Sergeant *Shee* has gone over to the government side, and also from the fact that the

Catholic Bishop of Athlone has pushed through the re-election of Keogh and that moreover the Catholic clergy has promoted the re-election of Fitzgerald. Wherever the lower ranks of the Catholic clergy have taken their "Irish patriotism" seriously and have stood up to those members of the Irish Brigade who deserted to the government, they have been rebuked by their bishops who are well aware of the diplomatic secret.

A protestant Tory newspaper bemoans the "complete congruity existing between Lord Palmerston and the Irish clergy. When Palmerston hands over Ireland to the priests, the priests will elect M.P.s who will hand over England to Lord Palmerston."

The Whigs use the Irish Brigade to dominate the English Parliament and they toss posts and salaries to the Brigade; the Catholic clergy permits the one to buy and the other to sell on condition that both acknowledge the power of the clergy and help to extend and strengthen it. It is, however, a very remarkable phenomenon that in the same measure as the Irish influence in the *political* sphere grows in England, the Celtic influence in the *social* sphere decreases in Ireland. Both the "Irish quarter" in Parliament and the Irish clergy seem to be equally unaware of the fact that behind their back the Irish society is being radically transformed by an Anglo-Saxon revolution. In the course of this revolution *the Irish agricultural system is being replaced by the English system, the system of small tenures by big tenures,* and *the modern capitalist* is taking the place of the old landowner.

The chief factors which prepared the ground for this transformation are: 1847, the year of famine,[53] which killed nearly one million Irishmen; emigration to America and Australia, which removed another million from the land and still carries off thousands; the unsuccessful insurrection of 1848,[54] which finally destroyed Ireland's faith in herself; and lastly the Act of Parliament which exposed the estates of the debt-ridden old Irish aristocrats to the hammer of the auctioneer or bailiff,[55] thus driving them from the land just as starvation drove away their small tenants, sub-tenants and cottagers.

Published in *Neue Oder-Zeitung*
No. 127, March 16, 1855

Translated from the German

Karl Marx

[FROM PARLIAMENT]

[EXCERPT]

For two years Parliament, as is well known, has been considering three bills designed to regulate the relations of Irish landowners and tenants. One of the bills determines the amount of compensation which the tenant should be entitled to claim for the improvements he made on the land, in the event of the landowner terminating the lease. Hitherto, all improvements made by Irish tenants—most of whom hold a temporary lease concluded for one year— have merely enabled the landowner to demand a higher rent on the expiration of the existing lease. Thus the tenant either loses the farm, if he does not wish to renew the lease under less favourable conditions, and with the farm he loses the capital he has invested in the improvements, or he is compelled to pay the landlord, in addition to the original rent, interest on the improvements made with his (the tenant's) capital. Support for the earlier mentioned bills was one of the arrangements with which the coalition cabinet purchased the votes of the Irish Brigade. In 1854, therefore, they were passed by the House of Commons, but the House of Lords with the connivance of the Ministers shelved them till the next session (in 1855) and then amended them in such a way that their point was blunted, sending them back to the House of Commons in this distorted form. There the main clause of the Compensation Bill was sacrificed on the altar of landed property last Thursday, and the Irish were astonished to see that the scales had been turned against them partly by the votes of members of the government and partly by the votes of those directly associated with them. Sergeant Shee's furious attack on Palmerston threatened to unleash a riot in the "Irish quarter" which at this moment could have serious consequences. Palmerston therefore negotiated with the help of Sadleir, an ex-member of the coalition and middleman of the Irish Brigade. He arranged for a deputation of 18 Irish M.P.s to visit him the day before yesterday to enquire whether he was willing to use his influence to have the parliamentary vote rescinded and to carry the clause through the House of Commons in another division. Palmerston, of course, is ready to promise anything in order to secure the support of the Irish Brigade during the vote on the no-confidence motion. The premature exposure of this intrigue in the House of Commons gave rise to one of those scandalous scenes typical of the decline of the oligarchic parliament. The Irish have 105 votes, but it became known that the majority of M.P.s had not authorised the deputation of 18. Altogether, Palmerston is no longer able to use the Irish during government crises in quite the same way as in O'Connell's time. Along with the dissolution of the old established parliamentary factions, the "Irish quarter", too, crumbles and disintegrates. In any case, the incident shows how Palmerston makes use of the time won to influence the various cliques. At the same time he waits for some favourable news from the theatre of war, some small incident which can be exploited in the parliamentary sphere, if not in the military. The submarine telegraph has wrenched the direction of the war from the hands of the generals and made it dependent on the amateurish astrological whims of Bonaparte and on parliamentary and diplomatic intrigues. Hence the inexplicable and quite unparalleled character of the second Crimean campaign.[56]

Published in *Neue Oder-Zeitung*   Translated from the
No. 325, July 16, 1855   German

Karl Marx

From LORD JOHN RUSSELL[57]

IV

London, August 4. With the outbreak of the anti-Jacobin war,[58] the influence of the Whigs in *England* began to ebb lower and lower. They therefore turned to Ireland, decided to throw *her* into the scales and wrote *Irish Emancipation* on their party standard. When they temporarily stepped into office in 1806, they introduced a small Irish Emancipation Bill and carried it through the second reading in the House of Commons, but then they withdrew it to flatter the bigot idiocy of George III. In 1812, they sought, though in vain, to thrust themselves on the Prince Regent (later George IV) on the ground that only they could bring about a reconciliation with Ireland. Before and during the Reform agitation they fawned upon O'Connell, and the "hopes raised in Ireland" were for them merely powerful instruments to be used for party purposes. Yet their first act at the first meeting of the reformed parliament was a declaration of war against Ireland, a "brutal and bloody measure", the Irish Coercion Bill,[59] which imposed martial law on Ireland.* The Whigs fulfilled their old promise with "fire, imprisonment, transportation and even with death". O'Connell was prosecuted for sedition.[60] But they introduced and carried the Coercion Bill only on the express stipulation that they would bring in an *Irish Church Bill*, with a clause stipulating that a certain *portion* of the revenue of

---

* The last part of the sentence reads as follows in *The New-York Daily Tribune*: "a declaration of civil war against Ireland, a 'brutal and bloody measure', the Irish Coercion 'Red-Coat Tribunal Bill', according to which men were to be tried in Ireland by military officers, instead of by Judges and Juries."—*Ed.*

the Established Church in Ireland should be placed at the disposal of Parliament and that Parliament was to use it for the benefit of Ireland. The importance of this clause flows from the fact that it acknowledges the principle that Parliament has the power to expropriate the Established Church, a principle Lord John Russell certainly ought to be convinced of since the whole immense property of his family consists of church plunder. As soon as the Coercion Bill had been passed, the Whigs, though they had engaged to stand or fall by the Church Bill, hastened—on the ground of avoiding a collision with the Lords—to take out that very clause, the only part of the Bill of any value at all. They then voted against and defeated their own measure. This happened in 1834. But towards the end of the year the Whigs' sympathies for the Irish were aroused again as if by an electric shock. For Sir Robert Peel came into office in the autumn of 1834 and the Whigs had to retire to the Opposition benches. And immediately we see our John Russell busily engaged in working on reconciliation with Ireland. He was the principal agent in bringing about, in January 1835, the *Lichfield-House Compact*,[61] through which the Whigs surrendered to O'Connell the Irish patronage (the right to distribute offices, etc.), and O'Connell secured to them the Irish votes, both inside Parliament and out of it. But a *pretext* for ejecting the Tories from Downing Street was needed. With characteristic "impudence", Russell chose the *Ecclesiastical Revenues* of Ireland as the battlefield and used the very *clause*—it became notorious under the name "*appropriation clause*"—which he and his colleagues in the Reform Ministry had withdrawn and abandoned a short time ago, as a war-cry. Under the slogan of the "appropriation clause" Peel was defeated. The Melbourne Cabinet was formed, and Lord John Russell became Home Secretary and Leader of the House of Commons. He now began to boast on the one hand of his mental firmness, for although now in office he still adhered to his *views* on the appropriation clause, and on the other hand of his moral moderation in not *acting* upon those views. He never acted upon them. In 1846, when he was Prime Minister, his moral moderation overcame his mental firmness to such an extent that he renounced his "views" too. He professed that he could not conceive of a more fatal measure than those endangering the revenues, the essential root of the Established Church.

In February 1833, John Russell as a member of the Reform Ministry denounced Irish *Repeal*, and stated in the House of Commons that the real object of the agitation was

"to overturn at once the United Parliament, and to establish, in place of King, Lords, and Commons of the United Kingdom, some parliament of which Mr. O'Connell was to be the leader and the chief".

In February 1834, the Repeal agitation was again denounced in the King's Speech, and the Reform Ministry proposed an address

"to record in the most solemn manner the fixed determination of Parliament to maintain unimpaired and undisturbed the legislative union".

Immediately on being shifted to the Opposition benches, the very same John Russell declared that

"with respect to the *Repeal of the Union*, the subject was open to amendment or question, like any other act of the legislature",

that is, neither more nor less than any Bill dealing with beer.

In March 1846, Lord John Russell in alliance with the Tories, then burning with the passion to punish Peel for the repeal of the Corn Laws,[62] broke up Peel's administration. The pretext was Peel's Irish *Arms Bill*, against which the morally outraged Lord John Russell resolutely protested. He became Premier, and his first act was an attempt to renew that same Bill. But he exposed himself to ridicule without achieving any result. O'Connell had just conjured up huge protest meetings against Peel's Bill and had obtained 50,000 signatures to petitions—he was in Dublin where he brought all the means of agitation into play. King Dan (as Daniel O'Connell was generally called) would have lost empire and revenue if at this moment he appeared as Russell's accomplice. He therefore angrily told the little man to withdraw his Arms Bill immediately. Russell withdrew it. Despite his secret association with the Whigs, O'Connell added humiliation to the defeat, and in this he was a past master. To make it quite obvious *who had ordered* the retreat, O'Connell announced the withdrawal of the Arms Bill to the Repealers at the Conciliation Hall in Dublin on August 17, that is, on the *very day* John Russell announced it in the House of Commons. In 1844, Russell had charged Sir Robert Peel with "having filled Ireland with troops, and with not governing but militarily occupying that country". In 1848, Russell occupied Ireland militarily, passed the Felony Acts, proclaimed the suspension of the Habeas Corpus Act,[63] and bragged about the "vigorous measures" of Clarendon.[64] This display of energy, too, was a pretence. In Ireland there were on the one side the O'Connellites and priests acting in collusion with the Whigs, and on the other side Smith O'Brien and his followers.[65] The latter were simply dupes who took the Repeal game seriously, and hence came to a comic end. The "vigorous measures" taken by the Russell Government and the brutality employed were thus not demanded by the circumstances. Their aim was, not the maintenance of the English rule in Ireland, but the prolongation of the Whig regime in England.

Published in *Neue Oder-Zeitung* No. 365, August 8, 1855. A shortened English version was published in *The New-York Daily Tribune* of August 28, 1855

Translated from the German, but analogous passages from the English version have been used

May 23, 1856

Dear Marx,

During our tour in Ireland[66] we came from Dublin to Galway on the west coast, then twenty miles north inland, then to Limerick, down the Shannon to Tarbert, Tralee, Killarney and back to Dublin—a total of about 450 to 500 English miles inside the country itself, so that we have seen about two-thirds of the whole of it. With the exception of Dublin, which bears the same relation to London as Düsseldorf does to Berlin and has quite the character of a small one-time capital, all English-built, too, the look of the entire country, and especially of the towns, is as if one were in France or Northern Italy. Gendarmes, priests, lawyers, bureaucrats, country squires in pleasing profusion and a total absence of any industry at all, so that it would be difficult to understand what all these parasitic growths live on if the distress of the peasants did not supply the other half of the picture. "Strong measures" are visible in every corner of the country, the government meddles with everything, of so-called self-government there is not a trace. Ireland may be regarded as England's first colony and as one which, because of its proximity, is still governed exactly in the old way, and one can already notice here that the so-called liberty of English citizens is based on the oppression of the colonies. I have never seen so many gendarmes in any country, and the sodden look of the bibulous Prussian gendarme is developed to its highest perfection here among the constabulary, who are armed with carbines, bayonets and handcuffs.

Characteristic of this country are its ruins, the oldest dating from the fifth and sixth centuries, the latest from

the nineteenth—with every intervening period. The most
ancient are all churches; after 1100, churches and castles;
after 1800, houses of peasants. The whole of the west,
especially in the neighbourhood of Galway, is covered with
ruined peasant houses, most of which have only been
deserted since 1846. I never thought that famine could have
such tangible reality.[67] Whole villages are devastated, and
there among them lie the splendid parks of the lesser
landlords, who are almost the only people still living there,
mostly lawyers. Famine, emigration and clearances together
have accomplished this. There are not even cattle to be
seen in the fields. The land is an utter desert which nobody
wants. In County Clare, south of Galway, it is somewhat
better. Here there are at least cattle, and the hills towards
Limerick are excellently cultivated, mostly by Scottish
farmers, the ruins have been cleared away and the country
has a bourgeois appearance. In the south-west there are a
lot of mountains and bogs but there is also wonderfully
luxuriant forest land; beyond that again fine pastures,
especially in Tipperary, and towards Dublin there is land
which, one can see, is gradually coming into the hands of
big farmers.

The country was completely ruined by the English wars
of conquest from 1100[68] to 1850 (for in reality both the
wars and the state of siege lasted as long as that). It has
been established as a fact that most of the ruins were produced by destruction during the wars. The people itself has
got its peculiar character from this, and for all their national
Irish fanaticism the fellows feel that they are no longer
at home in their own country. Ireland for the Saxon! That
is now being realised. The Irishman knows that he cannot
compete with the Englishman, who comes equipped with
means superior in every respect; emigration will go on until
the predominantly, indeed almost exclusively, Celtic
character of the population is gone to the dogs. How often
have the Irish started out to achieve something, and
every time they have been crushed, politically and industrially. By consistent oppression they have been artificially
converted into an utterly impoverished nation and now, as
everyone knows, fulfil the function of supplying England,
America, Australia, etc., with prostitutes, casual labourers,

pimps, pickpockets, swindlers, beggars and other rabble. Impoverishment characterises the aristocracy too. The land-owners, who everywhere else have become bourgeoisified, are here reduced to complete poverty. Their country-seats are surrounded by enormous, amazingly beautiful parks, but all around is waste land, and where the money is to come from it is impossible to see. These fellows are droll enough to make your sides burst with laughing. Of mixed blood, mostly tall, strong, handsome chaps, they all wear enormous moustaches under colossal Roman noses, give themselves the false military airs of retired colonels, travel around the country after all sorts of pleasures, and if one makes an inquiry, they haven't a penny, are laden with debts, and live in dread of the Encumbered Estates Court.[69]

Concerning the ways and means by which England rules this country—repression and corruption—long before Bonaparte attempted this, I shall write shortly if you won't come over soon. How about it?

Yours,

F. E.

Published in *Der Briefwechsel zwischen F. Engels und K. Marx*, Bd. II, Stuttgart, 1913

Translated from the German

Karl Marx

From THE QUESTION OF THE IONIAN ISLANDS[70]

According to his oracle in Printing-House Square,[71] he grasps after colonies only in order to educate them in the principles of public liberty; but, if we adhere to facts, the Ionian Islands, like India and Ireland, prove only that to be free at home, John Bull must enslave abroad. Thus, at this very moment, while giving vent to his virtuous indignation against Bonaparte's spy system at Paris, he is himself introducing it at Dublin.

Published in *The New-York Daily Tribune* No. 5526, January 6, 1859

Printed according to the text of the newspaper

THE EXCITEMENT IN IRELAND

London, Dec. 24, 1858

A Government, representing, like the present British Ministry, a party in decay, will always better succeed in getting rid of its old principles, than of its old connections. When installing himself at Downing street, Lord Derby, doubtless, made up his mind to atone for the blunders which in times past had converted his name into a byword in Ireland; and his versatile Attorney-General for Ireland, Mr. Whiteside, would not one moment hesitate flinging to the wind the oaths that bound him to the Orange Lodges.[72] But, then, Lord Derby's advent to power gave, simultaneously, the signal for one coterie of the governing class to rush in and fill the posts just vacated by the forcible ejection of the other coterie. The formation of the Derby Cabinet involved the consequence that all Government places should be divided among a motley crew still united by a party name which has become meaningless, and still marching under a banner torn to tatters, but in fact having nothing in common save reminiscences of the past, club intrigues, and, above all, the firm resolution to share together the loaves and fishes of office. Thus, Lord Eglinton, the Don Quixote who wanted to resuscitate the tournaments of chivalry in money-mongering England, was to be enthroned Lord Lieutenant at Dublin Castle,[73] and Lord Naas, notorious as a reckless partisan of Irish landlordism, was to be made his First Minister. The worthy couple, *arcades ambo*, on leaving London, were, of course, seriously enjoined by their superiors to have done with their crotchets, to behave properly, and by no capricious pranks to upset their own employers. Lord Eglinton's path across the channel was, we do not doubt,

paved with good intentions, the vista of the Vice-royal
baubles dancing before his childish mind; while Lord Naas,
on his arrival at Dublin Castle, was determined to satisfy
himself that the wholesale clearance of estates, the burning
down of cottages, and the merciless unhousing of their poor
inmates were proceeding at the proper ratio. Yet as party
necessities had forced Lord Derby to instal wrong men in
the wrong place, party necessities falsified at once the position
of those men, whatever their individual intentions might be.
Orangeism had been officially snubbed for its intruding
loyalty, the Government itself had been compelled to
denounce this organisation as illegal, and very unceremoniously it was told that it was no longer good for any earthly
purpose, and that it must vanish. The mere advent of a
Tory Government, the mere occupancy of Dublin Castle
by an Eglinton and a Naas revived the hopes of the chop-
fallen Orangemen. The sun shone again on the "true blues";
they would again lord it over the land as in the days of
Castlereagh, and the day for taking their revenge had
visibly dawned. Step by step, they led the bungling, weak,
and, therefore, temerarious representatives of Downing
street from one false position to the other, until one fine
morning at last, the world was startled by a proclamation
of the Lord Lieutenant, placing Ireland (so to say) in a
state of siege, and turning, through the means of £100 and
£50 rewards, the trade of the spy, the informer, the perjur-
er, and the *agent provocateur* into the most profitable trade
in Green Erin. The placards announcing rewards for the
detection of secret societies were hardly posted, when an
infamous fellow, named O'Sullivan, an apothecary's apprentice at Killarney, denounced his own father and some boys
of Killarney, Kenmare, Bantry, Skibbereen, as members of
a formidable conspiracy which, in secret understanding with
filibusters from the other side of the Atlantic, intended not
only, like Mr. Bright, to "Americanise English institutions",
but to annex Ireland to the model Republic. Consequently,
detectives busied themselves in the Counties of Kerry and
Cork, nocturnal arrests took place, mysterious informations
went on; from the south-west the conspiracy hunting spread
to the north-east, farcical scenes occurred in the County
of Monaghan, and alarmed Belfast saw some dozen of

schoolmasters, attorneys' clerks and merchants' clerks paraded through the streets and locked up in the jails. What rendered the thing worse was the veil of mystery thrown over the judicial proceedings. Bail was declined in all cases, midnight surprises became the order of the day, all the inquisitions were kept secret, copies of the informations on which the arbitrary arrests had been made were regularly refused, the stipendiary magistrates were whirling up and down from their judicial seats to the ante-chambers of Dublin Castle, and of all Ireland might be said, what Mr. Rea, the counsel for the defendants at Belfast, remarked with respect to that place, "I believe the British Constitution has left Belfast this last week."

Now, through all this hubbub and all this mystery, there transpires more and more the anxiety of the Government, that had given way to the pressure of its credulous Irish agents, who, in their turn, were mere playthings in the hands of the Orangemen, how to get out of the awkward fix without losing at once their reputation and their places. At first, it was pretended that the dangerous conspiracy, extending its ramifications from the south-west to the north-east over the whole surface of Ireland, issued from the Americanising Phoenix Club.[74] Then it was a revival of Ribbonism[75]; but now it is something quite new, quite unknown, and the more awful for all that. The shifts the Government is driven to may be judged from the manoeuvres of *The Dublin Daily Express*, the Government organ, which day by day treats its readers to false rumours of murders committed, armed men marauding, and midnight meetings taking place. To its intense disgust, the men killed return from their graves, and protest in its own columns against being so disposed of by the editor.

There may exist such a thing as a Phoenix Club, but at all events, it is a very small affair, since the Government itself has thought fit to stifle this Phoenix in its own ashes. As to Ribbonism, its existence never depended upon secret conspirators. When, at the end of the eighteenth century, the Protestant Peep-o'-Day boys combined to wage war against the Catholics in the north of Ireland, the opposing society of the Defenders[76] sprang up. When, in 1791, the Peep-o'-Day boys merged into Orangeism, the Defenders transformed themselves into Ribbonmen. When, at last, in our own days, the British Government disavowed Orangeism, the Ribbon Society, having lost its condition of life, dissolved itself voluntarily. The extraordinary steps taken by Lord Eglinton may, in fact, revive Ribbonism, as may the present attempts of the Dublin Orangemen to place English officers at the head of the Irish Constabulary, and fill its inferior ranks with their own partisans. At present there exist no secret societies in Ireland except agrarian societies. To accuse Ireland of producing such societies would be as judicious as to accuse woodland of producing mushrooms. The landlords of Ireland are confederated for a fiendish war of extermination against the cotters; or, as they call it, they combine for the economical experiment of clearing the land of useless mouths. The small native tenants are to be disposed of with no more ado than vermin is by the housemaid. The despairing wretches, on their part, attempt a feeble resistance by the formation of secret societies, scattered over the land, and powerless for effecting anything beyond demonstrations of individual vengeance.

But if the conspiracy hunted after in Ireland is a mere invention of Orangeism, the premiums held out by the Government may succeed in giving shape and body to the airy nothing. The recruiting sergeant is no more sure to press with his shilling and his gin some of the Queen's mob into the Queen's service, than a reward for the detection of Irish secret societies is sure to create the societies to be detected. From the entrails of every county there rise immediately blacklegs who, transforming themselves into revolutionary delegates, travel through the rural districts, enrol members, administer oaths, denounce the victims, swear them to the gallows, and pocket the blood-money. To characterise this race of Irish informers and the effect on them of Government rewards, it will suffice to quote one passage from a speech delivered by Sir Robert Peel in the House of Commons:

"When I was Chief Secretary of Ireland, a murder was committed between Carrick-on-Suir and Clonmel. A Mr.——had a deadly revenge toward a Mr. ——, and he employed four men at two guineas each to murder him. There was a road on each side of the River Suir, from Carrick to Clonmel; and placing two men on each road, the escape of his victim was impossible. He was, therefore, foully murdered, and the country was so shocked by this heinous crime, that the Government offered a reward of £500 for the discovery of each of the murderers. And can it be believed, the miscreant who bribed the four murderers was the very man who came and gave the information which led to their execution, and with these hands I paid in my office in Dublin Castle the sum of £2,000 to that monster in human shape."

Published in *The New-York Daily Tribune* No. 5530, January 11, 1859

Printed according to the text of the newspaper

Karl Marx

From POPULATION, CRIME AND PAUPERISM

There must be something rotten in the very core of a social system which increases its wealth without diminishing its misery, and increases in crimes even more rapidly than in numbers. It is true enough that, if we compare the year 1855 with the preceding years, there seems to have occurred a sensible decrease of crime from 1855 to 1858. The total number of people committed for trial, which in 1854 amounted to 29,359, had sunk down to 17,855 in 1858; and the number of convicted had also greatly fallen off, if not quite in the same ratio. This apparent decrease of crime, however, since 1854, is to be exclusively attributed to some technical changes in British jurisdiction; to the Juvenile Offenders' Act[77] in the first instance, and, in the second instance, to the operation of the Criminal Justice Act of 1855, which authorises the Police Magistrates to pass sentences for short periods, with the consent of the prisoners. Violations of the law are generally the offspring of economical agencies beyond the control of the legislator, but, as the working of the Juvenile Offenders' Act testifies, it depends to some degree on official society to stamp certain violations of its rules as crimes or as transgressions only. This difference of nomenclature, so far from being indifferent, decides on the fate of thousands of men, and the moral tone of society. Law itself may not only punish crime, but improvise it, and the law of professional lawyers is very apt to work in this direction. Thus, it has been justly remarked by an eminent historian, that the Catholic clergy of the medieval times, with its dark views of human nature, introduced by its influence into criminal legislation, has created more crimes than forgiven sins.

Strange to say, the only part of the United Kingdom in which crime has seriously decreased, say by 50, and even by 75 per cent, is Ireland. How can we harmonise this fact with the public-opinion slang of England, according to which Irish nature, instead of British misrule, is responsible for Irish shortcomings? It is, again, no act on the part of the British ruler, but simply the consequence of a famine,[78] an exodus, and a general combination of circumstances favourable to the demand for Irish labour, that has worked this happy change in Irish nature. However that may be, the significance of the following tabular statements cannot be misunderstood:

I.—Crimes in Ireland.

—Committed for Trial—

Years	Males	Females	Total	Convicted
1844	14,799	4,649	19,448	8,042
1845	12,807	3,889	16,696	7,101
1846	14,204	4,288	18,492	8,639
1847	23,552	7,657	31,209	15,233
1848	28,765	9,757	38,522	18,206
1849	31,340	10,649	41,989	21,202
1850	22,682	3,644	31,326	17,108
1851	17,337	7,347	24,684	14,377
1852	12,444	5,234	17,678	10,454
1853	10,260	4,884	15,144	8,714
1854	7,937	3,851	11,788	7,051
1855	6,019	2,993	9,012	5,220
1856	5,097	2,002	7,099	4,024
1857	5,458	1,752	7,210	3,925
1858	4,708	1,600	6,308	3,350

II.—Paupers in Ireland.

Years	No. of Parishes	Paupers	Years	No. of Parishes	Paupers
1849	880	82,357	1854	883	78,929
1850	880	79,031	1855	883	79,887
1851	881	76,906	1856	883	79,973
1852	882	75,111	1857	883	79,217
1853	882	75,437	1858	883	79,199

Published in *The New-York Daily Tribune* No. 5741, September 16, 1859

Printed according to the text of the newspaper

Karl Marx

From THE CRISIS IN ENGLAND[79]

Today, as fifteen years ago, England faces a catastrophe which threatens to undermine the foundation of her entire economic system. *Potatoes* as is known were almost the only food of the Irish and of a considerable part of the English working population when the potato blight of 1845 and 1846 struck the Irish root of life with rot. The results of that big catastrophe are well known. The Irish population decreased by two millions, some of whom starved, while others fled across the Atlantic. At the same time, this enormous calamity promoted the victory of the English Free-Trade party; the English landed aristocracy was compelled to sacrifice one of its most profitable monopolies, and the Repeal of the Corn Laws[80] ensured a wider and sounder basis for the reproduction and maintenance of the working millions.

What the *potato* was to Irish agriculture, *cotton* is to the dominant branch of Great Britain's industry. On its processing depends the subsistence of a mass of the population which is greater than the whole population of Scotland or two-thirds of the present population of Ireland. According to the 1861 census, the population of Scotland was 3,061,117, and that of Ireland only 5,764,543, while more than four million people in England and Scotland live directly or indirectly on the cotton industry. True, the cotton plant has not contracted any disease. Neither is its production the monopoly of a few areas of the world. On the contrary, no other plant providing material for clothing thrives on such extensive areas in America, Asia and Africa. The cotton monopoly of the slave-owning states of the American Union is not natural, but historically shaped. It grew and developed simultaneously with the monopoly of the English cotton industry on the world market. ...

Suddenly the American Civil War threatens this mainstay of English industry. While the Union blockades the ports of the Southern States to prevent the export of this year's cotton harvest and thereby cut off the secessionists' main source of income, the Confederation imparts compulsive force to this blockade merely by its decision not to export a single bale of cotton voluntarily and, moreover, to force England to come and fetch cotton herself from the southern ports. England is to be driven to break through the blockade by force, to declare war on the Union, and thus to throw her sword on the scales in favour of the slave-owning states.

Published in *Die Presse* No. 305,   Translated from the
November 6, 1861   German

Karl Marx

From ENGLISH HUMANISM AND AMERICA

This time it is the ladies from New Orleans, yellow beauties, tastelessly adorned with jewels and comparable in a way to the wives of the old Mexicans, except that they do not eat up their slaves *in natura*, who provide the occasion for a display of British aristocratic humanism—formerly it was the ports of Charleston. The English women who starve in Lancashire (but then they are not ladies and own no slaves) have not so far set any parliamentarian lips in motion; the cries of distress of the Irish women, who, because of the progressive concentration of small tenancies in Green Erin, are thrown half-naked into the streets and chased from house and home as if there were a Tatar invasion, have as yet elicited a single echo from Lords, Commons and Her Majesty's Government—homilies on the absolute rights of landownership.[81] But the ladies of New Orleans! That is quite a different matter. These ladies were far too enlightened to take part in the turmoil of war like the goddesses of Olympus, or to throw themselves into the flames like the women of Saguntum.[82] They have invented a new and safe kind of heroism, a kind that could have been invented only by women slave-owners, and at that by women slave-owners in a land where the free portion of the population consists of shop-keepers, cotton or sugar or tobacco merchants, who do not keep slaves as the *cives* of antiquity did. When their husbands had run away from New Orleans or hidden in their back-rooms, these ladies ran out into the streets to spit in the face of the victorious Unionist troops,[83] to stick out their tongues at them or, generally, like Mephistopheles, to make "obscene gestures" accompanied with invectives. These megaeras thought they would be allowed to be insolent "with impunity".

That was their heroism. General Butler issued a proclamation in which he announced that they would be treated like street-walkers if they continued to behave like street-walkers. Butler, though a lawyer by profession, does not seem to have studied English Statute Law properly.[84] Otherwise, by analogy with the laws imposed on Ireland under Castlereagh,[85] he would have prohibited them to set foot in the streets at all. Butler's warning to the "ladies" of New Orleans has made the Earl of Carnarvon, Sir J. Walsh (who played a ridiculous and odious role in Ireland), and Mr. Gregory, who demanded the recognition of the Confederation already a year ago, so morally indignant, that the Earl in the Upper House, and the knight and man "without a handle to his name" in the Lower House have questioned the government on what steps it intended to take in the name of affronted "humanism". Both Russell and Palmerston chastised Butler, and both expected him to be disavowed by the Government of Washington, and the very sensitive Palmerston, who, for "humane" admiration only, recognised the coup d'état of December 1851[86] (on which occasion some "ladies" were even shot dead and others were raped by the Zouaves[87]) behind the Queen's back and without previous knowledge by his colleagues, that same sensitive Viscount declared that Butler's warning was an *"infamy"*. Indeed, ladies, moreover ladies who even own slaves, should not even be allowed to vent their wrath and their spite on ordinary Unionist troops, peasants, artisans, and other rabble with impunity! It is "infamous".

Published in *Die Presse* No. 168, June 20, 1862

Translated from the German

Karl Marx

From Chapter XXV of CAPITAL, Volume I[88]

SECTION 5. ILLUSTRATIONS OF THE GENERAL LAW OF CAPITALIST ACCUMULATION

(f.) Ireland

In concluding this section, we must travel for a moment to Ireland. First, the main facts of the case.

The population of Ireland had, in 1841, reached 8,222,664; in 1851, it had dwindled to 6,623,985; in 1861, to 5,850,309; in 1866, to 5½ millions, nearly to its level in 1801. The diminution began with the famine year, 1846, so that Ireland, in less than twenty years, lost more than $\frac{5}{16}$ ths of its people.*

Its total emigration from May, 1851, to July, 1865, numbered 1,591,487: the emigration during the years 1861-1865 was more than half-a-million. The number of inhabited houses fell, from 1851-1861, by 52,990. From 1851-1861, the number of holdings of 15 to 30 acres increased 61,000, that of holdings over 30 acres, 109,000, whilst the total number

Table A

LIVE STOCK

Year	Horses		Cattle		
	Total Number	Decrease	Total Number	Decrease	Increase
1860	619,811	—	3,606,374	—	—
1861	614,232	5,579	3,471,688	134,686	—
1862	602,894	11,338	3,254,890	216,798	—
1863	579,978	22,916	3,144,231	110,659	—
1864	562,158	17,820	3,262,294	—	118,063
1865	547,867	14,291	3,493,414	—	231,120

* Population of Ireland, 1801, 5,319,867 persons; 1811, 6,084,996; 1821, 6,869,544; 1831, 7,828,347; 1841, 8,222,664.

Continued

Year	Sheep			Pigs		
	Total Number	Decrease	Increase	Total Number	Decrease	Increase
1860	3,542,080	—	—	1,271,072	—	—
1861	3,556,050	—	13,970	1,102,042	169,030	—
1862	3,456,132	99,918	—	1,154,324	—	52,282
1863	3,308,204	147,928	—	1,067,458	86,866	—
1864	3,366,941	—	58,737	1,058,480	8,978	—
1865	3,688,742	—	321,801	1,299,893	—	241,413

of all farms fell 120,000, a fall, therefore, solely due to the suppression of farms under 15 acres—i.e., to their centralisation.

The decrease of the population was naturally accompanied by a decrease in the mass of products. For our purpose, it suffices to consider the 5 years from 1861-1865 during which over half-a-million emigrated and the absolute number of people sank by more than $\frac{1}{3}$ of a million.

From the above table it results:—

Horses	Cattle	Sheep	Pigs
Absolute Decrease	Absolute Decrease	Absolute Increase	Absolute Increase
71,944	112,960	146,662	28,821*

Let us now turn to agriculture, which yields the means of subsistence for cattle and for men. In the following table is calculated the decrease or increase for each separate year, as compared with its immediate predecessor. The Cereal

---

* The result would be found yet more unfavourable if we went further back. Thus: Sheep in 1865, 3,688,742, but in 1856, 3,694,294. Pigs in 1865, 1,299,893, but in 1858, 1,409,883.

Table B

INCREASE OR DECREASE IN THE AREA UNDER CROPS AND GRASS
IN ACREAGE

Year	Cereal Crops	Green Crops		Grass and Clover		Flax		Total Cultivated Land	
	De-crease	De-crease	In-crease	De-crease	In-crease	De-crease	In-crease	De-crease	In-crease
	Acres	Acres	Acres	Acres	Acres	Acres	Acres	Acres	Acres
861	15,701	36,974	—	47,969	—	—	19,271	81,373	—
862	72,734	74,785	—	—	6,623	—	2,055	138,841	—
863	144,719	19,358	—	—	7,724	—	63,922	92,431	—
864	122,437	2,317	—	—	47,486	—	87,761	—	10,493
865	72,450	—	25,241	—	68,970	50,159	—	28,398	—
861-65	428,041	108,193	—	—	82,834	—	122,850	330,550	—

Crops include wheat, oats, barley, rye, beans, and peas; the Green Crops, potatoes, turnips, mangolds, beet-root, cabbages, carrots, parsnips, vetches, &c.

In the year 1865, 127,470 additional acres came under the heading "grass land", chiefly because the area under the heading of "bog and waste unoccupied", decreased by 101,543 acres. If we compare 1865 with 1864, there is a decrease in cereals of 246,667 qrs., of which 48,999 were wheat, 166,605 oats, 29,892 barley, &c.: the decrease in potatoes was 446,398 tons, although the area of their cultivation increased in 1865. [See *Table C.*]

From the movement of population and the agricultural produce of Ireland, we pass to the movement in the purse of its landlords, larger farmers, and industrial capitalists. It is reflected in the rise and fall of the Income-tax. It may be remembered that Schedule D. (profits with the exception of those of farmers), includes also the so-called "professional" profits—i.e., the incomes of lawyers, doctors, &c.; and the Schedules C. and E., in which no special details are given, include the incomes of employés, officers, State sinecurists, State fundholders, &c.

INCREASE OR DECREASE IN THE AREA UNDER CULTIVATION, PRODUCT

Product	Acres of Cultivated Land		Increase or Decrease, 1865		Product per Acre		
	1864	1865				1864	1865
Wheat . . .	276,483	266,989		9,494	Wheat, cwt.	13.3	13.0
Oats	1,814,886	1,745,228		69,658	Oats, cwt.	12.1	12.3
Barley . . .	172,700	177,102	4,402		Barley, cwt.	15.9	14.9
Bere . . . ⎫ Rye . . . ⎭	8,894	10,091	1,197		⎧Bere, cwt. ⎩Rye, cwt.	16.4 8.5	14.8 10.4
Potatoes . .	1,039,724	1,066,260	26,536		Potatoes, tons	4.1	3.6
Turnips . .	337,355	334,212		3,143	Turnips, tons	10.3	9.9
Mangold-wurzel . .	14,073	14,389	316		Mangold-wurzel, tons	10.5	13.3
Cabbages . .	31,821	33,622	1,801		Cabbages, tons	9.3	10.4
Flax	301,693	251,433		50,260	Flax, st.	(14 lb) 34.2	25.2
Hay	1,609,569	1,678,493	68,924		Hay, tons	1.6	1.8

* The data of the text are put together from the materials of the "Agricultural Statistics, Ireland, General Abstracts, Dublin", for the years 1860, et seq., and "Agricultural Statistics, Ireland. Tables showing the estimated average produce, &c., Dublin, 1866". These statistics are official, and laid before Parliament annually. [Note to 2nd edition. The official statistics for the year 872 show, as compared with 1871, a decrease in area under cultivation of 134,915 acres. An increase occurred in the cultivation of green crops, turnips,

Table C

PER ACRE, AND TOTAL PRODUCT OF 1865 COMPARED WITH 1864

Increase or Decrease, 1865		Total Product		Increase or Decrease, 1865	
		1864	1865		
	0.3	875,782 Qrs.	826,783 Qrs.		48,999 qrs.
0.2		7,826,332 "	7,659,727 "		166,605 "
	1.0	761,909 "	732,017 "		29,892 "
	1.6	15,160 "	13,989 "		1,171 "
1.9		12,680 "	18,364 "	5,684 qrs.	
	0.5	4,312,388 ts.	3,865,990 ts.		446,398 ts.
	0.4	3,467,659 "	3,301,683 "		165,976 "
2.8		147,284 "	191,937 "	44,653 ts.	
1.1		297,375 "	350,252 "	52,877 "	
	9.0	64,506 st.	39,561 st.		24,945 st.*
0.2		2,607,153 ts.	3,068,707 ts.	461,554 "	

mangold-wurzel, and the like; a decrease in the area under cultivation for wheat of 16,000 acres; oats, 14,000; barley and rye, 4,000; potatoes, 66,632; flax, 34,667; grass, clover, vetches, rape-seed, 30,000. The soil under cultivation for wheat shows for the last 5 years the following stages of decrease : — 1868. 285,000 acres; 1869, 280,000; 1870, 259,000; 1871. 244,000; 1872, 228,000. For 1872 we find, in round numbers, an increase of 2,600 horses, 80,000 horned cattle, 68,609 sheep, and a decrease of 236,000 pigs.]

Table D

THE INCOME-TAX ON THE SUBJOINED INCOMES IN POUNDS STERLING

	1860	1861	1862	1863	1864	1865
Schedule A. Rent of Land	12,893,829	13,003,554	13,398,938	13,494,091	13,470,700	13,801,616
Schedule B. Farmers' Profits	2,765,387	2,77	2,937,899	2,938,923	2,930,874	2,946,072
Schedule D. Indust- rial, &c., Profits	4,891,652	4,83 3	4,858,800	4,846,497	4,546,147	4,850,199
Total Sche- dules A. to E.	22,962,885	22,998,394	23,597,574	23,658,631	23,236,298	23,930,340*

Under Schedule D. the average annual increase of income from 1853-1864 was only 0.93; whilst, in the same period, in Great Britain, it was 4.58. The following table shows

Table E

SCHEDULE D. INCOME FROM PROFITS (OVER £60) IN IRELAND

	1864 £		1865 £	
Total yearly in- come of . . .	4,368,610 divided	among 17,467 persons.	4,669,979 divided	among 18,081 persons.
Yearly income over £60 and under £100 . .	238,726 "	5,015 "	222,575 "	4,703 "
Of the yearly to- tal income . .	1,979,066 "	11,321 "	2,028,571 "	12,184 "
Remainder of the total yearly in- come	2,150,818 "	1,131 "	2,418,833 "	1,194 "
Of these . . {	1,073,906 "	1,010 "	1,097,927 "	1,044 "
	1,076,912 "	121 "	1,320,906 "	150 "
	430,535 "	95 "	584,458 "	122 "
	646,377 "	26 "	736,448 "	28 "
	262,819 "	3 "	274,528 "	3** "

* Tenth Report of the Commissioners of Inland Revenue. London,
1866.
** The total yearly income under Schedule D. is different in this table
from that which appears in the preceding ones, because of certain deduc-
tions allowed by law.

the distribution of the profits (with the exception of those of farmers) for the years 1864 and 1865:—[See *Table E.*]

England, a country with fully developed capitalist production, and pre-eminently industrial, would have bled to death with such a drain of population as Ireland has suffered. But Ireland is at present only an agricultural district of England, marked off by a wide channel from the country to which it yields corn, wool, cattle, industrial and military recruits.

The depopulation of Ireland has thrown much of the land out of cultivation, has greatly diminished the produce of the soil,* and, in spite of the greater area devoted to cattle-breeding, has brought about, in some of its branches, an absolute diminution, in others, an advance scarcely worthy of mention, and constantly interrupted by retrogressions. Nevertheless, with the fall in numbers of the population, rents and farmers' profits rose, although the latter not as steadily as the former. The reason of this is easily comprehensible. On the one hand, with the throwing of small holdings into large ones, and the change of arable into pasture land, a larger part of the whole produce was transformed into surplus-produce. The surplus-produce increased, although the total produce, of which it formed a fraction, decreased. On the other hand, the money-value of this surplus-produce increased yet more rapidly than its mass, in consequence of the rise in the English market-price of meat, wool, &c., during the last 20, and especially during the last 10, years.

The scattered means of production that serve the producers themselves as means of employment and of subsistence, without expanding their own value by the incorporation of the labour of others, are no more capital than a product consumed by its own producer is a commodity. If, with the mass of the population, that of the means of production employed in agriculture also diminished, the mass of the capital employed in agriculture increased, because a part of

---

* If the product also diminishes relatively per acre, it must not be forgotten that for a century and a half England has indirectly exported the soil of Ireland, without as much as allowing its cultivators the means for making up the constituents of the soil that had been exhausted,

the means of production that were formerly scattered, was concentrated and turned into capital.

The total capital of Ireland outside agriculture, employed in industry and trade, accumulated during the last two decades slowly, and with great and constantly recurring fluctuations; so much the more rapidly did the concentration of its individual constituents develop. And, however small its absolute increase, in proportion to the dwindling population it had increased largely.

Here, then, under our own eyes and on a large scale, a process is revealed, than which nothing more excellent could be wished for by orthodox economy for the support of its dogma: that misery springs from absolute surplus-population, and that equilibrium is re-established by depopulation. This is a far more important experiment than was the plague in the middle of the 14th century so belauded of Malthusians.[89] Note further: If only the naïveté of the schoolmaster could apply, to the conditions of production and population of the nineteenth century, the standard of the 14th, this naïveté, into the bargain, overlooked the fact that whilst, after the plague and the decimation that accompanied it, followed on this side of the Channel, in England, enfranchisement and enrichment of the agricultural population, on that side, in France, followed greater servitude and more misery.*

The Irish famine of 1846 killed more than 1,000,000 people, but it killed poor devils only. To the wealth of the country it did not the slightest damage. The exodus of the next 20 years, an exodus still constantly increasing, did not, as, e.g., the Thirty Years' War,[90] decimate, along with the human beings, their means of production. Irish genius discovered an altogether new way of spiriting a poor people thousands of miles away from the scene of its misery. The exiles transplanted to the United States, send home sums of money every year as travelling expenses for those left behind. Every

---

* As Ireland is regarded as the promised land of the "principle of population", Th. Sadler, before the publication of his work on population, issued his famous book, "Ireland, its Evils and their Remedies." 2nd edition, London, 1829. Here, by comparison of the statistics of the individual provinces, and of the individual counties in each province, he proves that the misery there is not as Malthus would have it, in proportion to the number of the population, but in inverse ratio to this,

troop that emigrates one year, draws another after it the next. Thus, instead of costing Ireland anything, emigration forms one of the most lucrative branches of its export trade. Finally, it is a systematic process, which does not simply make a passing gap in the population, but sucks out of it every year more people than are replaced by the births, so that the absolute level of the population falls year by year.*

What were the consequences for the Irish labourers left behind and freed from the surplus-population? That the relative surplus-population is to-day as great as before 1846; that wages are just as low, that the oppression of the labourers has increased, that misery is forcing the country towards a new crisis. The facts are simple. The revolution in agriculture has kept pace with emigration. The production of relative surplus-population has more than kept pace with the absolute depopulation. A glance at Table C shows that the change of arable to pasture land must work yet more acutely in Ireland than in England. In England the cultivation of green crops increases with the breeding of cattle; in Ireland, it decreases. Whilst a large number of acres, that were formerly tilled, lie idle or are turned permanently into grass-land, a great part of the waste land and peat bogs that were unused formerly, become of service for the extension of cattle-breeding. The smaller and medium farmers—I reckon among these all who do not cultivate more than 100 acres—still make up about $\frac{8}{10}$ ths of the whole number.**
They are, one after the other, and with a degree of force unknown before, crushed by the competition of an agriculture managed by capital, and therefore they continually furnish new recruits to the class of wage-labourers. The one great industry of Ireland, linen-manufacture, requires relatively few adult men and only employs altogether, in spite of its expansion since the price of cotton rose in 1861-1866, a comparatively insignificant part of the population. Like all other great modern industries, it constantly produces, by incessant fluctuations, a relative surplus-population within

---

* Between 1851 and 1874, the total number of emigrants amounted to 2,325,922.

** [Note to 2nd edition.] According to a table in Murphy's "Ireland Industrial, Political and Social", 1870, 94.6 per cent. of the holdings do not reach 100 acres, 5.4 exceed 100 acres.

its own sphere, even with an absolute increase in the mass of human beings absorbed by it. The misery of the agricultural population forms the pedestal for gigantic shirt-factories, whose armies of labourers are, for the most part, scattered over the country. Here, we encounter again the system described above of domestic industry,[91] which in underpayment and over-work, possesses its own systematic means for creating supernumerary labourers. Finally, although the depopulation has not such destructive consequences as would result in a country with fully developed capitalistic production, it does not go on without constant reaction upon the home-market. The gap which emigration causes here, limits not only the local demand for labour, but also the incomes of small shopkeepers, artisans, tradespeople generally. Hence the diminution in incomes between £60 and £100 in Table E.

A clear statement of the condition of the agricultural labourers in Ireland is to be found in the Reports of the Irish Poor Law Inspectors (1870).* Officials of a government which is maintained only by bayonets and by a state of siege, now open, now disguised, they have to observe all the precautions of language that their colleagues in England disdain. In spite of this, however, they do not let their government cradle itself in illusions. According to them the rate of wages in the country, still very low, has within the last 20 years risen 50-60 per cent., and stands now, on the average, at 6s. to 9s. per week. But behind this apparent rise, is hidden an actual fall in wages, for it does not correspond at all to the rise in price of the necessary means of subsistence that has taken place in the meantime. For proof, the following extract from the official accounts of an Irish workhouse.

AVERAGE WEEKLY COST PER HEAD

Year ended	Provisions and Necessaries	Clothing	Total
29th Sept., 1849	1s. 3¼ d.	3d.	1s. 6¼ d.
„ 1869	2s. 7¼ d.	6d.	3s. 1¼ d.

---

* "Reports from the Poor Law Inspectors on the Wages of Agricultural Labourers in Ireland", Dublin, 1870. See also "Agricultural Labourers (Ireland). Return, etc.", 8th March, 1861.

The price of the necessary means of subsistence is therefore fully twice, and that of clothing exactly twice, as much
as they were 20 years before.

Even apart from this disproportion, the mere comparison
of the rate of wages expressed in gold would give a result
far from accurate. Before the famine, the great mass of
agricultural wages were paid in kind, only the smallest part
in money; to-day, payment in money is the rule. From this
it follows that, whatever the amount of the real wage, its
money rate must rise.

"Previous to the famine, the labourer enjoyed his cabin ... with
a rood, or half-acre or acre of land, and facilities for ... a crop of
potatoes. He was able to rear his pig and keep fowl.... But they now
have to buy bread, and they have no refuse upon which they can
feed a pig or fowl, and they have consequently no benefit from the
sale of a pig, fowl, or eggs."*

In fact, formerly, the agricultural labourers were but the
smallest of the small farmers, and formed for the most part
a kind of rearguard of the medium and large farms on which
they found employment. Only since the catastrophe of 1846
have they begun to form a fraction of the class of purely
wage-labourers, a special class, connected with its wage-
masters only by monetary relations.

We know what were the conditions of their dwellings in
1846. Since then they have grown yet worse. A part of the
agricultural labourers which, however, grows less day by
day, dwells still on the holdings of the farmers in overcrowded huts, whose hideousness far surpasses the worst
that the English agricultural labourers offered us in this way.
And this holds generally with the exception of certain
tracts of Ulster; in the south, in the counties of Cork, Limerick, Kilkenny, &c.; in the east, in Wicklow, Wexford, &c.;
in the centre of Ireland, in King's and Queen's County,
Dublin, &c.; in the north, in Down, Antrim, Tyrone, &c.;
in the west, in Sligo, Roscommon, Mayo, Galway, &c.
"The agricultural labourers' huts," an inspector cries out,
"are a disgrace to the Christianity and to the civilisation
of this country."** In order to increase the attractions of

---

* *l.c.*, pp. 29, 1.
** *l.c.*, p. 12.

these holes for the labourers, the pieces of land belonging thereto from time immemorial, are systematically confiscated.

"The mere sense that they exist subject to this species of ban, on the part of the landlords and their agents, has ... given birth in the minds of the labourers to corresponding sentiments of antagonism and dissatisfaction towards those by whom they are thus led to regard themselves as being treated as ... a proscribed race."*

The first act of the agricultural revolution was to sweep away the huts situated on the field of labour. This was done on the largest scale, and as if in obedience to a command from on high. Thus many labourers were compelled to seek shelter in villages and towns. There they were thrown like refuse into garrets, holes, cellars and corners, in the worst back slums. Thousands of Irish families, who according to the testimony of the English, eaten up as these are with national prejudice, are notable for their rare attachment to the domestic hearth, for their gaiety and the purity of their home-life, found themselves suddenly transplanted into hot-beds of vice. The men are now obliged to seek work of the neighbouring farmers and are only hired by the day, and therefore under the most precarious form of wage. Hence

"they sometimes have long distances to go to and from work, often get wet, and suffer much hardship, not unfrequently ending in sickness, disease and want".**

"The towns have had to receive from year to year what was deemed to be the surplus-labour of the rural division"***; and then people still wonder "there is still a surplus of labour in the towns and villages, and either a scarcity or a threatened scarcity in some of the country divisions".**** The truth is that this want only becomes perceptible "in harvest-time, or during spring, or at such times as agricultural operations are carried on with activity; at other periods of the year many hands are idle"*****; that "from the dig-

---

* *l.c.*, p. 12.
** *l.c.*, p. 25.
*** *l.c.*, p. 27.
**** *l.c.*, p. 26.
***** *l.c.*, p. 1.

ging out of the main crop of potatoes in October until the early spring following ... there is no employment for them"*; and further, that during the active times they "are subject to broken days and to all kinds of interruptions".**

These results of the agricultural revolution—*i.e.*, the change of arable into pasture land, the use of machinery, the most rigorous economy of labour, &c., are still further aggravated by the model landlords, who, instead of spending their rents in other countries, condescend to live in Ireland on their demesnes. In order that the law of supply and demand may not be broken, these gentlemen draw their

"labour-supply ... chiefly from their small tenants, who are obliged to attend when required to do the landlord's work, at rates of wages, in many instances, considerably under the current rates paid to ordinary labourers, and without regard to the inconvenience or loss to the tenant of being obliged to neglect his own business at critical periods of sowing or reaping".***

The uncertainty and irregularity of employment, the constant return and long duration of gluts of labour, all these symptoms of a relative surplus-population, figure therefore in the reports of the Poor Law administration, as so many hardships of the agricultural proletariat. It will be remembered that we met, in the English agricultural proletariat, with a similar spectacle. But the difference is that in England, an industrial country, the industrial reserve recruits itself from the country districts, whilst in Ireland, an agricultural country, the agricultural reserve recruits itself from the towns, the cities of refuge of the expelled agricultural labourers. In the former, the supernumeraries of agriculture are transformed into factory operatives; in the latter, those forced into the towns, whilst at the same time they press on the wages in towns, remain agricultural labourers, and are constantly sent back to the country districts in search of work.

---

* *l.c.*, pp. 31, 32.
** *l.c.*, p. 25.
*** *l.c.*, p. 30.

The official inspectors sum up the material condition of the agricultural labourer as follows:

"Though living with the strictest frugality, his own wages are barely sufficient to provide food for an ordinary family and pay his rent, and he depends upon other sources for the means of clothing himself, his wife, and children.... The atmosphere of these cabins, combined with the other privations they are subjected to, has made this class particularly susceptible to low fever and pulmonary consumption."*

After this, it is no wonder that, according to the unanimous testimony of the inspectors, a sombre discontent runs through the ranks of this class, that they long for the return of the past, loathe the present, despair of the future, give themselves up "to the evil influence of agitators", and have only one fixed idea, to emigrate to America. This is the land of Cockaigne, into which the great Malthusian panacea, depopulation, has transformed Green Erin.

What a happy life the Irish factory operative leads, one example will show:

"On my recent visit to the North of Ireland," says the English Factory Inspector, Robert Baker, "I met with the following evidence of effort in an Irish skilled workman to afford education to his children; and I give his evidence verbatim, as I took it from his mouth. That he was a skilled factory hand, may be understood when I say that he was employed on goods for the Manchester market. 'Johnson.— I am a beetler and work from 6 in the morning till 11 at night, from Monday to Friday. Saturday we leave off at 6 p.m., and get three hours of it (for meals and rest). I have five children in all. For this work I get 10s. 6d. a week; my wife works here also, and gets 5s. a week. The oldest girl who is 12, minds the house. She is also cook, and all the servant we have. She gets the young ones ready for school. A girl going past the house wakes me at half past five in the morning. My wife gets up and goes along with me. We get nothing (to eat) before we come to work. The child of 12 takes care of the little children all the day, and we get nothing till breakfast at eight. At eight we go home. We get tea once a week; at other times we get stirabout, sometimes of oat-meal, sometimes of Indian meal, as we are able to get it. In the winter we get a little sugar and water to our Indian meal. In the summer we get a few potatoes, planting a small patch ourselves; and when they are done we get back to stirabout. Sometimes we get a little milk as it may be. So we go on from day to day,

---

* *l.c.*, pp. 21, 13.

Sunday and week day, always the same the year round. I am always very much tired when I have done at night. We may see a bit of flesh meat sometimes, but very seldom. Three of our children attend school, for whom we pay 1d. a week a head. Our rent is 9d. a week. Peat for firing costs 1s. 6d. a fortnight at the very lowest.' "*

Such are Irish wages, such is Irish life!

In fact the misery of Ireland is again the topic of the day in England. At the end of 1866 and the beginning of 1867, one of the Irish land magnates, Lord Dufferin, set about its solution in *The Times.* "Wie menschlich von solch grossem Herrn!"

From Table E we saw that, during 1864, of £4,368,610 of total profits, three surplus-value makers pocketed only £262,819; that in 1865, however, out of £4,669,979 total profits, the same three virtuosi of "abstinence" pocketed £274,528; in 1864, 26 surplus-value makers reached to £646,377; in 1865, 28 surplus-value makers reached to £736,448; in 1864, 121 surplus-value makers, £1,076,912; in 1865, 150 surplus-value makers, £1,320,906; in 1864, 1,131 surplus-value makers, £2,150,818, nearly half of the total annual profit; in 1865, 1,194 surplus-value makers, £2,418,833, more than half of the total annual profit. But the lion's share, which an inconceivably small number of land magnates in England, Scotland and Ireland swallow up of the yearly national rental, is so monstrous that the wisdom of the English State does not think fit to afford the same statistical materials about the distribution of rents as about the distribution of profits. Lord Dufferin is one of those land magnates. That rent-rolls and profits can ever be "excessive", or that their plethora is in any way connected with plethora of the people's misery is, of course, an idea as "disreputable" as "unsound". He keeps to facts. The fact is that, as the Irish population diminishes, the Irish rent-rolls swell; that depopulation benefits the landlords, therefore also benefits the soil, and, therefore, the people, that mere accessory of the soil. He declares, therefore, that Ireland is still over-populated, and the stream of emigration still flows too lazily. To be perfectly happy, Ireland must get rid of at

---

* "Rept of Insp. of Fact., 31st Oct., 1866", p. 96.

least one-third of a million of labouring men. Let no man imagine that this lord, poetic into the bargain, is a physician of the school of Sangrado, who as often as he did not find his patient better, ordered phlebotomy and again phlebotomy, until the patient lost his sickness at the same time as his blood. Lord Dufferin demands a new bloodletting of one-third of a million only, instead of about two millions; in fact, without the getting rid of these, the millennium in Erin is not to be. The proof is easily given.

NUMBER AND EXTENT OF FARMS IN IRELAND IN 1864

(1) Farms not over 1 acre.		(2) Farms over 1, not over 5 acres.		(3) Farms over 5, not over 15 acres.		(4) Farms over 15, not over 30 acres.	
No.	Acres.	No.	Acres.	No.	Acres.	No.	Acres.
48,653	25,394	82,037	288,916	176,368	1,836,310	136,578	3,051,343

(5) Farms over 30, not over 50 acres.		(6) Farms over 50, not over 100 acres.		(7) Farms over 100 acres.		(8) Total area.
No.	Acres.	No.	Acres.	No.	Acres.	Acres.
71,961	2,906,274	54,247	3,983,880	31,927	8,227,807	20,319,924*

Centralisation has from 1851 to 1861 destroyed principally farms of the first three categories, under 1 and not over 15 acres. These above all must disappear. This gives 307,058 "supernumerary" farmers, and reckoning the families the low average of 4 persons, 1,228,232 persons. On the extravagant supposition that, after the agricultural revolution is complete one-fourth of these are again absorbable, there remain for emigration 921,174 persons. Categories 4, 5, 6, of over 15 and not over 100 acres, are, as was known

---

* The total area includes also peat bogs and waste land.

long since in England, too small for capitalistic cultivation of corn, and for sheep-breeding are almost vanishing quantities. On the same supposition as before, therefore, there are further 788,761 persons to emigrate; total, 1,709,532. And as l'appétit vient en mangeant, Rent-roll's eyes will soon discover that Ireland, with 3½ millions, is still always miserable, and miserable because she is over-populated. Therefore her depopulation must go yet further, that thus she may fulfil her true destiny, that of an English sheep-walk and cattle-pasture.*

Like all good things in this bad world, this profitable method has its drawbacks. With the accumulation of rents in Ireland, the accumulation of the Irish in America keeps pace. The Irishman, banished by sheep and ox, re-appears on the other side of the ocean as a Fenian,[92] and face to face

---

* How the famine and its consequences have been deliberately made the most of, both by the individual landlords and by the English legislature, to forcibly carry out the agricultural revolution and to thin the population of Ireland down to the proportion satisfactory to the landlords, I shall show more fully in Vol. III. of this work, in the section on landed property. There also I return to the condition of the small farmers and the agricultural labourers. At present, only one quotation. Nassau W. Senior says, with other things, in his posthumous work, "Journals, Conversations and Essays relating to Ireland", 2 vols. London, 1868; Vol. II., p. 282. "Well," said Dr. G., "we have got our Poor Law and it is a great instrument for giving the victory to the landlords. Another, and a still more powerful instrument is emigration.... No friend to Ireland can wish the war to be prolonged [between the landlords and the small Celtic farmers]—still less, that it should end by the victory of the tenants. The sooner it is over—the sooner Ireland becomes a grazing country, with the comparatively thin population which a grazing country requires, the better for all classes." The English Corn Laws of 1815 secured Ireland the monopoly of the free importation of corn into Great Britain. They favoured artificially, therefore, the cultivation of corn. With the abolition of the Corn Laws in 1846, this monopoly was suddenly removed. Apart from all other circumstances, this event alone was sufficient to give a great impulse to the turning of Irish arable into pasture land, to the concentration of farms, and to the eviction of small cultivators. After the fruitfulness of the Irish soil had been praised from 1815 to 1846, and proclaimed loudly as by Nature herself destined for the cultivation of wheat, English agronomists, economists, politicians, discover suddenly that it is good for nothing but to produce forage. M. Léonce de Lavergne has hastened to repeat this on the other side of the Channel. It takes a "serious" man, à la Lavergne, to be caught by such childishness.

with the old queen of the seas rises, threatening and more threatening, the young giant Republic:

Acerba fata Romanos agunt
Scelusque fraternae necis.

Published in the book: Karl Marx, *Das Kapital. Kritik der politischen Oekonomie.* Erster Band. Hamburg, 1867

Printed according to the text of the English edition of *Capital*, Vol. I, Moscow, 1965, which follows the fourth German edition of 1890

Karl Marx

From Chapter XXXVII of CAPITAL, Volume III[93]

We are not speaking now of conditions in which ground-rent, the manner of expressing landed property in the capitalist mode of production, formally exists without the existence of the capitalist mode of production itself, i.e., without the tenant himself being an industrial capitalist, nor the type of his management being a capitalist one. Such is the case, e.g., in *Ireland*. The tenant there is generally a small farmer. What he pays to the landlord in the form of rent frequently absorbs not merely a part of his profit, that is, his own surplus-labour (to which he is entitled as possessor of his own instruments of labour), but also a part of his normal wage, which he would otherwise receive for the same amount of labour. Besides, the landlord, who does nothing at all for the improvement of the land, also expropriates his small capital, which the tenant for the most part incorporates in the land through his own labour. This is precisely what a usurer would do under similar circumstances, with just the difference that the usurer would at least risk his own capital in the operation. This continual plunder is the core of the dispute over the Irish Tenancy Rights Bill. The main purpose of this Bill is to compel the landlord when ordering his tenant off the land to indemnify the latter for his improvements on the land, or for his capital incorporated in the land.[94] Palmerston used to wave this demand aside with the cynical answer:

"The House of Commons is a house of landed proprietors."

Published in the book: Karl Marx, *Das Kapital. Kritik der politischen Oekonomie*. Dritter Band. Herausgegeben von Friedrich Engels

Printed according to the text of the English edition of *Capital*, Vol. III, Moscow, 1966

Karl Marx

THE FENIAN PRISONERS AT MANCHESTER AND THE INTERNATIONAL WORKING MEN'S ASSOCIATION[95]

At a special meeting of the General Council of the I. W. A. held at the office 16, Castle Street, East, W., on Wednesday evening the following memorial was adopted:

"Memorial of the General Council of the International Working Men's Association.

"To the Right Hon. Gathorne-Hardy, her Majesty's Secretary of State.

"The memorial of the undersigned, representing working men's associations in all parts of Europe, showeth:

"That the execution of the Irish prisoners condemned to death at Manchester will greatly impair the moral influence of England upon the European Continent. The Execution of the four prisoners resting upon the same evidence and the same verdict which, by the free pardon of Maguire, have been officially declared, the one false, the other erroneous, will bear the stamp not of a judicial act, but of political revenge. But even if the verdict of the Manchester jury and the evidence it rests upon had not been tainted by the British Government itself, the latter would now have to choose between the blood-handed practices of old Europe and the magnanimous humanity of the young Transatlantic Republic.[96]

"The commutation of the sentence for which we

pray will be an act not only of justice, but of political wisdom.

"By order of the General Council of the I. W. Association,

"*JOHN WESTON*, Chairman
R. SHAW, Secretary for America
EUGÈNE DUPONT, Secretary for France
KARL MARX, Secretary for Germany
HERMANN JUNG, Secretary for Switzerland
P. LAFARGUE, Secretary for Spain
ZABICKI, Secretary for Poland
DERKINDEREN, Secretary for Holland
BESSON, Secretary for Belgium
G. ECCARIUS, General Secretary"

20 November, 1867

Published in *Le courrier français* No. 163, November 24, 1867

Printed according to the text of the book *The General Council of the First International. 1866-1868. Minutes,* Moscow

Karl Marx

[NOTES FOR AN UNDELIVERED SPEECH
ON IRELAND][97]

I. EXORDIUM. THE EXECUTION

Since our last meeting the object of our discussion, Fenianism, has entered a new phase. It has been baptised in blood by the English Government. The Political Executions at Manchester remind us of the fate of John Brown at Harpers Ferry.* They open a new period in the struggle between Ireland and England. The whole Parliament and liberal press responsible. Gladstone.

Reason: to keep up the hypocrisy that this was no political, but a common criminal affair. The effect produced upon Europe quite the contrary. They seem anxious to keep up the Act of the Long Parliament.[98] English [have] a divine right to fight the Irish on their native soil, but every Irish fighting against the British Government in England to be treated as an outlaw. Suspension of the Habeas Corpus Act.[99] State of siege. Facts from the *Chronicle*. Governmental organisation of "Assassination and Violence".[100] Case of Bonaparte.[101]

II. THE QUESTION

What is Fenianism?

---

* Here the following text is crossed out in the manuscript: "But the slaveholders have at least treated John Brown as a rebel, not as common felon."—*Ed.*

III. THE LAND QUESTION
Decrease of Population

1846
1841: $\left.\begin{array}{l} 8,222,664 \\ \underline{5,571,971} \\ 2,650,693 \end{array}\right\}$ in 25 Jahren* *1801:* 5,319,867
1866: 2,650,693

$\left.\begin{array}{l} 1855: \quad 6,604,665 \\ 1866: \quad \underline{5,571,971} \\ \quad 1,032,694 \end{array}\right\}$ in 11 years
1,032,694

Population not only decreased, but the number of the deaf-mutes, the blind, the decrepit, the lunatic, and idiotic increased relatively to the numbers of the population.

Increase of Live-Stock from 1855 to 1866

In the same period from 1855 to 1866 [the] number of the live-stock increased as follows: cattle by 178,532, sheep by 667,675, pigs by 315,918. If we take into account the simultaneous decrease of horses by 20,656, and equalise 8 sheep to 1 horse *total increase of live-stock*: 996,877, about one million.

Thus 1,032,694 Irishmen have been displaced by about one million cattle, pigs, and sheep. What has become of them? The *emigration list* answers.

Emigration

From 1st May 1851 to 31 December 1866: 1,730,189. Character of that emigration.

The process has been brought about and is still functioning upon an always enlarging scale by the *throwing together* or *consolidation of farms* (eviction) and by the simultaneous conversion of tillage into pasture.

From 1851-1861 [the] total number of farms decreased by 120,000, while simultaneously the number of farms of 15-30 acres increased by 61,000, that of 30 acres by 109,000 (together 170,000). The decrease was almost exclusively owed to the extinction of farms from less than one to less

---

* Years.—*Ed.*

than 15 acres. Lord Dufferin.* The increase means only
that amongst the decreased number of farms there is a
larger portion of farms of large dimension.

How the Process Works

a) *The People.*
The situation of the *mass of the people* has deteriorated,
and their state is verging to a crisis similar to that of 1846.
The relative surplus population now as great as before the
famine.

Wages have not risen more than 20%, since the potato
famine. The price of potatoes has risen nearly 200%; the
necessary means of life on an average by 100%. *Professor
Cliffe Leslie*, in the London *Economist* dated February 9,
1867, says:

> "After a loss of $^2/_5$ of the population in 21 years, throughout most
> of the island, the rate of wages is now only 1s. a day; a shilling does
> not go further than 6d. did 21 years ago. Owing to this rise in his
> ordinary food the labourer is worse off than he was 10 years ago."

b) *The Land.*
1) *Decrease of land under crops.*

Decrease in cereal crops:	*Decrease in green crops:*
1861-66: 470,917 acres	1861-66: 128,061 acres

2) *Decrease per statute acre of every crop.* There has
been decrease of *yield* in wheat, but greater *1847 to 1865*
per cent; the exact decrease: oats 16.3, flax 47.9, turnips 36.1, potatoes 50%. Some years would show a greater
decrease, but on the whole it has been *gradual* since 1847.

Since the exodus, the land has been underfed and overworked, partly from the injudicious consolidation of farms,
and, partly, because, under the corn-acre system,[102] the
farmer in a great measure trusted to his labourers to manure
the land for him. Rents and profits may increase, although
the profit of the soil decreases. The total produce may diminish, but that part of it, which is converted into surplus

---

* See pp. 113-14.—*Ed.*

produce, falling to landlord and greater farmers, instead of to the labourer. And the price of the surplus produce has risen.

So result: gradual expulsion of the natives, gradual deterioration and exhaustion of the source of national life, the soil.

Process of Consolidation

This process has only begun; it is going on in rapid strides. The consolidation has first attacked the farms of under one to under 15 acres. It will be far from having reached the English point of consolidation, if all farms under 100 acres have disappeared. Now the state was this in 1864:

The *total area of Ireland*, including *bogs and waste land*: 20,319,924 acres. *Of those* $^3/_5$,$=12,092,117$ *acres*, form still farms from *under 1 to under 100 acres*, and are in the hands of *569,844 farmers*; $^2/_5$,$=8,227,807$, form farms *from 100 till over 500 acres*, and are in the hands of 31,927 persons. Thus to be cleared off 2,847,220, if we number only the farmers and their families.

This system [is a] natural offspring of the famine of 1846, accelerated by the abolition of corn-laws,[103] and the rise in the price of meat and wool, now systematic.

Clearing of the estate of Ireland, transforming it in an English agricultural district, minus its resident lords and their retainers, separated from England by a broad water ditch.

Change of Character of the English Rule in Ireland

State only tool of the landlords. *Eviction*, also employed as means of political punishment. (*Lord Abercorn.** England. *Gaels: in the Highlands of Scotland.*[104]) Former English policy: displacing the Irish by English (Elizabeth), roundheads[105] (Cromwell). Since Anne, 18th-century politico-economical character only again in the protectionist measures of England against her own Irish colony; within that colony making *religion* a proprietary title. After the *Union*[106] [the] system of rack-renting and middlemen, but

* See pp. 143-44.—*Ed.*

left the Irish, however ground to the dust, holder of their native soil. Present system, quiet business-like extinction, and government only instrument of landlords (and usurers).

From this altered state:

1) *Distinguishing character of Fenianism: Socialist, lower-class movement.*

2) *Not Catholic movement.*

Priests leaders as long as Catholic Emancipation and their leader, Daniel O'Connell, remained leader of the Irish movement. Ridiculous Popishism of the English. High Catholic priests against Fenianism.

3) *No representative leader in the British Parliament.* Character of O'Connell's physical force movement.[107] Extinction of Irish party in Parliament.

4) *Nationality.* Influence of European movement, and English phraseology.

5) *America, Ireland, England*—three fields of action, leadership of America.

6) *Republican*, because America republic.

I have now given the characteristics of Fenianism.

IV. THE ENGLISH PEOPLE

A cause of humanity and right, but above all a specific English question.

a) *Aristocracy and Church and Army.* (*France*, Algiers.)

b) *Irish in England.* Influence on wages, etc. Lowering the character of the English and Irish. *The Irish Character.* Chastity of Irishmen. Attempts at education in Ireland. Diminution of crimes.

	Convicted in Ireland *Committed for trial:*	*Convicted:*
1852	17,678	10,454
1866	4,326	2,418

The decrease in the numbers of persons committed for trial in England and Wales, since 1855, is partly due to the *Criminal Justice Act of 1855*, authorising Justices to pass sentences for short periods with the consent of the prisoners, instead of committing for trial to the sessions.

Birmingham. Progress of the English people. Infamy of the English press.

c) *The Foreign Policy*. Poland, etc. Castlereagh. Palmerston.[108]

V. THE REMEDY

Foolishness of the minor parliamentary propositions. Error of the Reform League.[109]

Repeal as one of the articles of the English Democratic Party.

Published in: Marx and Engels, *Collected Works*, second Russian ed., Vol. 16, Moscow, 1960

Printed according to the text of the book *The General Council of the First International. 1866-1868. Minutes*, Moscow

Karl Marx

[OUTLINE OF A REPORT ON THE IRISH QUESTION TO THE COMMUNIST EDUCATIONAL ASSOCIATION OF GERMAN WORKERS IN LONDON

DECEMBER 16, 1867[110]]

I

What is distinctive of Fenianism? Actually, it originates from the Irish Americans. They are the initiators and leaders. But in Ireland the movement took root (and is still really rooted) only in the mass of the people, the lower orders. That is what *characterises* it. In all earlier Irish movements the people followed the aristocracy or middle-class men, and always the Catholic churchmen. The Anglo-Irish chiefs and the priests during the rising against Cromwell; even James II, King of England, in the war against William III; the Protestant Republicans of Ulster (Wolfe Tone, Lord Fitzgerald)[111] in the 1798 revolution and, finally, in this century the bourgeois O'Connell supported by the Catholic clergy, which also played a leading role in all earlier movements excepting 1798. The Catholic clergy decreed a ban on Fenianism, which it did not lift until it realised that its attitude would deprive it of all influence on the Irish masses.

II

Here is what baffles the English: they find the present regime mild compared with England's former oppression of Ireland. So why this most determined and irreconcilable form of opposition now? What I want to show—and what even those Englishmen who side with the Irish, who concede them the right to secession, do not see—is that the regime since 1846, though less barbarian in form, is in effect destructive, leaving no alternative but Ireland's voluntary emancipation by England or life-and-death struggle.

III

Concerning past history the facts are available in any history book. Hence, I shall give only a few, firstly, to clarify the difference between the present and past and, secondly, to bring out a few points about the character of those who are now called the Irish people.

a) The English in Ireland Before the Protestant Reformation

1172. Henry II conquered less than $^1/_3$ of Ireland. It was a nominal conquest. A gift from Pope Adrian IV, the Englishman. Some 400 years later another Pope (in Elizabethan times, 1576), Gregory XIII, took back the present from the English (Elizabeth).[112] The *"English Pale"*.[113] Capital: *Dublin.* Mixing of English common colonists with Irish, and of Anglo-Norman nobles with Irish chiefs. Otherwise, the war of conquest was conducted (originally) as against Red Indians. No English reinforcements sent to Ireland until 1565 (Elizabeth).

b) Protestant Epoch. Elizabeth. James I. Charles I. Cromwell. Colonisation Plan (16th and 17th Centuries)

Elizabeth. The plan was to exterminate the Irish at least up to the river Shannon, to take their land and settle English colonists in their place, etc. In battles against Elizabeth the still Catholic Anglo-Irish fought the English alongside natives. The avowed plan of the English:
Clearing the island of the natives, and stocking it with loyal Englishmen. They succeeded only to plant a landowning aristocracy. English *Protestant "adventurers"* (merchants, usurers), who obtained from the English crown the confiscated lands, and "gentlemen undertakers" who were to plant the ceded estates with native English families.
James I. Ulster. (Jacobite plantation, 1609-12.) British undertakers, "to stock the confiscated, stolen lands with Irish". Not until *1613.* are Irish considered English *subjects;* previously they were looked upon as "outlaws" and "ene-

mies". The *Irish Parliament*[114] governed only the *Pale*.
Persecution of Catholics.

Elizabeth settled *Munster, James I Ulster*, but *Leinster*
and *Connaught* have not yet been purged. *Charles I* tried
to purge *Connaught*.

Cromwell. First national revolt of Ireland, its 2nd Complete Conquest. Partial Re-colonisation. (1641-60.)

Irish Revolution of *1641. August 1649 Cromwell landing
in Dublin.* (Followed by Ireton, Lambert, Fleetwood, Henry
Cromwell.)

In *1652 the 2nd Complete Conquest of Ireland completed.
Division of spoils*: the Government itself, the "*adventurers*" who had lent £360,000 for the 11 years of war, the
officers and soldiers, by the *Acts of the English Parliament,
12 August, 1652*, and *26 September, 1653*.[115] Smite the
Amalekites of the Irish Nation hip and thigh, and replant
the re-devastated land with new colonies of brand-new
Puritan English.—Bloodshed, devastation, depopulation of
entire counties, removal of their inhabitants to other regions,
sale of many Irish into slavery in the West Indies.

By engaging in the conquest of Ireland, Cromwell threw
the English Republic out the window.

Thence the Irish mistrust of the *English people's party*.

c) Restoration of the Stuarts. William III.
Second Irish Revolt, and the Capitulation
on Terms[116]

1660-1692.*
The British were then more numerous in Ireland than
at any other time. Never higher than ³/₁₁, never lower than
²/₁₁ of the Irish population.

1684. Charles II begins to favour the Catholic interest of
Ireland, and to enlist a Catholic army.

1685. James II gives full rein to the Catholics of Ireland. Catholic army increased and favoured. The Catholics
soon began to declare that the Acts of Settlement must be
repealed and the proprietors of 1641 re-established. James
calls some Irish regiments to England.

---

* Followed by "(1701) (Anne)" in the manuscript.—*Ed*

1689. William III in England. *12 March, 1689*: James landed at *Kinsale* at the head of Irish soldiers. *Limerick capitulates to William III, 1691.* Shameful violation of the treaty, already under William III, still more under Anne.

d) Ireland Defrauded and Humbled to the Dust. 1692-July 4, 1776

α) All notions of *"planting"* the country with English and Scotch yeomen or tenant farmers were discarded. Settling German and French Protestants attempted. French Protestants in the towns (woollen manufacturers) flee the English protectionist and mercantile system.

1698. The Anglo-Irish Parliament (like obedient colonists) passed, on the command of the mother country, *a prohibitory tax on Irish woollen goods export* to foreign countries.

1698. In the same year, the English Parliament laid a *heavy tax* on the import of the home manufactures in England and Wales, and *absolutely prohibited their export to other countries.* She struck down the manufactures of Ireland, depopulated her cities and threw the people back upon the land.

The *Williamite* (imported lords) *absentees.*[117] Cry *against absentee landlords since 1692.*

Similar *legislation of England against Irish Cattle.*

1698: Molyneux pamphlet for the independence of the Irish Parliament (i.e., the *English Colony in Ireland*) against the English.[118] Thus began the *struggle of the English Colony in Ireland and the English Nation. Simultaneously, struggle between the Anglo-Irish Colony and the Irish Nation.* William III resisted the shameful attempts of the English and *Anglo-Irish Parliaments to violate the treaties of Limerick and Galway.*

β) *Queen Anne. (1701-13; George until 1776.)*

Penal Code[119] *built up by the Anglo-Irish Parliament with assent of the English Parliament.* Most infamous means to make *Protestant Proselytes* amongst the Irish Catholics by regulations of *"Property".* A code for the *transfer of "Property"* from Catholics to Protestants, or to make *"Anglicanism"* a *proprietary title. (Education. Personal disabilities.)* (No Catholic able to be a private soldier.) To teach the

Catholic religion was a transportable felony, to convert a Protestant to Catholicism an act of treason. To be a Catholic Archbishop—banishment, if returning from banishment—act of high treason; hanged, disembowelled alive, and afterwards quartered.

Experiment to coerce the mass of the Irish nation into the Anglican religion. Catholics deprived of vote for members of Parliament.[120]

This *Penal Code intensified the hold of the Catholic Priesthood upon the Irish people.*

The poor people fell into habits of indolence.

During the palmy days of Protestant ascendancy and Catholic degradation, the Protestants did not encroach upon the Catholics in numbers.

e) 1776-1801. Time of Transition

α) Before dealing with this transition period, what was the result of English terrorism?

English incomers absorbed into the Irish people and Catholicised.

The towns founded by the English Irish.

No English colony (except *Ulster* Scotch) *but English landowners.*

The North American Revolution forms the first turning-point in Irish history.

β) *1777* the British army surrendered at Saratoga Springs to the American "rebels". *British cabinet forced to make concessions to the Nationalist (English) party in Ireland.*

1778. Roman Catholic Relief Bill (passed by *the Anglo-Irish Parliament*). (Catholics were still excluded from acquiring *by purchase, or as tenants, any freeholds*[121] *interest.*)

1779. Free Trade with Great Britain. Almost all restraints put upon Irish industry swept away.

1782. The Penal Code still further released. The Roman Catholics allowed to acquire *freehold property for life,* or *in fee simple,* and—to *open schools.*

1783. Equal rights of the *Anglo-Irish Parliament.*

Winter 1792-93. After the French Government had annexed Belgium and England resolved upon French war, another portion of the Penal Code was released. Irish could

become Colonels in Army, *elective franchise for Irish Par-
liament, etc.*
Rebellion of 1798. Belfast Republicans (Wolfe Tone, Lord
Fitzgerald). Irish peasants not ripe.
*Anglo-Irish House of Commons voted for the Act of
Union passed in 1800.* By the Legislature and Customs
Union of Britain and Ireland *closed the struggle between
the Anglo-Irish and the English.* The colony itself protested
against the illegal Act of Union.

1801-1846

a) *1801-1831.* At this time (after the end of the war[122]) a
movement for emancipation of Catholics under way among
Irish and English (1829).
From *1783 legislative independence of Ireland,* shortly
after which duties were imposed *on various articles of
foreign* manufacture, avowedly with the intention of en-
abling some of her people to employ some of their surplus
labour, etc.
The natural consequence was that Irish manufactures
gradually disappeared as the Act of Union came into effect.

Dublin

Master woollen manufacturers . .	*1800*	...	91	*1840* .·.	12
Hands employed	"	...	4,918	"	... 602
Master woolcombers	"	...	30	*1834* ...	5
Hands employed	"	...	230	"	... 66
Carpet manufacturers	"	...	13	*1841* ...	1
Hands employed.	"	...	720	"	... 0
Silk-loom weavers at work . . .	"	...	2,500	*1840* ... 250	

Kilkenny

Blanket manufacturers	*1800*	...	56	*1822* ...	42
Hands employed	"	...	3,000	"	... 925

Balbriggan

Calico-looms at work	*1799* ...	2,500	*1841* ...	226

Wicklow

Handlooms at work	*1800* ...	1,000	*1841* ...	0

Cork

Braid weavers	1800	... 1,000	1834	... 40
Worsted weavers	"	... 2,000	"	... 90
Hosiers	"	... 300	"	... 28
Woolcombers	"	... 700	"	... 110
Cottonweavers,	"	... 2,000	"	... 220

etc. The linen industry (Ulster) did not compensate for this.

"The *cotton manufacture of Dublin*, which employed 14,000 operatives, has been destroyed; the 3,400 silk looms have been destroyed; the serge manufacture, which employed 1,491 operatives, has been destroyed; the flannel manufacture of Rathdrum, the blanket manufacture of Kilkenny, the camlet trade of Bandon, the worsted manufactures of Waterford, the ratteen and frieze manufactures of Carrick-on-Suir have been destroyed. One business alone survives!... That fortunate business—which the Union Act has not struck down—that favoured, and privileged, and patronised business is the Irish coffin-maker's." (*Speech of T. F. Meagher, 1847.*)

Every time Ireland was about to develop industrially, she was crushed and reconverted into a purely agricultural land. After the latest General Census of *1861*:

Agricultural Population of Ireland (including all cottiers[123] and farm labourers with their families)	4,286,019
In the 798 towns (of which many were in fact small market towns)	1,512,948
	5,798,967.

Therefore (1861) approximately $\frac{4}{5}$ purely agricultural, and actually perhaps $\frac{6}{7}$ if market towns are also counted.

Ireland is therefore purely agricultural: "Land is life" (*Justice Blackburne*). Land became the great object of pursuit. The people had now before them the choice between the occupation of land, *at any rent*, or *starvation*. System of *rack-renting*.

"The lord of the land was thus enabled to dictate his own terms, and therefore it has been that we have heard of the payment of £5, 6, 8, and even as much as £10 per acre. Enormous rents, low wages, farms of an enormous extent, let by rapacious and indolent proprietors to *monopolising landjobbers*, to be relet by intermediate oppressors, for five times their value, among the wretched starvers on potatoes and water."

State of popular starvation.

Corn Laws in England create a monopoly to a certain extent for the export of Irish corn to England. The average export of grain in the *first 3 years* following the passage of the Act of Union about 300,000 qrs,

1820 over 1 million qrs,

1834 yearly average of 2½ million qrs.

Amount to pay *rent to absentees*, and interest to mortgagees (1834), over 30 million dollars (*or 7 million pounds sterling*). Middlemen accumulated fortunes that they *would* not invest in the improvement of land, and *could* not, under the system which prostrated manufactures, invest in machinery, etc. All their accumulations were sent therefore to England for investment. An official document published by the British Government shows that the transfers of British securities from England to Ireland, i.e., the investment of Irish capital in England, in the 13 years following the adoption of free trade in 1821, amounted to as many millions of pounds sterling, and thus was Ireland forced to contribute cheap labour and cheap capital to building up "the great works of Britain".

Many pigs and export of same.

1831-1841. Accretion of Ireland's population from *7,767,401 to 8,175,238*

In 10 years	407,837
In the same period there *emigrated* (somewhat more than 40,000 per year)	450,873
The total being	858,710.

O'Connell. Repeal Movement. Lichfield-House Contract with Whigs.[124] Partial famines. *Insurrection Acts, Arms Acts, Coercion Acts.*

IV

THE PERIOD OF THE LAST 20 YEARS (FROM 1846). CLEARING OF THE ESTATE OF IRELAND

Earlier, repeated cases of partial famine. Now famine was general.

This new period was ushered in by the potato blight (1846-47), starvation and the consequent exodus.

Over one million die, partly from hunger, partly from
diseases, etc. (caused by hunger). In nine years, 1847-55,
1,656,044 left the country.

The revolution of the old agricultural system was but a
natural result of the barren fields. People fled. (Families
clubbed together to send away the youngest and most
enterprising.) Hence, of course, the pooling of small lease-
holds and substitution of pasturage for crop farming.

However, soon circumstances arose whereby this became
a conscious and deliberate system.

Firstly, the chief factor: Repeal of the Corn Laws was
one of the direct consequences of the Irish disaster. As a
result, Irish corn lost its monopoly on the English market
in the ordinary years. Corn prices dropped. Rents could
no longer be paid. In the meantime, the price of meat, wool
and other animal products increased steadily in the preceding 20 years. Tremendous growth of the wool industry in
England. Pig-raising was partly connected with the old
system. Now, chiefly sheep and horned cattle. Deprived of
the English market now, as by the Act of Union of her own.

Contributing circumstances that made this systematic:

Secondly: Reorganisation of agriculture in England. Cari-
cature of same in Ireland.

Thirdly: The despairing flight of starving Irish to England
filled basements, hovels, workhouses in Liverpool, Manchester, Birmingham, Glasgow with men, women, children in a
state almost of starvation.

Act of Parliament passed (1847-48) that Irish landlords
had to support their own paupers. (The English Pauper
Law is extended to Ireland.) Hence, the Irish (especially
English) landlords, mostly deep in debt, try to get rid of
the people and clear their estates.

Fourthly: Encumbered Estates Act (*1853*?).

"The landlord was ruined, for he could collect no rents, and he
was at the same time liable for the payment of enormous taxes for
the maintenance of his poor neighbours. His land was encumbered with
mortgages and settlements, created when food was high, and he could
pay no interest; and now a law was passed, by aid of which property
could be summarily disposed of at a public sale, and the proceeds distributed among those who had legal claims upon it."

Absentee Proprietors. (English capitalists, insurance societies, etc., thereby multiplied, equally former middlemen, etc.,

who wanted to run their farms on modern economic lines.)
Eviction of farmers partly by friendly agreement
terminating tenure. But much more *eviction* en masse
(forcibly by *crowbar* brigades, beginning with the destruction of roofs), forcible ejection. (Also used as political
retribution.) This has continued since 1847 to this day.
(*Abercorn*, Viceroy of Ireland.) African razzias (razzias of
the little African kings). (People driven from the land. The
starving population of the towns largely increased.)

"The tenantry are turned out of the cottages by scores at a time....
Land agents direct the operation. The work is done by a large force
of police and soldiery. Under the protection of the latter, the 'crowbar
brigade' advances to the devoted township, takes possession of the
houses.... The sun that rose on a village sets on a desert." (*Galway
Paper*, 1852.) (*Abercorn.**)

Let us now see how this system affected the land in
Ireland, where conditions are quite different from those in
England.

Decrease of Cultivated Land. 1861-66

Decrease in cereal crops		*Decrease in green crops*
1861-65	428,041 acres	107,984 acres
1866	42,876	20,077
Total decrease	470,917	128,061

Decrease of Yield per Statute Acre of Every Crop

1847-1865 per cent: the exact decrease: *oats* 16.3, *flax*
47.9, *turnips* 36.1, *potatoes* 50. Some years would show a
greater decrease, but on the whole it has been gradual since
1847.

ESTIMATED AVERAGE PRODUCE PER
STATUTE ACRE

	Wheat cwts	*Potatoes tons*	*Flax stones (14 lbs.)*
1851	12.5	5.1	38.6
1866	11.3	2.9	24.9

* See pp. 143-44.—*Ed*.

Though Ireland exported considerable quantities of wheat in the past, it is now said to be good only for cultivating *oats* (the yield of which per acre also continuously decreases).

In fact: *1866* Ireland shipped out only 13,250 qrs of wheat against 48,589 qrs shipped in (that is, almost fourfold). Meanwhile, it shipped out approximately one million qrs of oats (for £1,201,737).

Since the exodus, the land has been underfed and overworked, partly from the injudicious consolidation of farms, and partly because, under the corn-acre system,[125] the farmer in a great measure trusted to his labourers to manure the land for him. Rents and profits (where the farmer is no peasant farmer) may increase, although the produce of the soil decreases. The total produce may diminish, and still greater part of it be converted into surplus produce, falling to the landlord and (great) farmer. And the price of the surplus produce has risen.

Hence, *sterilisation* (gradual) of the land, as in *Sicily by the ancient Romans* (ditto in *Egypt*).

We shall speak of the *livestock*, but first about the *population*.

Decrease of the Population

1801: 5,319,867; *1841*: 8,222,664; *1851*: 6,515,794; *1861*: 5,764,543. If the trend continues, there will be 5,300,000 in *1871*, that is, less than in 1801. I shall now show, however, that the population will be lower still in 1871, even though the emigration rate remains constant.

Emigration

Emigration accounts naturally for part of the decrease. In *1845-66* there emigrated 1,990,244, or approximately *2,000,000 Irish*. (Unheard of.) (About $2/5$ of the *total emigration* from the *United Kingdom in 1845-66*, which was 4,657,588.) In 1831-41 emigration approximately equalled *half the accretion* of population during the decade, and after 1847 it was considerably higher than the accretion.

However, emigration alone does not account for the decrease of the population since 1847.

Decrease of the Natural Annual Accretion of the Population

The accretion (annual) in *1831-41* was 1.1 per cent, or about $1^1/_{10}$ per cent a year. If the population had increased in the same proportion in *1841-51*, it would have been 9,074,514 in *1851*. In fact, however, it was only 6,515,794. Consequently, the deficit was 2,558,720. Out of this figure, emigration accounted for 1,274,213. That leaves 1,284,507 unaccounted for. Over a million, but not the whole deficit of 1,284,507, died in the famine. Hence, evidently, natural population growth decreased in 1841-51.

This is borne out by the *decade of 1851-61*. No famine. The population decreased from 6,515,794 to 5,764,543. Absolute decrease: 751,251. Yet emigration in this period claimed over 1,210,000. Hence, there was an accretion of nearly 460,000 during the ten years. Because 751,251 + 460,000=the number of emigrants=1,211,251. Emigration claimed almost triple the accretion. The rate of accretion was 0.7 per cent per year, hence considerably lower than the 1.1 per cent of 1831-41.

The explanation is very simple. The increase of a population by births must principally depend on the proportion which those between 20 and 35 bear to the rest of the community. Now the proportion of persons between the ages of 20 and 35 in the population of the United Kingdom is about 1:3.98 or 25.06 per cent, while their proportion in the emigration even of the present day is about 1:1.89 or 52.76 per cent. And probably still greater in Ireland.

Physical Deterioration of the Population

In *1806*, with a total population of 5,574,107, there was an excess of males over females by 50,469, whilst in 1867, with a total population of 5,557,196, there is an excess of the females over males. At the same time *not only a relative*, but an *absolute increase* in the number of deaf-mutes, blind, insane, idiotic, and decrepit inhabitants. Contrasting 1851 with 1861, whilst the population had decreased enormously, the number of deaf-mutes had increased by 473, on their former total of 5,180; the lame and decrepit by 225, on their former total of 4,375; the blind by 1,092, on their former total of 5,767; the lunatic and idiotic, by the immense number of 4,118, on their former total of 9,980; mounting up, in 1861, notwithstanding the decrease in the population, to 14,098.

Wages

Wages have not risen more than 20 per cent since the potato famine. The price of potatoes has risen nearly 200 per cent, and 100 per cent on an average of essential food products.

Professor Cliffe Leslie, in the *Economist of February 9, 1867*, says:

"After a loss of two-fifths of the population in 21 years, throughout most of the island the rate of wages is now only 1s. a day; a shilling does not go farther than 6d. did 21 years ago. Owing to this rise in the ordinary food the labourer is worse off than he was ten years ago."

Partial famines especially in Munster and Connaught.

Bankruptcy of shopkeepers is permanent. Market towns, etc., fall to ruin.

The Results of This Process

In *1855-66*, 1,032,694 Irishmen replaced by *996,877* head of livestock (cattle, sheep and pigs). That, in fact, was the accretion of livestock during that period, with the decrease of horses (20,656) compensated by eight sheep (to one horse), which are therefore subtracted from the accretion.

Consolidation of Farms

From *1851 to 1861* the total decrease of farms was 120,000. (Though the number of 15-30 acre farms and farms of 30 acres and over increased.) Thus, the decrease affected particularly farms of one to under 15 acres.

In 1861 about $3/5$ of the area (Ireland's total area: 20,319,924 acres) or 12,000,000 acres was held by *569,844 tenants* who worked plots of one up to less than 100 acres, and about $2/5$ (8 million acres) by tenants with over 100 and 500 acres and over (*31,927 tenants*).

The process of consolidation in full gear. Ulster. (Cultivation of flax; Scottish Protestant tenants.)

The Times, etc., officially congratulates Abercorn as

Viceroy on this system. He, too, is one of these devastators. Lord Dufferin: over-population, etc.*
In sum, it is a question of life and death.
Meagher, Hennessy,** *Irishman.*[126]

DECREASE OF CRIME IN IRELAND

	Committed for trial	Convicted
1852	17,678	10,454
1866	4,326	2,418

V
UNITED STATES AND FENIANISM

Published in:
Marx and Engels,
Collected Works, second
Russian ed., Vol. 16,
Moscow, 1960

Printed according to the manuscript in German and English
Part of the manuscript translated from the German

---

* See pp. 113-14.—*Ed.*
** See p. 148.—*Ed.*

On December 16, Karl Marx delivered a lecture to the London German Workers' Educational Association on the conditions in Ireland, in which he proved that all attempts of the English government to anglicise the Irish population in past centuries had ended in failure. The English, including aristocrats, who immigrated before the Reformation[127] were transformed into Irishmen by their Irish wives and their descendants fought against England. The brutalities of the war against the Irish under Queen Elizabeth, the destruction of crops and the displacement of the population from one area to another to make place for English colonists did not change anything. At that time, gentlemen and merchant adventurers received large plots of land on condition that they would be colonised by English people. In Cromwell's time, the descendants of these colonists fought with the Irish against the English. Cromwell sold many of them as slaves in the West Indies. Under the Restoration,[128] Ireland received many favours. Under William III, a class came to power which only wanted to make money, and Irish industry was suppressed in order to force the Irish to sell their raw materials to England at any price. With the help of the Protestant Penal Laws, the new aristocrats received freedom of action under Queen Anne. The Irish Parliament[129] was a means of oppression. Those who were Catholic were not allowed to hold an official post, could not be landowners, were not allowed to make wills, could not claim an inheritance; to be a Catholic bishop was high treason. All these were means for robbing the Irish of their land; yet over 50 per cent of the English descendants in Ulster have remained Catholic. The people were driven into the arms of the Catholic clergy, who thus became powerful. All that the English government succeeded in doing was to plant an aristocracy in Ireland. The towns built by the English have become Irish. That is why there are so many English names among the Fenians.

During the American War of Independence the reins were loosened a little. Further concessions had to be granted during the French revolution. Ireland rose so quickly that her people threatened to outstrip the English. The English government drove them to rebellion and achieved the Union by bribery. The Union delivered the death blow to reviving Irish industry. On one occasion Meagher said: all Irish branches of industry have been destroyed, all we have been left is the making of coffins. It became a vital necessity to have land; the big landowners leased their lands to speculators; land passed through four or five lease stages before it reached the peasant, and this made prices disproportionately high. The agrarian population lived on potatoes and water; wheat and meat were sent to England; the rent was eaten up in London, Paris and Florence. In 1836, £7,000,000 was sent abroad to absent landowners. Fertilisers were exported with the produce and rent, and the soil was exhausted. Famine often set in here and there, and owing to the potato blight there was a general famine in 1846. A million people died of starvation. The potato blight resulted from the exhaustion of the soil, it was a product of English rule.

Through the repeal of the Corn Laws Ireland lost her monopoly position on the English market, the old rent could no longer be paid. High prices for meat and the bankruptcy of the still remaining small landowners further contributed to the eviction of the small peasants and the transformation of their land into sheep pastures. Over half a million acres of arable land have not been tilled since 1860. The yield per acre has dropped: oats by 16 per cent, flax by 36 per cent, potatoes by 50 per cent. At present only oats are cultivated for the English market, and wheat is imported.

With the exhaustion of the soil, the population has deteriorated physically. There has been an absolute increase in the number of lame, blind, deaf and dumb, and insane in the decreasing population.

Over 1,100,000 people have been replaced with 9,600,000 sheep. This is a thing unheard of in Europe. The Russians replace evicted Poles with Russians, not with sheep. Only under the Mongols in China was there once a discussion whether or not to destroy towns to make place for sheep.

The Irish question is therefore not simply a nationality question, but a question of land and existence. Ruin or revolution is the watchword; all the Irish are convinced that if anything is to happen at all it must happen quickly. The English should demand separation and leave it to the Irish themselves to decide the question of landownership. Everything else would be useless. If that does not happen soon the Irish emigration will lead to a war with America. The domination over Ireland at present amounts to collecting rent for the English aristocracy.

Published in:
Marx and Engels,
Collected Works, second
Russian ed., Vol. 16,
Moscow, 1960

Translated from the
German

Karl Marx and Frederick Engels

EXCERPTS FROM LETTERS ON IRELAND WRITTEN BETWEEN 1867 AND 1868

MARX TO LUDWIG KUGELMANN

October 11, 1867

Ernest Jones was to speak in Ireland to *Irish people* as a representative of the party, and, since big landownership there is identical with *England's ownership of Ireland*, he was to speak *against* big landownership. You should never look for general principles in the hustings speeches of English politicians but only for what is useful for the *immediate* aim.

MARX TO ENGELS

November 2, 1867

The proceedings against the Fenians in Manchester were every inch what could be expected.[130] You will have seen what a row "our people" kicked up in the Reform League. I have sought in every way to provoke this manifestation of the English workers in support of Fenianism.[131]
Greetings.

Yours,
K. M.

Previously I thought Ireland's separation from England impossible. Now I think it inevitable, although after separation there may come *federation*. How the English carry on is evidenced by the *Agricultural Statistics* for the current year,[132] which appeared a few days ago. Furthermore, the form of the eviction. The Irish Viceroy, Lord Abicorn*

---

* Lord Abercorn.—*Ed.*

(that *seems* to be his name), "cleared" his estate in the last few weeks by forcibly evicting thousands of people. Among them were prosperous tenants, whose improvements and investments were thus confiscated! In no other European country did foreign rule adopt this form of direct expropriation of the stock population. The Russians confiscate solely on political grounds; the Prussians in Western Prussia buy out.

ENGELS TO MARX

November 5, 1867

How low the English judges have sunk was demonstrated yesterday by Blackburne when he asked witness Beck (who first swore to *William* Martin but later said that it was *John* M.): "Then you swore to William and you *meant* to swear to John?" I think the whole prosecution will fall to pieces more and more with each new batch of accused; perjury for a reward of £200 is simply incredible.

Can you tell me where I can read in greater detail about Lord Abercorn's evictions?

MARX TO ENGELS

November 7, 1867

There was a detailed description of the Abercorn evictions about a fortnight ago in *The Irishman* (Dublin). I may manage to get again the issue that was lent to me for only 24 hours.

At the meeting, at which Colonel Dickson presided and Bradlaugh made a speech about Ireland, our old Weston, seconded by Fox and Cremer, tabled a resolution for the Fenians which was passed unanimously. Last Tuesday, too, there was a stormy demonstration for the Fenians[133] during Acland's lecture on the Reform Bill in Cleveland Hall (above our heads, we had our meeting down in the coffee room, which is in the basement). This business stirs the feelings of the intelligent part of the working class here.

ENGELS TO LUDWIG KUGELMANN

November 8, 1867

The Irish, too, are a very substantial ferment in this business, and the London proletarians declare every day more openly for the Fenians and, hence—an unheard-of and splendid thing here—for, first, a violent and, secondly, an anti-English movement.

ENGELS TO MARX

November 24, 1867

Dear Moor,

I am returning the encl. letters.

So yesterday morning the Tories, by the hand of Mr. Colcraft, accomplished the final act of separation between England and Ireland. The *only thing* that the Fenians still lacked were martyrs. They have been provided with these by Derby and G. Hardy. Only the execution of the three[*] has made the liberation of Kelly and Deasy the heroic deed as which it will now be sung to every Irish babe in the cradle in Ireland, England and America. The Irish women will do that just as well as the Polish women.

To my knowledge, the only time that anybody has been executed for a similar matter in a civilised country was the case of John Brown at Harpers Ferry. The Fenians could not have wished for a better precedent. The Southerners had at least the decency to treat J. Brown as a *rebel*, whereas here everything is being done to transform a political attempt into a common crime.

ENGELS TO MARX

November 29, 1867

As regards the Fenians you are quite right.[134] The beastliness of the English must not make us forget that the leaders of this sect are mostly asses and partly exploiters and we

---

[*] Michael Larkin, William Allen and Michael O'Brien.—*Ed.*

cannot in any way make ourselves responsible for the stupidities which occur in every conspiracy. And they are certain to happen.

I need not tell you that black and green predominate in my home too.[135] The English press has once again behaved most meanly. Larkin is said to have fainted and the others* to have looked pale and confused. The Catholic priests who were there declare that this is a lie. Larkin, they say, *stumbled* on a rough spot and the three of them showed great courage. The Catholic bishop of Salford complained bitterly that Allen would not repent of his deed, saying he had nothing to repent of and were he at liberty he would do the same again. By the way, the Catholic priests were very insolent—on Sunday it was given out from the pulpit in all churches that these three men had been *murdered.*

MARX TO ENGELS

November 30, 1867

If you read the papers you will have seen that 1) the Memorial of the International Council for the Fenians** was sent to Hardy, and that 2) the debate on Fenianism was public (last Tuesday*** week) and reported in *The Times.*[136] Reporters of the Dublin *Irishman* and *Nation* were among those present. I came very late (I ran a temperature for about a fortnight and the fever passed only two days ago) and really did not intend to speak, firstly because of my troublesome physical condition, and secondly because of the ticklish situation. However Weston, who was in the chair, tried to force me to, so I moved for an adjournment, which obliged me to speak last Tuesday.**** As a matter of fact what I had prepared for Tuesday last was not a speech but the points of a speech.***** But the Irish reporters failed to come, and waiting for them it had become 9 o'clock, while the establishment was at our disposal only till 10.30. Fox

---

* William Allen and Michael O'Brien.—*Ed.*
** See pp. 118-19.—*Ed.*
*** The 19th of November. See pp. 368-72.—*Ed.*
**** November 26th.—*Ed.*
***** See pp. 120-25.—*Ed.*

(because of the quarrel in the Council he had not shown himself for the past 2 weeks, and had moreover sent in his resignation as member of the Council, containing rude attacks on Jung[137]) had, at my request, prepared a long speech. After the opening of the sitting I therefore stated I would yield the floor to Fox on account of the belated hour. Actually— because of the Manchester executions that had taken place in the meantime—our subject, Fenianism, was liable to inflame the passions to such heat that *I* (but not the abstract *Fox*) would have been forced to hurl revolutionary thunderbolts instead of soberly analysing the state of affairs and the movement as I had intended. The Irish reporters therefore, by staying away and delaying the opening of the meeting, did signal service for me. I don't like to mix with a crowd like Roberts, Stephens, and the rest.

Fox's speech was good, for one thing because it was delivered by an *Englishman* and for another because it concerned only the political and international aspects. For that very reason he just skimmed along the surface of things. The resolution he handed up was absurd and inane. I objected to it and had it referred to the Standing Committee.[138]

What the English do not yet know is that since 1846 the economic content and therefore also the political aim of English domination in Ireland have entered into an entirely new phase, and that, precisely because of this, Fenianism is characterised by a socialistic tendency (in a negative sense, directed against the appropriation of the soil) and by being a lower orders movement. What can be more ridiculous than to confuse the barbarities of Elizabeth or Cromwell, who wanted to supplant the Irish by English colonists (in the Roman sense), with the present system, which wants to supplant them by sheep, pigs and oxen! The system of 1801-46, with its rack-rents and middlemen, collapsed in 1846. (During that period evictions were exceptional, occurring mainly in Leinster where the land is especially good for cattle-raising.) The repeal of the Corn Laws, partly the result of or at any rate hastened by the Irish famine, deprived Ireland of its *monopoly* of England's corn supply in normal times. Wool and meat became the slogan, hence conversion of tillage into pasture. Hence from then onwards systematic consolidation of farms. The Encumbered Estates

Act, which turned a mass of previously enriched middle-men into landlords, hastened the process. *Clearing of the Estate of Ireland!* is now the one purpose of English rule in Ireland. The *stupid* English government in London knows nothing of course itself of this immense change since 1846. But the Irish know it. From *Meagher's Proclamation* (1848) down to the *election manifesto of Hennessy* (Tory and Urquhartite) (1866), the Irish have expressed their conscious-ness of it in the clearest and most forcible manner.

The question now is, what shall *we* advise the *English* workers? In my opinion they must make the *Repeal of the Union* (in short, the *affair of 1783*, only democratised and adapted to the conditions of the time) an article of their *pronunziamento.*[139] This is the only *legal* and therefore only possible form of Irish emancipation which can be admitted in the programme of an *English* party. Experience must show later whether a mere personal union can continue to subsist between the two countries. I half think it can if it takes place in time.

What the Irish need is:

1) Self-government and independence from England.

2) An agrarian revolution. With the best intentions in the world the English cannot accomplish this for them, but they can give them the legal means of accomplishing it for them-selves.

3) *Protective tariffs against England.* Between 1783 and 1801 every branch of Irish industry flourished. The Union, which overthrew the protective tariffs established by the Irish Parliament, destroyed all industrial life in Ireland. The bit of linen industry is no compensation whatever. The Union of 1801 had just the same effect on Irish industry as the measures for the suppression of the Irish woollen in-dustry, etc., taken by the English Parliament under Anne, George II, and others. Once the Irish are independent, neces-sity will turn them into protectionists, as it did Canada, Australia, etc.

Before I present my views in the Central Council (next Tuesday, this time fortunately *without* reporters),[140] I would like you to give me your opinion in a few lines.

MARX TO ENGELS

December 14, 1867

Dear Fred,

The last exploit of the Fenians in Clerkenwell[141] was a very stupid thing. The London masses, who have shown great sympathy for Ireland, will be made wild by it and driven into the arms of the government party. One cannot expect the London proletarians to allow themselves to be blown up in honour of the Fenian emissaries. There is always a kind of fatality about such a secret, melodramatic sort of conspiracy.

ENGELS TO MARX

December 19, 1867

The stupid affair in Clerkenwell was obviously the work of a few specialised fanatics; it is the misfortune of all conspiracies that they lead to such stupidities, because "after all something must happen, after all something must be done". In particular, there has been a lot of bluster in America about this blowing up and arson business, and then a few asses come and instigate such nonsense. Moreover, these cannibals are generally the greatest cowards, like this Allen, who seems to have already turned Queen's evidence, and then the idea of liberating Ireland by setting a London tailor's shop on fire!

MARX TO ENGELS

March 16, 1868

The present way in which the English treat political prisoners in Ireland, and also suspects, or even those sentenced to ordinary prison terms (like Pigott of *The Irishman* and Sullivan of the *News*)[142] is really worse than anything happening on the Continent, except in Russia. What dogs!

MARX TO LUDWIG KUGELMANN

April 6, 1868

The Irish question predominates here just now. It has been exploited by Gladstone and company, of course, only in order to get into office again, and, above all, to have an *electoral cry* at the next elections, which will be based on household suffrage.[143] *For the moment* this turn of events is bad for the workers' party; the intriguers among the workers, such as Odger and Potter, who want to get into the next Parliament, have now a new *excuse* for attaching themselves to the bourgeois Liberals.

However, this is only a *penalty* which England—and consequently also the English working class—is paying for the great crime she has been committing for many centuries against Ireland. And in the long run it will benefit the English working class itself. You see, the *English* Established *Church in Ireland*—or what they use to call here the *Irish Church*—is the religious bulwark of *English landlordism* in Ireland, and at the same time the outpost of the Established Church in England herself. (I am speaking here of the Established Church as a *landowner*.) The overthrow of the Established Church in Ireland will mean its downfall in England and the two will be followed by the doom of landlordism—first in Ireland and then in England. I have, however, been convinced from the first that the social revolution must begin *seriously* from the bottom, that is, from landownership.

Karl Marx

[ON THE REFUSAL BY THE ENGLISH PRESS TO TAKE NOTICE OF THE GROWTH OF SYMPATHY WITH IRELAND AMONG ENGLISH WORKERS AND ON THE OPENING OF THE DEBATE ON THE IRISH QUESTION

(Record of the Speech and Content of the Letter. From the Minutes of the General Council Meetings of October 26 and November 9, 1869)]

I

Cit. *Marx* said the principal thing was whatever was passed would be suppressed by the London press. The main feature of the demonstration[144] had been ignored, it was that at least a part of the English working class had lost their prejudice against the Irish. This might be put in writing and addressed to somebody, not the government. He thought it a good opportunity to do something. . . .

II

The *Secretary* reported from the Sub-Committee that it had been agreed not to proceed with an address on the Irish question[145] because if the views of the Council were properly set forth, the government and the press would turn them against the prisoners.

Cit. *Jung* read a letter from Cit. Marx in support of the report and, if adopted, Cit. Marx proposed the discussion of the following questions: (1) The attitude of the British Government on the Irish question; (2) The attitude of the English working class towards the Irish. Cit. Marx volunteered to open the debate.

The report was adopted and the questions ordered to be put on the order of the day.

Published in the book *The General Council of the First International. 1868-1870. Minutes,* Moscow

Printed according to the text of the book

Karl Marx

[ON THE POLICY OF THE BRITISH GOVERNMENT WITH RESPECT TO THE IRISH PRISONERS

(Record of the Speech and Draft Resolution.
From the Minutes of the General Council Meeting
of November 16, 1869)]

Cit. *Marx* then opened the debate on the attitude of the British Government on the Irish question. He said political amnesty proceeds from two sources: 1. When a government is strong enough by force of arms and public opinion, when the enemy accepts the defeat, as was the case in America,[146] then amnesty is given. 2. When misgovernment is the cause of quarrel and the opposition gains its point, as was the case in Austria and Hungary.[147] Such ought to have been the case in Ireland.

Both Disraeli and Gladstone have said that the government ought to do for Ireland what in other countries a revolution would do. Bright asserted repeatedly that Ireland would always be rife for revolution unless a radical change was made. During the election Gladstone justified the Fenian insurrection and said that every other nation would have revolted under similar circumstances. When taunted in the House he equivocated his fiery declarations against the "policy of conquest"[148] implied that "Ireland ought to be ruled according to Irish ideas". To put an end to the "policy of conquest" he ought to have begun like America and Austria by an amnesty as soon as he became minister. He did nothing. Then the amnesty movement in Ireland by the municipalities. When a deputation was about to start with a petition containing 200,000 signatures for the release of the prisoners he anticipated it by releasing some to prevent the appearance of giving way to Irish pressure. The petition came, it was not got up by Fenians, but he gave no answer. Then it was mooted in the House that the prisoners were infamously treated. In this at least the English Government is impartial; it treats Irish and English alike; there is no country in Europe where political prisoners are treated like in England and Russia. Bruce was obliged to admit the fact. Moore wanted an inquiry; it was refused. Then commenced the popular amnesty movement at Limerick. A meeting was held at which 30,000 people were present and a memorial for the unconditional release was adopted. Meetings were held in all the towns in the North. Then the great meeting was announced in Dublin where 200,000 people attended. It was announced weeks beforehand for the 10th October. The trade societies wanted to go in procession. On the 8th proclamations were issued prohibiting the procession to go through certain streets. Isaac Butt interpreted it as a prohibition of the procession. They went to Fortescue to ask but he was not at home, his Secretary Burke did not know. A letter was left to be replied to; he equivocated. The government wanted a collision. The procession was abandoned and it was found afterwards that the soldiers had been supplied with 40 rounds of shot for the occasion.

After that Gladstone answered the Limerick memorial of August in a roundabout way.[149] He says the proceedings varied much. There were loyal people and others who used bad language demanding as a right what could only be an act of clemency.

It is an act of presumption on the part of a paid public servant to teach a public meeting how to speak.

The next objection is that the prisoners have not abandoned their designs which were cut short by their imprisonment.

How does Gladstone know what their designs were and that they still entertain them? Has he tortured them into a confession? He wants them to renounce their principles, to degrade them morally. Napoleon did [not] ask people to renounce their republican principles before he gave an amnesty and Prussia attached no such conditions.

Then he says the conspiracy still exists in England and America.

If it did, Scotland Yard would soon be down upon it. It is only "disaffection of 700 years' standing". The Irish have declared they would receive unconditional freedom as an act of conciliation. Gladstone cannot quell the Fenian conspiracy in America, his conduct promotes it, one paper calls him the Head Centre.[150] He finds fault with the press. He has not the courage to prosecute the press; he wants to make the prisoners responsible. Does he want to keep them as hostages for the good behaviour of the people outside? He says "it has been our desire to carry leniency to the utmost point". This then is the utmost point.

When Mountjoy was crowded with untried prisoners, Dr. M'Donnell wrote letter after letter to Joseph Murray about their treatment. Lord Mayo said afterwards that Murray had suppressed them. M'Donnell then wrote to the inspector of prisons, to a higher official. He was afterwards dismissed and Murray was promoted.*

He then says: we have advised the minor offenders to be released; the principal leaders and organisers we could not set free.

This is a positive lie. There were two Americans amongst them who had 15 years each. It was fear for America that made him set them free. Carey was sentenced in 1865 to 5 years, he is in the lunatic asylum, his family wanted him home, he could not upset the government.

He further says: to rise in revolt against the public order has ever been a crime in this country. Only in this country. Jefferson Davis's revolt was right because it was not against the English, the government.[151] He continues, the administration can have no interest except the punishment of crimes.

The administration are the servants of the oppressors of Ireland. He wants the Irish to fall on their knees because an enlightened sovereign and Parliament have done a great act of justice. They were the criminals before the Irish people. But the Irish was the only question upon which Gladstone and Bright could become ministers and catch the dissenters and give the Irish place-hunters an excuse of selling themselves.[152] The church was only the badge of conquest. The badge is removed, but the servitude remains. He states that the government is resolved to continue to remove any grievance, but that they are determined to give security to life and property and maintain the integrity of the empire.

---

* See p. 167.—*Ed.*

Life and property are endangered by the English aristocracy. Canada makes her own laws[153] without impairing the integrity of the empire, but the Irish know nothing of their own affairs, they must leave them to Parliament, the same power that has landed them where they are. It is the greatest stupidity to think that the prisoners out of prison could be more dangerous than insulting a whole nation. The old English leaven of the conqueror comes out in the statement: we will grant but you must ask.

In his letter to Isaac Butt he says:

> "You remind me that I once pleaded for foreigners. Can the two cases correspond? The Fenians were tried according to lawful custom and found guilty by a jury of their countrymen. The prisoners of Naples were arrested and not tried and when they were tried they were tried by exceptional tribunals and sentenced by judges who depended upon the government for bread."

If a poacher is tried by a jury of country squires he is tried by his countrymen. It is notorious that the Irish juries are made up of purveyors to the castle whose bread depends upon their verdict. Oppression is always a lawful custom. In England the judges can be independent, in Ireland they cannot. Their promotion depends upon how they serve the government. Sullivan the prosecutor has been made master of the rolls.

To the Ancient Order of Foresters in Dublin he answered that he was not aware that he had given a pledge that Ireland was to be governed according to Irish ideas.[154] And after all this he comes to Guild-Hall and complains that he is inadequate for the task.

The upshot is that all the tenant right meetings are broken up; they want the prisoners [released]. They have broken with the clerical party. They now demand that Ireland is to govern herself. Moore and Butt have declared for it.* They have resolved to liberate O'Donovan Rossa by electing him a member of Parliament.[155]

Cit. Marx ended by proposing the following resolution:
Resolved,

---

* This sentence was inserted between the lines of the Minute Book. —*Ed.*

That in his reply to the Irish demands for the release of the imprisoned Irish patriots (in a reply contained in his letter to Mr. O'Shea d.d. Oct. 18, 1869, and to Mr. Isaac Butt d.d. Oct. 23, 1869) Mr. Gladstone has deliberately insulted the Irish nation;

That he clogs political amnesty with conditions alike degrading to the victims of misgovernment and the people they belong to;

That having in the teeth of his responsible position publicly and enthusiastically cheered on the American slaveholders' rebellion, he now steps in to preach to the Irish people the doctrine of passive obedience;

That his whole proceedings with reference to the Irish amnesty question are the true and genuine offspring of that *"policy of conquest"* by the fiery denunciation of which Mr. Gladstone ousted his Tory rivals from office;

That the *General Council of the International Working Men's Association* express their admiration of the spirited and high-souled manner in which the Irish people carry on their amnesty movement;

That this resolution be communicated to all the branches of, and working men's bodies connected with, the *International Working Men's Association* in Europe and the United States.

The resolution published in November-December in a number of the International's papers. Record of the speech published in: Marx and Engels, *Collected Works*, second Russian ed., Vol. 16, Moscow, 1960

Printed according to the text of the book *The General Council of the First International. 1868-1870. Minutes,* Moscow

Karl Marx

[ON THE POLICY OF THE BRITISH GOVERNMENT WITH RESPECT TO THE IRISH PRISONERS

(Record of the Speeches in Support of the General Council Resolution. From the Minutes of the General Council Meetings of November 23 and 30, 1869)]

I

Cit. *Marx.* Cit. Mottershead has given a history of Gladstone. I could give another, but that has nothing to do with the question before us. The petitions which were adopted at the meetings were quite civil, but he found fault with the speeches by which they were supported. Castlereagh was as good a man as Gladstone and I found today in the *Political Register* that he used the same words against the Irish as Gladstone, and Cobbett made the same reply as I have done.[156]

When the electoral tour commenced all the Irish candidates spouted about amnesty, but Gladstone did nothing till the Irish municipalities moved.

I have not spoken of the people killed abroad, because you cannot compare the Hungarian war with the Fenian insurrection.[157] We might compare it with 1798[158] and then the comparison would not be favourable to the English.

I repeat that political prisoners are not treated anywhere so bad as in England.

Cit. Mottershead is not going to tell us his opinion of the Irish; if he wants to know what other people think of the English let him read Ledru-Rollin[159] and other Continental writers. I have always defended the English and do so still.

These resolutions are not to be passed to release the prisoners, the Irish themselves have abandoned that.

It is a resolution of sympathy with the Irish and a review of the conduct of the government, it may bring the English and the Irish together. Gladstone has to contend with the opposition of the *Times,* the *Saturday Review,* etc., if we speak out boldly; on the other side, we may support him against an opposition to which he might otherwise have to succumb. He was in office during the Civil War and was responsible for what the government did and if the North was low when he made his declaration, so much the worse for his patriotism.

Cit. Odger is right, if we wanted the prisoners released, this would not be the way to do it, but it is more important to make a concession to the Irish people than to Gladstone. . . .

Cit. *Marx* had no objection to leave out the word "deliberately", as a Prime Minister must necessarily be considered to do everything deliberately.[160]

II

Cit. *Marx* said if Odger's suggestions were followed the Council would put themselves on an English party standpoint.[161] They could not do that. The Council must show the Irish that they understood the question and the Continent that they showed no favour to the British Government. The Council must treat the Irish like the English would treat the Polish.

Published in: Marx and Engels, *Collected Works,* second Russian ed., Vol. 16, Moscow, 1960, and in the book *The General Council of the First International. 1868-1870. Minutes,* Moscow

Printed according to the text of the book *The General Council of the First International.   1868-1870. Minutes,* Moscow

Karl Marx

[ON THE SIGNIFICANCE OF THE IRISH QUESTION

*(Record of the Speech. From the Minutes of
the General Council Meeting of December 14, 1869)*]

Cit. *Marx* proposed that the Council at its rising should adjourn to January 4th. He said it would not be advisable to discuss the Irish during the holiday weeks when the attendance of members might be small.[162] He considered the solution of the Irish question as the solution of the English, and the English as the solution of the European.

The proposition was agreed to.

Published in the book *The General
Council of the First International.
1868-1870. Minutes*, Moscow

Printed according to the
text of the book

Karl Marx

From CONFIDENTIAL COMMUNICATION[163]

4) *Question of separating the General Council from the Federal Council for England.*

Long before the foundation of *L'Égalité*, this proposition used to be made periodically inside the General Council by one or two of its English members.[164] It was always rejected almost unanimously.

Although revolutionary *initiative* will probably come from France, England alone can serve as the *lever* for a serious *economic* revolution. It is the only country where there are no more peasants and where land property is concentrated in a few hands. It is the only country where the *capitalist form*, i.e., combined labour on a large scale under capitalist masters, embraces virtually the whole of production. It is the only country *where the great majority of the population consists of wages labourers*. It is the only country where the class struggle and organisation of the working class by the *Trades Unions* have *acquired* a certain degree of maturity and universality. It is the only country where, because of its domination on the world market, every revolution in economic matters must immediately affect the whole world. If landlordism and capitalism are classical examples in England, on the other hand, the *material conditions* for their *destruction* are the most mature here. The General Council now being in the *happy position of having its hand directly on this great lever of proletarian revolution*, what folly, we might say even what a crime, to let this lever fall into purely English hands!

The English have all the *material* prerequisites necessary for the social revolution. What they lack is the *spirit of ge-*

neralisation and *revolutionary fervour*. Only the General Council can provide them with this, can thus accelerate the truly revolutionary movement here, and in consequence, *everywhere*. The great effect we have already had is attested to by the most intelligent and influential of the newspapers of the ruling classes, as, e.g., *Pall Mall Gazette, Saturday Review, Spectator* and *Fortnightly Review*, not to speak of the so-called radicals in the *Commons* and the *Lords* who a little while ago still exerted a big influence on the leaders of the English workers. They accuse us publicly of having poisoned and almost extinguished the *English spirit* of the working class and of having pushed it into revolutionary socialism.

The only way to bring about this change is to agitate like the *General Council* of the *International Association*. As the General Council we can initiate measures (e.g., the founding of the *Land and Labour League*[165]) which as a result of their execution will later appear to the public as spontaneous movements of the English working class.

If a *Regional Council* were formed outside of the *General Council*, what would be the immediate effects?

Placed between the *General Council* and the *General Trades Union Council*, the *Regional Council* would have no authority. On the other hand, the *General Council of the International* would lose *this great lever*. If we preferred the showman's chatter to serious action behind the scenes, we would perhaps commit the mistake of replying publicly to *L'Égalité's* question, why the *General Council* permits "such a burdensome combination of functions".

England cannot be treated simply as a country along with other countries. She must be treated as the *metropolis of capital*.

5) *Question of the General Council Resolution on the Irish Amnesty*.

If England is the bulwark of landlordism and European capitalism, the only point where one can hit official England really hard is *Ireland*.

In the first place, Ireland is the *bulwark* of English land-lordism. If it fell in Ireland it would fall in England. In Ireland this is a hundred times easier since *the economic struggle there is concentrated exclusively on landed property*, since this struggle is at the same time national, and since the people there are more revolutionary and exasperated than in England. Landlordism in Ireland is maintained solely by the *English army*. The moment the *forced union* between the two countries ends, a social revolution will immediately break out in Ireland, though in outmoded forms. English landlordism would not only lose a great source of wealth, but also its *greatest moral force*, i.e., that of *representing the domination of England over Ireland*. On the other hand, by maintaining the power of their landlords in Ireland, the English proletariat makes them invulnerable in England itself.

In the second place, the English bourgeoisie has not only exploited the Irish poverty to keep down the working class in England by *forced immigration* of poor Irishmen, but it has also divided the proletariat into two hostile camps. The revolutionary fire of the Celtic worker does not go well with the nature of the Anglo-Saxon worker, solid, but slow. On the contrary, in all *the big industrial centres in England* there is profound antagonism between the Irish proletariat and the English proletariat. The average English worker hates the Irish worker as a competitor who lowers wages and the *standard of life*. He feels national and religious antipathies for him. He regards him somewhat like the *poor whites* of the Southern States of North America regard their black slaves. This antagonism among the proletarians of England is artificially nourished and supported by the bourgeoisie. It knows that this scission is the true secret of maintaining its power.

This antagonism is reproduced on the other side of the Atlantic. The Irish, chased from their native soil by the *bulls* and the sheep, reassemble in North America where they constitute a huge, ever-growing section of the population. Their only thought, their only passion, is hatred for England. The English and American governments (or the classes they represent) play on these feelings in order to perpetuate the covert struggle between the United States and England. They thereby prevent a sincere and lasting alliance between the workers on both sides of the Atlantic, and consequently, their emancipation.

Furthermore, Ireland is the only pretext the English Government has for retaining a *big standing army*, which, if need be, as has happened before, can be used against the English workers after having done its military training in Ireland.

Lastly, England today is seeing a repetition of what happened on a monstrous scale in Ancient Rome. Any nation that oppresses another forges its own chains.

Thus, the attitude of the International Association to the Irish question is very clear. Its first need is to encourage the social revolution in England. To this end a great blow must be struck in Ireland.

The General Council's resolutions on the Irish amnesty serve only as an introduction to other resolutions which will affirm that, quite apart from international justice, it is a *precondition to the emancipation of the English working class* to transform the present *forced union* (i.e., the enslavement of Ireland) into *equal and free confederation* if possible, into *complete separation* if need be.[166]

Written by Marx about March 28, 1870

Published in the journal
Die Neue Zeit,
Bd. 2, Nr. 15, 1902

Printed according to the text of the document "The General Council to the Federal Council of Romance Switzerland" in *The General Council of the First International. 1868-1870. Minutes,* Moscow
Translated from the French

Karl Marx

THE ENGLISH GOVERNMENT AND THE FENIAN PRISONERS[167]

London, February 21, 1870

I

The silence which is observed in the European press concerning the disgraceful acts committed by this oligarchical bourgeois government is due to a variety of reasons. Firstly, the English Government is *rich* and the press, as you know, is *immaculate*. Moreover, the English Government is the model government, recognised as such by the landlords, by the capitalists on the Continent and even by Garibaldi (see his book[168]): consequently we should not revile this ideal, government. Finally, the French Republicans are narrow-minded and selfish enough to reserve all their anger for the Empire. It would be an insult to free speech to inform their fellow countrymen that in the *land of bourgeois freedom* sentences of 20 years hard labour are given for offences which are punished by 6 months in prison in the *land of barracks*. The following information on the treatment of Fenian prisoners has been taken from English journals:

Mulcahy, sub-editor of the newspaper *The Irish People*,[169] sentenced for taking part in the Fenian conspiracy, was harnessed to a cart loaded with stones with a metal band round his neck at Dartmoor.

O'Donovan Rossa, owner of *The Irish People*, was shut up for 35 days in a pitch-black dungeon with his hands tied behind his back day and night. They were not even untied to allow him to eat the miserable slops which were left for him on the earthen floor.

Kickham, one of the editors of *The Irish People*, although he was unable to use his right arm because of an abscess, was forced to sit with his fellow prisoners on a heap of rubble in the November cold and fog and break up stones and bricks with his left hand. He returned to his cell at night and had nothing to eat but 6 ounces of bread and a pint of hot water.

O'Leary, an old man of sixty or seventy who was sent to prison, was put on bread and water for three weeks because he would not renounce *paganism* (this, apparently, is what a jailer called free thinking) and become either Papist, Protestant, Presbyterian or even Quaker, or take up one of the many religions which the prison governor offered to the heathen Irish.

Martin H. Carey is incarcerated in a lunatic asylum at Millbank. The silence and the other bad treatment which he has received have made him lose his reason.

Colonel *Richard Burke* is in no better condition. One of his friends writes that his mind is affected, he has lost his memory and his behaviour, manners and speech are those of a madman.

The political prisoners are dragged from one prison to the next as if they were wild animals. They are forced to keep company with the vilest knaves; they are obliged to clean the pans used by these wretches, to wear the shirts and flannels which have previously been worn by these criminals, many of whom are suffering from the foulest diseases, and to wash in the same water. Before the arrival of the Fenians at Portland all the criminals were allowed to talk with their visitors. A visiting cage was installed for the Fenian prisoners. It consists of three compartments divided by partitions of thick iron bars; the jailer occupies the central compartment and the prisoner and his friends can only see each other through this double row of bars.

In the docks you can find prisoners who eat all sorts of slugs, and frogs are considered dainties at Chatham. General Thomas Burke said he was not surprised to find a dead mouse floating in the soup. The convicts say that it was a bad day for them when the Fenians were sent to the prisons. (The prison regime has become much more severe.)

———

I should like to add a few words to these extracts.

Last year *Mr. Bruce*, the Home Secretary, a great liberal, great policeman and great mine owner in Wales who cruelly exploits his workers, was questioned on the bad treatment of Fenian prisoners and O'Donovan Rossa in particular. At first he denied everything, but was later compelled to confess. Following this Mr. Moore, an Irish member in the House of Commons, demanded an enquiry into the facts. This was flatly refused by the *radical ministry* of which the head is that demigod Mr. Gladstone (he has been compared to Jesus Christ publicly) and that old bourgeois demagogue, John Bright, is one of the most influential members.

The recent wave of reports concerning the bad treatment of the Fenians led several members of Parliament to request Mr. Bruce for permission to visit the prisoners *in order to be able to verify the falseness of these rumours*. Mr. Bruce refused this permission on the grounds that the prison governors were afraid that the prisoners would be too excited by visits of this kind.

Last week the Home Secretary was again submitted to questioning. He was asked whether it was true that O'Donovan Rossa received corporal punishment (i.e., whipping) after his election to Parliament as the member for Tipperary.[170] The Minister confirmed that he had not received such treatment since 1868 (which is tantamount to saying that the political prisoner had been given the whip over a period of two to three years).

I am also sending you extracts (which we are going to publish in our next issue) concerning the case of Michael Terbert, a Fenian sentenced as such to forced labour, who was serving his sentence at Spike Island Convict Prison in the county of Cork, Ireland. You will see that the coroner himself attributes this man's death to the torture which was inflicted on him. This investigation was held last week.

In the course of two years *more than twenty* Fenian workers have died or gone insane thanks to the philanthropic natures of these honest bourgeois souls, backed by the honest landlords.

You are probably aware that the English press professes a chaste distaste for the dreadful general security laws which grace "la belle France". With the exception of a few short intervals, it is security laws which formed the Irish

Charter. Since 1793 the English Government has taken advantage of any pretext to suspend the Habeas Corpus Act (which guarantees the liberty of the individual)[171] regularly and periodically, in fact all laws, except that of brute force. In this way thousands of people have been arrested in Ireland on *being suspected of Fenianism* without ever having been tried, brought before a judge or court, or even charged. Not content with depriving them of their liberty, the English Government has had them tortured in the most savage way imaginable. The following is but one example.

One of the prisons where persons suspected of being Fenians were buried alive is *Mountjoy Prison* in Dublin. The prison inspector, Murray, is a despicable brute who maltreated the prisoners so cruelly that some of them went mad. The prison doctor, an excellent man called M'Donnell (who also played a creditable part in the enquiry into Michael Terbert's death), spent several months writing letters of protest which he addressed in the first instance to Murray himself. When Murray did not reply he sent accusing letters to higher authorities, but being an expert jailer Murray intercepted these letters.

Finally M'Donnell wrote directly to Lord Mayo who was then Viceroy of Ireland. This was during the period when the Tories were in power (Derby and Disraeli). What effect did his actions have? The documents relating to the case were published by order of Parliament and . . . Dr. M'Donnell was dismissed from his post!!! Whereas Murray retained his.

Then the so-called radical government of Gladstone came to power, the warm-hearted, unctuous, magnanimous Gladstone who had wept so passionately and so sincerely before the eyes of the whole of Europe over the fate of Poerio and other members of the bourgeoisie who were badly treated by King Bomba.[172] What did this idol of the progressive bourgeoisie do? While insulting the Irish by his insolent replies to their demands for an amnesty, he not only confirmed the monster Murray in his post, but endowed the position of chief jailer with a nice fat sinecure as a token of his personal satisfaction! There's the apostle of the philanthropic bourgeoisie for you!

But something had to be done to pull the wool over the eyes of the public. It was essential to appear to be doing

something for Ireland, and the Irish Land Bill[173] was proclaimed with a great song and dance. All this is nothing but
a pose with the ultimate aim of deceiving Europe, winning
over the Irish judges and advocates with the prospect of endless disputes between landlords and farmers, conciliating
the landlords with the promise of financial aid from the
state and deluding the more prosperous farmers with a few
mild concessions.

In the long introduction to his grandiloquent and confused
speech Gladstone admits that even the "benevolent" laws
which liberal England bestowed on Ireland over the last
hundred years have always led to the country's further decline.[174] And after this naive confession the same man per-
sists in torturing those who want to put an end to this harmful and stupid legislation.

II

The following is an account taken from an English newspaper of the results of an enquiry into the death of Michael
Terbert, a Fenian prisoner who died at Spike Island Prison
due to the bad treatment which he had received.

"On Thursday last Mr. John Moore, *Coroner* of the Middleton district,
held an inquest at Spike Island Convict Prison, on the body of a convict ... named Michael Terbert, who had died in hospital.

"Peter Hay, governor of the prison, was called first. He deposed—
'The deceased, Michael Terbert, came to this prison in June, 1866; I
can't say how his health was at the time; he had been convicted on the
12th of January, 1866, and his sentence was seven years' penal servitude;
he appeared delicate for some time past, as will appear from one of the
prison books, which states that he was removed on the recommendation
of medical officers, as being unfit for cellular discipline.' Witness then
went into a detail of the frequent punishments inflicted on the deceased
for breach of discipline, many of them for the use 'of disrespectful
language to the medical officer'.

"Jeremiah Hubert Kelly deposed—'I remember when Michael Terbert
came here from Mountjoy Prison; it was then stated that he was unfit
for cellular discipline—that means being always confined to a cell;
certificate to the effect was signed by Dr. M'Donnell; ... I found him,
however, to be in good health, and I sent him to work; I find by the
record that he was in hospital from the 31st January, 1869, until the 6th
February, 1869; he suffered then from increased affection of the heart,
and from that time he did not work on the public works, but in-doors, at
oakum; from the 19th March, 1869, until the 24th March, 1869, he was
in hospital, suffering from the same affection of heart; from the 24th

April till the 5th May he was also in hospital from spitting of blood; from the 19th May till the 1st June he was in hospital for heart disease; from the 21st June till the 22nd June he was under hospital treatment for the same; he was also in hospital from the 22nd July till the 15th August, for the same—from 9th November till the 13th December for debility, and from 20th December to the 8th February, when he died from acute dropsy; on the 13th November he first appeared to suffer from dropsy, and it was then dissipated; I visit the cells every day, and I must have seen him when under punishment from time to time; it is my duty to remit, by recommendation, that punishment, if I consider the prisoner is not fit to bear it; I think I did so twice in his case.'

" 'As a medical man, did you consider that five days on bread and water per day was excessive punishment for him, notwithstanding his state of health in Mountjoy and here?'—'I did not; the deceased had a good appetite; I don't think that the treatment induced acute dropsy, of which he died'....

"Martin O'Connell, resident apothecary of Spike Island, was next examined—Witness mentioned to Dr. Kelly last July that while the deceased was labouring under heart disease, he should not have been punished; ... he was of opinion that such punishment as the deceased got was prejudicial to his health, considering that he was an invalid for the past twelve months ... he could not say that invalids were so punished, as he only attended cells in Dr. Kelly's absence; he was certain, considering the state of the deceased man's health, that five days continuously in cells would be injurious to his health; ... The *Coroner* then ... dealt forcibly with the treatment which the prisoner had received ... alternating between the hospital and the punishment cell.

"The jury returned the following verdict: 'We find that Michael Terbert died in hospital at Spike Island Convict Prison, on the 8th of February, 1870, of dropsy; he was twenty-five years of age, and unmarried. We have also to express in the strongest terms our total disapproval of the frequent punishment he suffered in cells on bread and water for several days in succession during his imprisonment in Spike Island, where he had been sent in June, 1866, from Mountjoy Prison, for the reason that in Dr. M'Donnell's opinion he was unfit for cellular discipline at Mountjoy; and we express our condemnation of such treatment.' "[175]

Published in the newspaper
L'Internationale Nos. 59 and 60,
February 27 and March 6, 1870

Translated from the
French

[RECORD OF KARL MARX'S SPEECH CONCERNING THE "BEE-HIVE" NEWSPAPER

(From the Minutes of the General Council Meeting of April 26, 1870)][176]

Cit. *Marx* proposed that the Council should cut off all connections with the *Bee-Hive*. He said it had suppressed our resolutions and mutilated our reports and delayed them so that the dates had been falsified, even the mention that certain questions respecting the Irish prisoners were being discussed had been suppressed.

Next to that, the tone of the *Bee-Hive* was contrary to the Rules and platform of the Association. It preached harmony with the capitalists, and the Association had declared war against the capitalists' rule.

Besides this, our branches abroad complained that by sending our reports to the *Bee-Hive* we gave it a moral support and led people to believe that we endorsed its policy. We would be better without its publicity than with it.

On the Irish Coercion Bill[177] it had not said a word against the government.

Published in:
Marx and Engels,
Collected Works, second
Russian ed., Vol. 16,
Moscow, 1960

Printed according to the text of the book
The General Council of the First International. 1868-1870. Minutes, Moscow

Frederick Engels

HISTORY OF IRELAND[178]

NATURAL CONDITIONS

At the north-western corner of Europe lies the land whose history will occupy us, an island of 1,530 German or 32,500 English square miles. But another island, three times as large, lies obliquely interposed between Ireland and the rest of Europe. For the sake of brevity we usually call this island England; it blocks Ireland off completely towards the north, east and south-east, and allows a free view only in the direction of Spain, Western France and America.

The channel between the two islands, 50-70 English miles wide at the narrowest points in the south, 13 miles wide at one point in the north and 22 miles wide at another, allowed the Irish Scots to emigrate from the north to the neighbouring island and to found the Kingdom of Scotland even before the fifth century. In the south it was too wide for Irish and British boats and a serious obstacle even for the flat-bottomed coastal vessels of the Romans. But when the Frisians, Angles and Saxons, and after them the Scandinavians, were able to venture beyond the sight of land on the open seas in their keeled vessels, this channel was an obstacle no longer; Ireland fell a victim to the raiding expeditions of the Scandinavians, and presented an easy booty for the English. As soon as the Normans had built up a powerful, unified government in England, the influence of the larger island made itself felt—in those times this meant a war of conquest.[179]

If during the war a period set in when England gained control of the sea, this precluded the possibility of successful foreign intervention.

When the larger island finally became unified into one state, the latter had to strive to assimilate Ireland completely.

If this assimilation had been successful, its whole course would have become a matter of history. It would be subject to its judgement but could never be reversed. But if after 700 years of fighting this assimilation has *not* succeeded; if instead each new wave of invaders flooding Ireland is assimilated by the *Irish*; if, even today, the Irish are as far from being English, or West Britons, as they say, as the Poles are from being West Russians after only 100 years of oppression; if the fighting is not yet over and there is no prospect that it can be ended in any other way than by the extermination of the oppressed race—then, all the geographical pretexts in the world are not enough to prove that it is England's mission to conquer Ireland.

———

To understand the nature of the soil of present-day Ireland we have to return to the distant epoch when the so-called Carboniferous System was formed.[*]

The centre of Ireland, to the north and south of a line from Dublin to Galway, forms a wide plain rising to 100-300 feet above sea-level. This plain, the foundation so to say of the whole of Ireland, consists of the massive bed of limestone (carboniferous limestone), which forms the middle layer of the Carboniferous System, and immediately above which lie the coal-measures of England and other places.

In the south and the north, this plain is encircled by a mountain chain which extends mainly along the coast, and consists almost entirely of older rock-formations which have broken through the limestone. These older rock-formations contain granite, mica-slate, Cambrian, Cambro-Silurian, Upper-Silurian, Devonian, together with argillaceous slate and sandstone, rich in copper and lead, found in the lowest layer of the Carboniferous System; apart from this they

———

[*] Unless otherwise stated, all the geological data given here is from J. Beete Jukes, *The Student's Manual of Geology*. New Edition. Edinburgh, 1862. Jukes was the local superior during the geological survey of Ireland and therefore the prime authority on this territory, which he treats in special detail.

contain a little gold, silver, tin, zinc, iron, cobalt, antimony and manganese.

The limestone itself rises to mountains only in a few places: it reaches 600 feet in the centre of the plain, in Queen's County,[180] and a little over 1,000 feet in the west, on the southern shore of Galway Bay (Burren Hills).

At several points in the southern half of the limestone plain there are to be found isolated coal-bearing mountain ridges of considerable extent and from 700 to 1,000 feet above sea-level. These rise from depressions in the limestone plain as plateaus with rather steep escarpments.

"The escarpments in these widely separated tracts of coal-measures are so similar, and the beds composing them so precisely alike, that it is impossible to suppose otherwise than that they originally formed continuous sheets of rock, although they are now separated by sixty or eighty miles.... This belief is strongly confirmed by the fact that there are often, between the two larger areas, several little outlying patches in which the coal-measures are found capping the summits of small hills, and that wherever the undulation of the limestone is such as to bring its upper beds down beneath the level of the present surface of the ground, we invariably find some of the lower beds of the coal-measures coming in upon them." (Jukes, pp. 285-86.)

Other circumstances, which are too detailed for us here and can be found in Jukes, pages 286-89, contribute to the certainty that the whole Irish central plain arose through denudation, as Jukes says, so that the lower layers of limestone were exposed after the coal-measures and the high limestone deposits—of an average thickness of at least 2,000-3,000 and possibly 5,000-6,000 feet of stone—had been washed away. Jukes even found another small coal-measure on the highest ridge of the Burren Hills, County Clare, which are pure limestone and 1,000 feet high (p. 513).

Some fairly considerable areas containing coal-measures have survived in Southern Ireland; but only a few of these contain enough coal to justify mining. Moreover, the coal itself is anthracite, that is, it contains little hydrogen and cannot be used for all industrial purposes without some addition.

There are also several not very extensive coal-fields in Northern Ireland in which the coal is bituminous, that is, ordinary coal rich in hydrogen. Their stratification does not coincide exactly with that of the southern coal deposits. But a similar washing away process did occur even here. This is shown by the fact that large fragments of coal, as well as sandstone and blue clay belonging to the same formation, are to be found on the surface of limestone valleys to the south-east of such a coal-field in the direction of Belturbet and Mohill. Large blocks of coal have been discovered by well-sinkers in this area of the drift; and in some cases the quantity of coal was so considerable that it was thought that deeper shafts must lead to a coal-bed. (Kane, *The Industrial Resources of Ireland*, 2nd edition, Dublin, 1845, p. 265.)

It is obvious that Ireland's misfortune is of ancient origin; it begins directly after the carboniferous strata were deposited. A country whose coal deposits are eroded, placed near a larger country rich in coal, is condemned by nature to remain for a long time the farming country for the larger country when the latter is industrialised. That sentence, pronounced millions of years ago, was carried out in this century. We shall see later, moreover, how the English assisted nature by crushing almost every seed of Irish industry as soon as it appeared.

More recent Secondary and Tertiary layers[181] occur almost exclusively in the north-east; amongst these we are interested chiefly in the beds of red marl in the vicinity of Belfast, which contain almost pure rock-salt to a thickness of 200 feet (Jukes, p. 554), and the chalk overlaid with a layer of basalt which covers the whole of County Antrim. Generally speaking, there are no important geological developments in Ireland between the end of the Carboniferous Period and the Ice Age.

It is known that after the Tertiary Epoch there was an era in which the low-lying lands of the medium latitudes of Europe were submerged by the sea, and in which such a low temperature prevailed in Europe that the valleys between the protruding island mountain tops were filled with glaciers which extended down to the sea. Icebergs used to separate themselves from these glaciers and carry rocks of all sizes which had been detached from the mountains, out to sea. When the ice melted, the rocks and other debris were deposited—a process still daily occurring on coasts of the polar regions.

During the Ice Age, Ireland too, with the exception of the mountain tops, was submerged by the sea. The degree of submergence may not have been the same everywhere, but an average of 1,000 feet below the present level can be accepted; the granite mountain chains south of Dublin must have been submerged by over 1,200 feet.

If Ireland had been submerged by only 500 feet, only the mountain chains would have remained exposed. These would then have formed two semi-circular groups of islands around a wide strait extending from Dublin to Galway. A still greater submergence would have made these islands smaller and decreased their number, until, at a submergence of 2,000 feet, only the most extreme tips would have risen above the water.*

As the submersion slowly proceeded, the limestone plains and mountain slopes must have been swept clean of much of the older rock covering them; subsequently there followed the depositing of the drift peculiar to the Ice Age on the whole of the area covered by water. Pieces of rock eroded from the mountain islands and fine fragments of rock scraped away by the glaciers as they pushed their way slowly and powerfully through the valleys—earth, sand, gravel, stones, rocks, worn smooth within the ice but sharp-edged above it—all this was carried out to sea and gradually deposited on the sea-bed by icebergs which were detaching themselves from the shore. The layer formed in this way varies according to circumstances and contains loam (originating from argillaceous slate), sand (originating from quartz and granite), limestone gravel (derived from limestone formations), marl (where finely-crumbled limestone mixes with loam) or mixtures of all these components; but it always contains a mass of stones of all sizes, sometimes rounded, sometimes sharp, ranging up to colossal erratic boulders, which are commoner in Ireland than in the North-German Plain or between the Alps and the Jura.

During the subsequent re-emergence of the land from the sea, this newly-formed surface was given roughly its present

---

* Ireland has an area of 32,509 English square miles. 13,243 square miles are 0-250 feet above sea-level; 11,797 are 251-500 feet above sea-level; 5,798 are 501-1,000 feet above sea-level; 1,589 are 1,001-2,000 feet above sea-level; 82 square miles are over 2,001 feet above sea-level.

structure. In Ireland, little washing away appears to have taken place then; with few exceptions varying thicknesses of drift cover all the plains, extend into all the valleys, and are also often found high up on the mountain slopes. Limestone is the most frequently occurring stone in them, and for this reason the whole stratum is usually called limestone gravel here. Big blocks of limestone are also extensively strewn over all the lowlands, one or more in nearly every field; apart from limestone, a lot of other local rocks, especially granite, are naturally to be found near the mountains they originated from. From the northern side of Galway Bay granite appears commonly in the plain extending south-east as far as the Galty Mountains and more rarely as far as Mallow (County Cork).

The north of the country is covered with drift to the same height above sea-level as the central plain; a similar deposit, originating from the local, mainly Silurian rocks, is to be found between the various more or less parallel mountain chains running through the south. This appears plentifully in Flesk and Laune valley near Killarney.

The glacier tracks on the mountain slopes and valley bottoms are common and unmistakable, particularly in the south-west of Ireland. Only in Oberhasli and here and there in Sweden do I remember seeing more sharply-stamped ice-trails than in Killarney (in the Black Valley and the Gap of Dunloe).

The emergence of the land during or after the Ice Age seems to have been so considerable that Britain was for a time connected by dry land not only with the Continent, but also with Ireland. At least this seems the only way the similarity between the fauna of these lands can be explained. Ireland has the following extinct large mammals in common with the Continent: the mammoth, the Irish giant stag, the cave-bear, a kind of reindeer, and so on. In fact, an emergence of less than 240 feet over the present level would be enough to connect Ireland with Scotland, and one of less than 360 feet would join Ireland and Wales with wide bridges of land.* The fact that Ireland emerged to a higher

---

* See Map 15a in Stielers Handatlas, 1868.[182] This map, as well as No. 15d, specially of Ireland, picture the ground structure very clearly.

level after the Ice Age than at present is proved by the underwater peat bogs with upright tree trunks and roots which occur all around the coast, and which are identical in every detail with the lowest layers of the neighbouring inland peat bogs.

From an agricultural point of view, Ireland's soil is almost entirely formed from the drift of the Ice Age, which here, thanks to its slate and limestone origin, is not the barren sand with which the Scottish, Scandinavian and Finnish granites have covered such a large part of North Germany, but an extremely fertile, light loam. The variety in the rocks, whose decomposition contributed and is still contributing to this soil, provides it with a corresponding variety of the mineral elements required for vegetable life; and if one of these, say lime, is greatly lacking in the soil, plenty of pieces of limestone of all sizes are to be found everywhere—quite apart from the underlying limestone bed—so it can be added quite easily.

When the well-known English agronomist, *Arthur Young*, toured Ireland in the 1770s, he did not know what amazed him more: the natural fertility of the soil or the barbaric manner in which the peasants cultivated it. "A light, dry, soft, sandy, loam soil" prevails where the land is good at all. In the "Golden Vale" of Tipperary and also elsewhere he found:

"the same sort of sandy reddish loam I have already described, incomparable land for tillage". From there, in the direction of Clonmel, "the whole way through the same rich vein of red sandy loam I have so often mentioned: I examined it in several fields, and found it to be of an extraordinary fertility, and as fine turnip land as ever I saw".

Further:

"The rich land reaches from Charleville, at the foot of the mountains, to Tipperary, by Kilfenning, a line of twenty-five miles, and across from Ardpatrick to within four miles of Limerick, sixteen miles." "The richest in the country is the Corcasses on the Maag, about Adair, a tract of five miles long, and two broad, down to the Shannon. . . . When they break this land up, they sow first oats, and get 20 barrels an acre, or 40 common barrels, and do not reckon that an extra crop; they take ten or twelve in succession, upon one ploughing, till the crops grow poor, and then they sow one of horse beans, which refreshes the land enough to take ten crops of oats more; the beans are very good.... Were such barbarians ever heard of?"

Further, near Castle Oliver, County Limerick,

"the finest soil in the country is upon the roots of mountains; it is a rich, mellow, crumbling, putrid, sandy loam, eighteen inches to three feet deep, the colour a reddish brown. It is dry sound land, and would do for turnips exceedingly well, for carrots, for cabbages, and in a word for everything. I think, upon the whole, it is the richest soil I ever saw, and such as is applicable to every purpose you can wish; it will fat the largest bullock, and at the same time do equally well for sheep, for tillage, for turnips, for wheat, for beans, and in a word, for every crop ... you must examine into the soil before you will believe that a country, which has so beggarly an appearance, can be so rich and fertile."

On the river Blackwater near Mallow,

"there are tracts of flat land in some places one quarter of a mile broad; the grass everywhere remarkably fine.... It is the finest sandy land I have anywhere seen, of a reddish-brown colour, would yield the greatest arable crops in the world, if in tillage; it is five feet deep, and has such a principle of adhesion, that it burns into good brick, yet it is a perfect sand.... The banks of this river, from its source to the sea, are equally remarkable for beauty of prospect, and fertility of soil." "Friable, sandy loams, dry but fertile, are very common, and they form the best soils in the kingdom, for tillage and sheep. Tipperary and Roscommon abound particularly in them. The most fertile of all are the bullock pastures of Limerick, and the banks of the Shannon in Clare, called the Corcasses.... Sand, which is so common in England, and yet more common through Spain, France, Germany, and Poland, quite from Gibraltar to Petersburg, is nowhere met with in Ireland, except for narrow slips of hillocks, upon the sea coast. Nor did I ever meet with, or hear of a chalky soil."*

Young's judgement on the soil of Ireland is summarised in the following sentences:

"If I was to name the characteristics of an excellent soil, I would say *that* upon which you may fat an ox and feed off a crop of turnips. By the way, I recollect little or no such land in England, yet it is not uncommon in Ireland." (Vol. 2, p. 271.)—"Natural fertility, acre for acre over the two kingdoms, is certainly in favour of

---

* Arthur Young, *A Tour in Ireland*, 3 vols. London, 177 ..., Vol. 2, pp. 28, 135, 143, 154, 165; Vol. 2, Part II, p. 4.

Ireland." (Vol. 2, Part II, p. 3.)—"As far as I can form a general idea of the soil of the two kingdoms, Ireland has much the advantage." (Vol. 2, Part II, p. 9.)

In 1808-10, Edward *Wakefield*, an Englishman likewise versed in agronomy, toured Ireland and recorded the result of his observations in a valuable work.* His remarks are better-ordered, more extensive and fuller than those in Young's travel-book; on the whole, both agree.

Wakefield found little disparity in the nature of the soil in Ireland on the whole. Sand occurs only on the coast (it is so seldom found inland that large quantities of sea sand are transported inland for improving the turf and loam soils); chalky soil is unknown (the chalk in Antrim is, as has already been mentioned, covered with a layer of basalt, the products of the decomposition of which produce a highly fertile soil. In England the chalky soils are the worst), "... tenacious clays, such as those found in Oxfordshire, in some parts of Essex, and throughout High Suffolk, I could never meet with...." The Irish call all loamy soils clay; there might be real clay in Ireland as well, but not on the surface as in several parts of England in any case. Limestone or limestone gravel is to be found everywhere. "The former is a useful production, and is converted into a source of wealth that will always be employed with advantage." Mountains and peat bogs certainly reduce the fertile surface considerably. There is little fertile land in the north; yet even here there are highly luxuriant valleys in every county, and Wakefield unexpectedly found a highly fertile tract even in furthest Donegal amongst the wildest mountains. The extensive cultivation of flax in the north is in itself sufficient proof of fertility, as this plant does not thrive in poor soil.

"A great portion of the soil in Ireland throws out a luxuriant herbage, springing up from a calcareous subsoil, without any considerable depth. I have seen bullocks of the weight of 180 stone, rapidly fattening on land incapable of receiving the print of a horse's foot, even in the wettest season, and where there were not many inches of soil. This is *one* species of the rich soil of Ireland, and is to be found throughout Roscommon, in some parts of Galway, Clare, and other districts. Some places exhibit the richest loam that I ever saw

---

* Edward Wakefield, *An Account of Ireland, Statistical and Political*, London, 1812, 2 vols.

turned up by a plough; this is the case throughout Meath in particular. Where such soil occurs, its fertility is so conspicuous, that it appears as if nature had determined to counteract the bad effects produced by the clumsy system of its cultivators. On the banks of the Fergus and Shannon, the land is of a different kind, but equally productive, though the surface presents the appearance of a marsh. These districts are called 'the caucasses' [so designated by Wakefield as distinct from Young]; the substratum is a blue silt, deposited by the sea, which seems to partake of the qualities of the upper stratum; for this land can be injured by no depth of ploughing.

"In the counties of Limerick and Tipperary there is another kind of rich land, consisting of a dark, friable, dry, sandy loam which, if preserved in a clean state, would throw out corn for several years in succession. It is equally well adapted to grazing and tillage, and I will venture to say, seldom experiences a season too wet, or a summer too dry. The richness of the land, in some of the vales, may be accounted for by the deposition of soil carried thither from the upper grounds by the rain. The subsoil is calcareous, so that the very richest manure is thus spread over the land below, without subjecting the farmer to any labour." (Vol. 1, pp. 79, 80.)

If a thinnish layer of heavy loam lies directly on limestone, the land is not suited to tillage and bears only a miserable crop of grain, but it makes excellent sheep-pastures. This improves it further by producing a thick grass mixed with white clover and. . . .* (Vol. 1, p. 80.)

Dr. Beaufort** states that there occur in the west, particularly in Mayo, many turloughs—shallow depressions of different sizes, which fill with water in the winter, although not visibly connected with streams or rivers. In the summer this drains away through underground fissures in the limestone, leaving luxurious firm grazing-ground.

"Independently of the caucasses," Wakefield continues, "the richest soil in Ireland is to be found in the counties of Tipperary, Limerick, Roscommon, Longford, and Meath. In Longford there is a farm called Granard Kill, which produced eight crops of potatoes without manure. Some parts of the County of Cork are uncommonly fertile, and upon the whole, Ireland may be considered as affording land of an excellent quality, though I am by no means prepared to go the length of many writers, who assert, that it is decidedly acre for acre richer than England." (Vol. 1, p. 81.)

---

* There is an omission in the manuscript. According to Wakefield it is "wild burnet".—*Ed.*

** Beaufort, Revd. Dr., *Memoir of a Map of Ireland,* 1792, pp. 75-76 Quoted in Wakefield, Vol. 1, p. 36.

The last observation, directed against Young, rests on a misunderstanding of Young's opinion, quoted above. Young does not say that Ireland's soil is more productive than England's, each taken in their present state of cultivation—which is naturally far higher in England; Young merely states that the *natural* fertility of the soil is greater in Ireland than in England. This does not contradict Wakefield.

After the last famine, in 1849, Sir* Robert Peel sent a Scottish agronomist, Mr. *Caird,* to Ireland to report on means of improving agriculture there. In a publication issued soon afterwards he said about the west of Ireland—the worst stricken part of the country apart from the extreme north-west:

"I was much surprised to find so great an extent of fine fertile land. The interior of the country is very level, and its general character stony and dry; the soil dry and friable. The humidity of the climate causes a very constant vegetation, which has both advantages and disadvantages. It is favourable for grass and green crops,** but renders it necessary to employ very vigorous and persevering efforts to extirpate weeds. The abundance of lime everywhere, both in the rock itself, and as sand and gravel beneath the surface, are of the greatest value."

Caird also confirms that County Westmeath consists of the finest pasture land. Of the region north of Lough Corrib (County Mayo) he writes:

"The greater part of this farm" (a farm of 500 acres) "is the finest feeding land for sheep and cattle—dry, friable, undulating land, all on limestone. The fields of rich old grass are superior to anything we have, except in small patches, in any part of Scotland I at present remember. The best of it *is too good for tillage,* but about one half of it might be profitably brought under the plough.... The rapidity with which the land on this limestone subsoil recovers itself, and, without any seeds being sown, reverts to good pasture, is very remarkable."***

---

* In the manuscript the word "Ministry" appears above the "Sir".
—*Ed.*

** "Green crops" embrace all cultivated fodder crops, as well as carrot, beetroot, turnip and potato, that is, everything except corn, grasses and garden plants.

*** Caird, *The Plantation Scheme, or the West of Ireland as a Field for Investment,* Edinburgh, 1850. He also wrote travel reports on the condition of agriculture in the main counties of England for *The Times* of 1850-51. The above quotations are found on pp. 6, 17-18, 121.

Finally we note a French authority*:

"Of the two divisions of Ireland, that of the north-west, embracing a fourth of the island, and comprehending the province of Connaught, with the adjacent counties of Donegal, Clare, and Kerry, resembles Wales, and even, in its worst parts, the Highlands of Scotland. Here again are two millions of unsightly hectares, the frightful aspect of which has given rise to the national proverb, 'Go to the devil or Connaught'.** The other, or south-east and much larger division, since it ... includes the provinces of Leinster, Ulster, and Munster, equal to about six millions of hectares, is *at least equal* in natural fertility to England proper. It is not all, however, equally good; the amount of humidity there is still greater than in England. Extensive bogs cover about a tenth of the surface; more than another tenth is occupied with mountains and lakes. In fact, five only out of eight millions of hectares in Ireland are cultivated [pp. 9, 10]. Even the English admit that Ireland, in point of soil, is superior to England.... Ireland contains eight millions of hectares. Rocks, lakes, and bogs occupy about two millions of these, and two millions more are indifferent land. The remainder— that is to say, about half the country—is rich land, with calcareous subsoil. What better could be conceived?" (P. 343.)

We see therefore that all authorities agree that Ireland's soil contains all the elements of fertility to an extraordinary degree. This, not only in its chemical ingredients but also in its structure. The two extremes of heavy impenetrable clay, completely impermeable, and loose sand, completely permeable, do not occur. But Ireland has another disadvantage. While the mountains are mainly along the coast, the watersheds between the inland river basins are mostly low-lying, and therefore the rivers are not capable of carrying all the rain water out to sea. Thus extensive peat bogs arise inland, especially on the watersheds. In the plain alone 1,576,000 acres are covered with peat bogs. These are largely depressions or troughs in the land, most of which were once shallow lake basins which were gradually overgrown with moss and marsh plants and were filled up with their decomposing remains. As with our north-German moors, their only use is for turf cutting. With the present system of agriculture cultivation can only gradually reclaim their edges. The soil in these former lake basins is mainly marl and its

---

* Léonce de Lavergne, *Rural Economy of England, Scotland and Ireland*. Translated from the French. Edinburgh, 1855.
** This expression, as will be seen later, owes its origin not to the dark mountains of Connaught, but to the darkest period in the entire history of Ireland.[183]

lime content (varying from 5 per cent to 90 per cent) is due to the shells of fresh-water mussels. Thus the material for their development into arable land exists within each of these peat bogs. Apart from this, most of them are rich in iron ore. Besides these low-lying peat bogs, there are 1,254,000 acres of mountain moor. These are the result of deforestation in a damp climate and are one of the peculiar beauties of the British Isles. Wherever flat or almost flat summits were deforested—and this occurred extensively in the 17th century and the first half of the 18th century to provide the iron works with charcoal—a layer of peat formed under the influence of rain and mist and gradually spread down the slopes where the conditions were favourable. Such moors cover the ridges of the mountain chain dividing Northern England from north to south almost as far as Derby; and are found in abundance wherever substantial mountain ranges are marked on the map of Ireland. Yet, the peat bogs of Ireland are by no means hopelessly lost to agriculture; on the contrary, in time we shall see what rich fruits some of these, and the two million hectares of the "indifferent land" contemptuously mentioned by Lavergne, can produce given correct management.

———

Ireland's climate is determined by her position. The Gulf Stream and the prevailing south-west winds provide warmth and make for mild winters and cool summers. In the southwest the summer lasts far into October which, according to Wakefield (Vol. 1, p. 221), is there regarded as the best month for sea bathing. Frost is rare and of short duration, snow usually melts immediately on the low-lying land. Spring weather prevails throughout the winter in the inlets of Kerry and Cork, which are open to the south-west and protected from the north; here, and in certain other places, myrtle thrives in the open (Wakefield mentions a country-residence where it grows into trees 16 feet high and is used to make stable-brooms, Vol. 1, p. 55), and laurel, arbutus and other evergreen plants grow into substantial trees. In Wakefield's time, the peasants in the south were still leaving their potatoes in the open all winter—and they had not been frost-bitten since 1740. On the other hand, Ireland also suffers the first powerful downpour of the heavy Atlantic rain clouds. Ireland's average rainfall is at least 35 inches, which is considerably more than England's average, yet is definitely lower than that of Lancashire and Cheshire and scarcely more than the average for the whole of the West of England. In spite of this the Irish climate is decidedly pleasanter than the English. The leaden sky which often causes days of continual drizzle in England is mostly replaced in Ireland by a continental April sky; the fresh sea-breezes bring on clouds quickly and unexpectedly, but drive them past equally quickly, if they do not come down immediately in sharp showers. And even when the rain lasts for days, as it does in late autumn, it does not have the chronic air it has in England. The weather, like the inhabitants, has a more acute character, it moves in sharper, more sudden contrasts; the sky is like an Irish woman's face: here also rain and sunshine succeed each other suddenly and unexpectedly and there is none of the grey English boredom.

The Roman, *Pomponius Mela*, gives us the oldest report on the Irish climate (in *De situ orbis*) in the first century of our era:

"Above Brittaine is Ireland, almost of like space but on both sides equall, with shores evelong, *of a evyll ayre to rypen things that are sown*, but *so aboundant of Grasse which is not onelie rancke but also sweete*, that the Cattell may in small parte of the daye fyll themselves, and if they bee not kept from feedying, they burste with grazing over-long."

"Coeli ad maturanda semina iniqui, verum adeo luxuriosa herbis non laetis modo sed etiam dulcibus!" We find this part amongst others translated into modern English by Mr. *Goldwin Smith*, Professor of History formerly of Oxford and now in Cornell University, America. He reports that it is difficult to gather in the harvest of wheat in a large part of Ireland and continues:

"Its [Ireland's] natural way to commercial prosperity seems to be to supply with the produce of its *grazing* and dairy farms the *population of England*."*

---

* Goldwin Smith, *Irish History and Irish Character*, Oxford and London, 1861.—What is more amazing in this work, which, under the mask of "objectivity", justifies English policy in Ireland, the ignorance of the professor of history, or the hypocrisy of the liberal bourgeois? We shall touch on both again later.

From Mela to Goldwin Smith and up to the present day, how often has this assertion been repeated—since 1846,[184] especially by a noisy chorus of Irish landowners—that Ireland is condemned by her climate to provide not Irishmen with bread but Englishmen with meat and butter, and that the destiny of the Irish people is, therefore, to be brought over the ocean to make room in Ireland for cows and sheep!

It can be seen that to establish the facts on the Irish climate is to unravel a topical political question. And indeed the climate only concerns us here insofar as it is important for agriculture. Rain measurements, at their present incomplete stage of observations, are only of secondary importance for our purpose; *how much* rain falls is not so important as *how* and *when* it falls. Here agronomical judgements are most important.

Arthur Young considers that Ireland is considerably damper than England; this is the cause of the amazing grass-bearing qualities of the soil. He speaks of cases when turnip- and stubble-land, left unploughed, produced a rich harvest of hay in the next summer, a thing of which there is no example in England. He further mentions that the Irish wheat is much lighter than that grown in drier lands; weeds and grass spring up in abundance under even the best management, and the harvests are so wet and so troublesome to bring in that revenue suffers greatly. (Young's *Tour*, Vol. 2, p. 100.)

At the same time, however, he points out that the soil in Ireland counteracts this dampness of the climate. It is generally stony, and for this reason lets the water through more easily.

"Harsh, tenacious, stoney, strong loams, difficult to work, are not uncommon [in Ireland]; but they are quite different from English clays. If as much rain fell upon the clays of England (a soil very rarely met with in Ireland, and never without much stone) as falls upon the rocks of her sister-island, those lands could not be cultivated. But the rocks here are cloathed with verdure;—those of limestone with only a thin covering of mold, have the softest and most beautiful turf imaginable." (Vol. 2, Part II, pp. 3-4.)

The limestone is known to be full of cracks and fissures which let the excess water through quickly.

Wakefield devotes to the climate a very comprehensive chapter in which he summarises all the earlier observations up to his own time. Dr. *Boate* (*Natural History of Ireland*, 1645)[185] describes the winters as mild, with three or four periods of frost every year, each of which usually lasts for only two or three days; the Liffey in Dublin freezes over scarcely once in 10 to 12 years. March is usually dry and fine, but then the weather becomes rainy; there are seldom more than two or three consecutive dry days in summer; and in the late autumn it is fine again. Very dry summers are rare, and dearth never occurs because of drought, but mostly because of too much rain. It seldom snows on the plains, so cattle remain in the open all the year round. Yet years of heavy snow do occur, as in 1635, when the people had difficulty in providing shelter for the cattle. (Wakefield, Vol. 1, p. 216 and following.)

In the beginning of the last century, Dr. *Rutty* (*Natural History of the County of Dublin*) made accurate meteorological observations which stretched over 50 years, from 1716 to 1765. During this whole period the proportion of south and west winds to north and east winds was 73:37 (10,878 south and west against 6,329 north and east). Prevailing winds were west and south-west, then came northwest and south-east, and most rarely north-east and east. In summer, autumn and winter west and south-west prevail. East is most frequent in spring and summer, when it occurs twice as frequently as in autumn and winter; north-east is most frequent in spring when, likewise, it is twice as frequent as in autumn and winter. As a result of this, the temperatures are more even, the winters milder and the summers cooler than in London, while on the other hand the air is damper. Even in summer, salt, sugar, flour, etc., soak dampness out of the air, and corn must be kiln-dried, a practice unknown in *some parts* of England. (Wakefield, Vol. 1. pp. 172-81.)

Rutty could at that time only compare Irish climate with that in London, which, as in all Eastern England, is drier, to be sure. If material on Western and especially North-Western England had been at his disposal, he would have found that his description of the Irish climate—distribution of winds over the year, wet summers, in which sugar, salt, etc., are ruined in unheated rooms—fits this area completely, except that Western England is colder in winter.

Rutty also kept data on the meteorological aspect of the seasons. In the fifty years referred to, there were 16 cold, late or too dry springs: a little more than in London; further, 22 hot and dry, 24 wet, and 4 changeable summers: a little damper than in London, where the number of dry and wet summers is equal; further, 16 fine, 12 wet, 22 changeable autumns: again a little damper and more changeable than in London; and 13 frosty, 14 wet and 23 mild winters: which is considerably damper and milder than in London.

According to measurements made in the Botanical Gardens in Dublin, the following total amount of rain fell each month in the ten years between 1802 and 1811 (in inches): December: 27.31; July: 24.15; November: 23.49; August: 22.47; September: 22.27; January: 21.67; October: 20.12; May: 19.50; March: 14.69; April: 13.54; February: 12.32; June: 12.07. Average for the year: 23.36 (Wakefield, Vol. 1, p. 191). These ten years were unusually dry. Kane (*Industrial Resources,* p. 73) gives an average of 30.87 inches for 6 years in Dublin and *Symons* (*English Rainfall*) puts it at 29.79 inches for 1860-62. Because of the fleeting nature of local showers in Ireland, such measurements mean very little unless they extend over many years and are undertaken at many stations. This is proved among other things by the fact that, of the three stations measuring rainfall in Dublin in 1862, the first recorded 24.63, the second 28.04, and the third 30.18 inches as the average. The average amount of rainfall recorded by 12 stations in different parts of Ireland in the years 1860-62, was not quite 39 inches according to Symons (individual averages varied from 25.45 to 51.44 inches).

In his book about Ireland's climate, Dr. *Patterson* says:

"The frequency of our showers, and not the amount of rainfall itself, has caused the popular notion about the wetness of our climate.... Sometimes the spring sowing is a little delayed because of wet weather, but our springs are so frequently cold and late that early sowing is not always advisable. If frequent summer and autumn showers make our hay and corn harvests risky, then vigilance and diligence would be just as successful in such exigencies as they are for the English in their 'catching' harvests, and improved cultivation would ensure that the seed-corn would aid the peasants' efforts."*

In Londonderry the number of rain-free days each year between 1791 and 1802 varied from 113 to 148—the average for the period was over 126. In Belfast the same average emerged. In Dublin it varied from 168 to 205, average 179 (Patterson, *ibid.*).

According to Wakefield, Irish harvests fall as follows: wheat mostly in September, more rarely in August, occasionally in October; barley usually a little later than wheat; and oats approximately a week after barley, therefore usually in October. After considerable research, Wakefield concluded that not nearly enough material existed for a *scientific* description of the Irish climate, but *nowhere* does he state that it provides a serious obstacle to the cultivation of corn. In fact he finds, as we shall see, that the losses incurred during wet harvest times are due to entirely different causes, and states so quite explicitly:

"The soil of Ireland is so fertile, and the *climate so favourable*, that under a proper system of agriculture, it will produce not only a *sufficiency* of corn for its own use, but a *superabundance* which may be ready at all times to relieve England when she may stand in need of assistance." (Vol. 2, p. 61.)

At that time, of course—1812—England was at war with the whole of Europe and America,[186] and it was much more difficult to import corn—corn was the primary need. Now America, Rumania, Russia and Germany deliver sufficient corn, and the question now is rather one of cheap *meat*. And because of this Ireland's climate is no longer suited to tillage.

Ireland has grown corn since ancient times. In her oldest laws, recorded long before the arrival of Englishmen, the "sack of wheat" is already a definite measurement of value. Fixed quantities of wheat, malt-barley and oatmeal are quite regularly mentioned in the tributes of inferiors to tribal and other chiefs.** After the English invasion, the cultiva-

---

* Dr. W. Patterson, *An Essay on the Climate of Ireland*, Dublin, 1804, p. 164.
** *Ancient Laws and Institutes of Ireland—Senchus Mor*. Two volumes. Dublin, printed for Her Majesty's Stationary Office, and

tion of corn diminished because of the continual battles, without ever ceasing completely; it increased between 1660 and 1725 and decreased again from 1725 to about 1780; more corn as well as a greater quantity of potatoes was again sown between 1780 and 1846, and since then they have both given way to the steadily advancing cattle pastures. If Ireland were not suited to the cultivation of corn, would it have been grown for over a thousand years?

Of course there are regions, in which because of the proximity of mountains the rainfall is always greater, and which are less suited to wheat-growing—notably in the south and west. Besides the good years, a series of wet summers will often occur there, as between 1860-62, which do great harm to the wheat. Wheat, however, is not Ireland's principal grain, and Wakefield even complains that too little of it is grown for lack of a market—the only one being the nearest mill. For the most part, barley is grown only for the secret distilleries (secret because of taxation). Ireland's principal grain was and still is oats. In 1810 no less than 10 times as much oats was grown as of all the other sorts of corn put together. As oats are harvested after wheat and barley, the harvest is usually in late September or October when the weather is usually fine, especially in the south. And in any case, oats can take a considerable amount of rain.

We have already seen that Ireland's climate, as far as the amount and distribution of rain throughout the year is concerned, corresponds almost entirely with that of the North-West of England. The rainfall is much greater in the mountains of Cumberland, Westmorland, and North Lancashire (in Coniston 96.03, in Windermere 75.02 inches, average in the years 1860-62), than in certain stations in Ireland known to me, and yet hay is made and oats are grown there. In the same years the rainfall varied in South Lancashire from 25.11 in Liverpool to 59.13 in Bolton, the average being about 40 inches; in Cheshire it varied from 33.02 to 43.40 inches, the average being approximately 37 inches. In Ireland, as we

published by Alexander Thom (London, Longmans) in 1865 and 1869.[187] See Vol. 2, pp. 239-51. The value of one sack of wheat was 1 screpall (denarius) or 20-24 grains of silver. The value of the screpall is fixed by Dr. *Petrie* in *Ecclesiastical Architecture of Ireland, anterior to the Anglo-Norman Invasion*, Dublin, 1845, 4°, pp. 212-19.

saw, it was not quite 39 inches in the same years. (All figures from Symons.) In both counties corn of all kinds, and in particular wheat, is cultivated; Cheshire carried on mainly cattle-rearing and dairy farming until the last epidemic of cattle-plague, but since most of the cattle perished the climate suddenly became quite admirably suited for wheat-growing. If there had been an epidemic of cattle-plague in Ireland causing devastation similar to that in Cheshire, instead of preaching that Ireland's natural occupation is cattle-raising, they would point to the place in Wakefield which says that Ireland is destined to be England's granary.

If one looks at the matter impartially and without being misled by the cries of the interested parties, the Irish landowners and the English bourgeois, one finds that Ireland, like all other places, has some parts which because of soil and climate are more suited to cattle-rearing, and others to tillage, and still others—the vast majority—which are suited to both. Compared with England, Ireland is more suited to cattle-rearing on the whole; but if England is compared with France, she too is more suited to cattle-rearing. Are we to conclude that the whole of England should be transformed into cattle pastures, and the whole agricultural population be sent into the factory towns or to America—except for a few herdsmen—to make room for cattle, which are to be exported to France in exchange for silk and wine? But that is exactly what the Irish landowners who want to put up their rents and the English bourgeoisie who want to decrease wages demand for Ireland: Goldwin Smith has said so plainly enough. And yet the social revolution inherent in this transformation from tillage to cattle-rearing would be far greater in Ireland than in England. In England, where large-scale agriculture prevails and where agricultural labourers have already been replaced by machinery to a large extent, it would mean the transplantation of at most one million; in Ireland, where small and even cottage-farming prevails, it would mean the transplantation of four million: the extermination of the Irish people.

It can be seen that even the facts of nature become points of national controversy between England and Ireland. It can also be seen, however, how the public opinion of the ruling class in England—and it is only this that is generally known on the Continent—changes with the fashion and in its own interests. Today England needs grain quickly and dependably—Ireland is just perfect for wheat-growing. Tomorrow England needs meat—Ireland is only fit for cattle pastures. The existence of five million Irish is in itself a smack in the eye to all the laws of political economy, they have to get out but whereto is their worry!

ANCIENT IRELAND

The writers of ancient Greece and Rome, and also the fathers of the Church, give very little information about Ireland.

Instead there still exists an abundant native literature, in spite of the many Irish manuscripts lost in the wars of the sixteenth and seventeenth centuries. It includes poems, grammars, glossaries, annals and other historical writings and law-books. With very few exceptions, however, this whole literature, which embraces the period at least from the eighth to the seventeenth centuries, exists only *in manuscript*. For the Irish language printing has existed only for a few years, only from the time when the language began to die out. Of this rich material, therefore, only a small part is available.

Amongst the most important of these annals are those of Abbot *Tigernach* (died 1088), those of *Ulster*, and above all, those of the *Four Masters*. These last were collected in 1632-36 in a monastery in Donegal under the direction of Michael O'Clery, a Franciscan monk, who was helped by three other Seanchaidhes (antiquarians), from materials which now are almost all lost. They were published in 1856 from the original Donegal manuscript which still exists, having been edited and provided with an English translation by O'Donovan.* The earlier editions by Dr. Charles O'Conor (the first part of the *Four Masters*, and the *Annals of Ulster*) are untrustworthy in text and translation.[188]

The beginning of most of these annals presents the mythical prehistory of Ireland. Its base was formed by old folk-

---

* *Annala Rioghachta Eireann. Annals of the Kingdom of Ireland by the Four Masters.* Edited, with an English Translation, by Dr. John O'Donovan. Second edition, Dublin, 1856, 7 volumes in 4°.

legends, which were spun out endlessly by poets in the 9th and 10th centuries and were then brought into suitable chronological order by the monk-chroniclers. The *Annals of the Four Masters* begins with the year of the world 2242, when Caesair, a granddaughter of Noah, landed in Ireland forty days before the Flood; other annals have the ancestors of the Scots, the last immigrants to Ireland, descend in direct line from Japheth and bring them into connection with Moses, the Egyptians and the Phoenicians, as the German chroniclers of the Middle Ages connected the ancestors of the Germans with Troy, Aeneas or Alexander the Great. The *Four Masters* devote only a few pages to this legend (in which the only valuable element, the original folk-legend, is not distinguishable even now); the *Annals of Ulster* leave it out altogether; and Tigernach, with a critical boldness wonderful for his time, explains that all the written records of the Scots before King Cimbaoth (approximately 300 B.C.) are uncertain. But when new national life awoke in Ireland at the end of the last century, and with it new interest in Irish literature and history, just these monks' legends were counted to be their most valuable constituent. With true Celtic enthusiasm and specifically Irish naïveté, belief in these stories was declared an intrinsic part of national patriotism, and this offered the supercunning world of English scholarship—whose own efforts in the field of philological and historical criticism are gloriously enough well known to the rest of the world—the desired pretext for throwing everything Irish aside as arrant nonsense.*

---

*One of the most naive products of that time is *The Chronicles of Eri, being the History of the Gaal Sciot Iber, or the Irish People, translated from the original manuscripts in the Phoenician dialect of the Scythian language by O'Connor*, London, 1822, 2 volumes. The Phoenician dialect of the Scythian language is naturally Celtic Irish, and the original manuscript is a verse chronicle chosen at will. The publisher is Arthur O'Connor, exile of 1798,[189] uncle of Feargus O'Connor who was later leader of the English Chartists, an ostensible descendant of the ancient O'Connors, Kings of Connaught, and, after a fashion, the Irish Pretender to the throne. His portrait appears in front of the title, a man with a handsome, jovial Irish face, strikingly resembling his nephew Feargus, grasping a crown with his right hand. Underneath is the caption: "O'Connor—cear-rige, head of his race, and O'Connor, chief of the prostrate people of his nation: 'Soumis, pas vaincus' [Subdued, not conquered]."

Since the thirties of this century a far more critical spirit has come into being in Ireland, especially through Petrie and O'Donovan. Petrie's already-mentioned researches prove that the most complete agreement exists between the oldest surviving inscriptions, which date from the 6th and 7th centuries, and the annals; and O'Donovan is of the opinion that these begin to report historical facts as early as the second and third centuries of our era. It makes little difference to us whether the credibility of the annals begins several hundred years earlier or later since, unfortunately, during that period they are almost wholly fruitless for our purpose. They contain short, dry notices of deaths, accessions to the throne, wars, battles, earthquakes, plagues, Scandinavian raiding expeditions, but little that has reference to the social life of the people. If the whole juridical literature of Ireland were published, the annals would acquire a completely different meaning; many a dry notice would obtain new life through explanations found in the law-books.

Almost all of these law-books, which are very numerous, still await the time when they will see the light of day. On the insistence of several Irish antiquarians, the English Government agreed in 1852 to appoint a commission for publishing the ancient laws and institutions of Ireland. But the commission consisted of three lords (who are never far away when there is state money to be spent), three lawyers of the highest rank, three Protestant clergymen, and Dr. Petrie and an official who is the chief surveyor in Ireland. Of these gentlemen only Dr. Petrie and two clergymen, Dr. Graves (now Protestant Bishop of Limerick) and Dr. Todd, could claim to understand anything at all about the tasks of the commission, and of these three Petrie and Todd have since died. The commission was instructed to arrange the transcription, translation and publication of the legal content of the ancient Irish manuscripts, and to employ the necessary people for that purpose. It employed the two best people that were to be had, Dr. O'Donovan and Professor O'Curry, who copied, and made a rough translation of, a large number of manuscripts; both died, however, before anything was ready for publication. Their successors, Dr. Hancock and Professor O'Mahony, then took up the work, so that up to the present the two volumes already cited, containing the *Senchus Mor,* have appeared. According to the publishers' acknowledgement only two of the members of the commission, Graves and Todd, have taken part in the work, through some annotations to the proofs. Sir Th. Larcom, a member of the commission, placed the original maps of the survey of Ireland at the disposal of the publishers for the verification of place names. Dr. Petrie soon died, and the other gentlemen confined their activities to drawing their salaries conscientiously for 18 years.

That is how public works are carried out in England, and even more so in English-ruled Ireland. Without jobbery,* they cannot begin. No public interest may be satisfied without a pretty sum or some fat sinecures being siphoned off for lords and government proteges. With the money that the wholly superfluous commission has wasted the entire unpublished historical literature could have been published in Germany—and better.

The *Senchus Mor* has until now been our main source for information about conditions in ancient Ireland. It is a collection of ancient legal decisions which, according to the later composed introduction, was compiled on the orders of St. Patrick, and with his assistance brought into harmony with Christianity, rapidly spreading in Ireland. The High King of Ireland, Laeghaire (428-458, according to the *Annals of the Four Masters*), the Vice-Kings, Corc of Munster and Daire, probably a prince of Ulster, and also three bishops: St. Patrick, St. Benignus and St. Cairnech, and three lawyers: Dubthach, Fergus and Rossa, are supposed to have formed the "commission" which compiled the book—and there is no doubt that they did their work more cheaply than the present commission, who only had to publish it. The *Four Masters* give 438 as the year in which the book was written.

The text itself is evidently based on very ancient heathen materials. The oldest legal formulas in it are written in verse with a precise metre and the so-called consonance, a kind of

---

* The using of public office to one's private advantage or to that of relations and friends, and likewise the using of public money for indirect bribery in the interests of a party, is called jobbery in England. An individual transaction is called a job. The English colony in Ireland is the main centre of jobbery.

alliteration or rather consonant-assonance, which is peculiar to Irish poetry and frequently goes over to full rhyme. As it is certain that old Irish law-books were translated in the fourteenth century from the so-called Fenian dialect (Bérla Feini), the language of the fifth century, into the then current Irish (Introduction (Vol. 1), p. xxxvi and following) it emerges that in the *Senchus Mor* too the metre has been more or less smoothed out in places; but it appears often enough along with occasional rhymes and marked consonance to give the text a definite rhythmical cadence. It is generally sufficient to read the translation in order to find out the verse forms. But then there are also throughout it, especially in the latter half, numerous pieces of undoubted prose; and, whereas the verse is certainly very ancient and has been handed down by tradition, these prose insertions seem to originate with the compilers of the book. At any rate, the *Senchus Mor* is quoted frequently in the glossary composed in the ninth or tenth century, and attributed to the King and Bishop of Cashel, Cormac, and it was certainly written long before the English invasion.

All the manuscripts (the oldest of which appears to date from the beginning of the 14th century or earlier) contain a series of mostly concordant annotations and longer commenting notes on this text. The annotations are in the spirit of old glossaries; quibbles take the place of etymology and the explanation of words, and comments are of varying quality, being often badly distorted or largely incomprehensible, at least without knowledge of the rest of the law-books. The age of the annotations and comments is uncertain. Most of them, however, probably date from after the English invasion. As at the same time they show only a very few traces of developments in the law outside the text itself, and these are only a more precise establishment of details, the greater part, which is purely explanatory, can certainly also be used with some discretion as a source concerning earlier times.

The *Senchus Mor* contains:

1. The law of distraint [*Pfändungsrecht*], that is to say, almost the whole judicial procedure;

2. The law of hostages, which during disputes were put up by people of different territories;

3. The law of Saerrath and Daerrath (see below)[190]; and
4. The law of the family.

From this we obtain much valuable information on the social life of that time, but, as long as many of the expressions are unexplained and the rest of the manuscripts is not published, much remains dark.

In addition to literature, the surviving architectural monuments, churches, round towers, fortifications and inscriptions also enlighten us about the condition of the people before the arrival of the English.

From foreign sources we need only mention a few passages about Ireland in the Scandinavian sagas and the life of St. Malachy by St. Bernard,[191] which are not fruitful sources, and then come immediately to the first Englishman to write about Ireland from his own experience.

Sylvester Gerald Barry, known as *Giraldus Cambrensis*, Archdeacon of Brecknock, was a grandchild of the amorous Nesta, daughter of Rhys ap Tewdwr, Prince of South Wales, and mistress of Henry I of England and the ancestor of almost all the Norman leaders who took part in the first conquest of Ireland. In 1185 he went with John (later "Lackland") to Ireland and in the following years wrote, first, the *Topographia Hibernica*, a description of the land and the inhabitants, and then the *Hibernia Expugnata*, a highly-coloured history of the first invasion. It is mainly the first work which concerns us here. Written in highly pretentious Latin and filled with the wildest belief in miracles and with all the church and national prejudices of the time and the race of its vain author, the book is nevertheless of great importance as the first at all detailed report by a foreigner.*

From here on, Anglo-Norman sources about Ireland naturally become more abundant; however, little knowledge is gained about the social circumstances of the part of the island that remained independent, and it is from this that conclusions regarding ancient conditions could be drawn. It

---

* *Giraldi Cambrensis Opera*, ed. J. S. Brewer, London, Longmans, 1863.[192]—A (weak) English translation of the historical works including the two works already mentioned was published in London by Bohn in 1863 (*The Historical Works of Giraldus Cambrensis*).

is only towards the end of the 16th century, when Ireland as a whole was first systematically subjugated, that we find more detailed reports about the actual living conditions of the Irish people, and these naturally contain a strong English bias. We shall find later that, in the course of the 400 years which elapsed since the first invasion, the condition of the people changed little, and not for the better. But, precisely because of this, these newer writings—Hanmer, Campion, Spenser, Davies, Camden, Moryson and others[193]— which we shall have to consult frequently, are one of our main sources of information on a period 500 years earlier, and a welcome and indispensable supplement to the poor original sources.

The mythical prehistory of Ireland tells of a series of immigrations which took place one after the other and mostly ended with the subduing of the island by the new immigrants. The three last ones are: that of the Firbolgs, that of the Tuatha-de-Dananns, and that of the Milesians or Scots, the last supposed to have come from Spain. Popular writing of history changed Firbolgs (fir—Irish fear, Latin vir, Gothic vair—man) into Belgian without further ado; the Tuatha-de-Dananns (tuatha—Irish people, tract of land, Gothic thiuda) into Greek Danai or German Danes as they felt the need. O'Donovan is of the opinion that something historical lies at the basis of at least the immigrations named above. According to the annals there occurred in the year 10 A.D. an insurrection of the aitheach tuatha (which Lynch, who is a good judge of the old language, translated in the seventeenth century as: *plebeiorum hominum gens*), that is, a plebeian revolution, in which the whole of the nobility (saorchlann) was slain. This points to the dominion of Scottish conquerors over the older inhabitants. O'Donovan draws the conclusion from the folk-tales that the Tuatha-de-Dananns, who were later transformed in folk-lore into elves of the mountain forest, survived up to the 2nd or 3rd century of our era in isolated mountain areas.

There is no doubt that the Irish were a mixed people even before large numbers of English settled among them. As early as the twelfth century, the predominant type was fair-haired as it still is. Giraldus (*Top. Hib.* III, 26) says of two strangers, that they had long *yellow* hair like the Irish. But there are also even now, especially in the west, two quite different types of black-haired people. The one is tall and well-built with fine facial features and curly hair, people whom one thinks that one has already met in the Italian Alps or Lombardy; this type occurs most frequently in the south-west. The other, thickset and short in build, with coarse, lank, black hair and flattened, almost negroid faces, is more frequent in Connaught. Huxley attributes this dark-haired element in the originally light-haired Celtic population to an Iberian (that is, Basque) admixture,[194] which would be correct in part at least. However, at the time when the Irish come clearly into the light of history, they have become a homogeneous people with Celtic speech and we do not find anywhere any other foreign elements, apart from the slaves acquired by conquest or barter, who were mostly Anglo-Saxons.

The reports of the classical writers of antiquity about that people do not sound very flattering. *Diodorus* recounts that those Britons who inhabit the island called Iris (or Irin? it is in the accusative, Ἴριν) eat people.[195] *Strabo* gives a more detailed report:

"Concerning this island [Jerne] I have nothing certain to tell, except that its inhabitants are more savage than the Britons, since they are man-eaters as well as heavy eaters [πολυφάγοι; according to another manner of reading ποηφάγοι —herbivorous], and since, further, they count it an honourable thing, when their fathers die, to devour them, and openly to have intercourse, not only with the other women, but also with their mothers and sisters."[196]

The patriotic Irish historians have been more than a little indignant over this alleged calumny. It was reserved to more recent investigation to prove that cannibalism, and especially the devouring of parents, was a stage in the development of probably all nations. Perhaps it will be a consolation to the Irish to know that the ancestors of the present Berliners were still honouring this custom a full thousand years later:

"*Aber Weletabi, die in Germania sizzent, tie wir Wilze heizên, die ne scament sih nieht ze chedenne daz sie iro parentes mit mêren* *rehte ezen sulin, danne die wurme."** (Notker, quoted in Jacob Grimm's *Rechtsaltertümer,* p. 488.)

And we shall see the consuming of human flesh reoccur more than once under English rule. As far as the phanerogamy (to use an expression of Fourier's[197]), which the Irish are reproached with, is concerned: such things occurred amongst all the barbarous peoples, and much more amongst the quite unusually gallant Celts. It is interesting to note that even then the island carried the present native name: Iris, Irin and Jerne are identical with Eire and Erinn; and how even Ptolemy already knew the present name of the capital, Dublin, Eblana (with the right accent   "Εβλανα).[198] This is all the more noteworthy since the Irish Celts have since ancient times given this city another name, Athcliath, and for them Duibhlinn—the black pool—is the name of a place on the River Liffey.

Moreover we also find the following passage in Pliny's *Historiae Naturalis,* IV, 16:

"The Britons travel there" (to Hibernia) "in boats of willow-branches across which animal-skins have been sewn together."

And later *Solinus* says of the Irish:

"They cross the sea between Hibernia and Britannia in boats of willow-branches, which they overlay with a cover of cattle-hide." (C. Jul. Solini, *Cosmographia,* Ch. 25.)

In the year 1810, Wakefield found that on the whole west coast of Ireland "no other boats occurred except ones which consisted of a wooden frame covered over with a horse-or ox-hide". The shape of these boats varies according to the district, but they are all distinguished by their extraordinary lightness, so that mishaps rarely occur on them. Naturally they are of no use on the open sea, for which reason fishing can only take place in the creeks and amongst the islands. Wakefield saw these boats in Malbay, County Clare. They were 15 feet long, 5 feet wide and 2 feet deep. Two cowhides with the hair on the inside and tarred on the outside

---

* "But the Weletabi who reside in Germany, which we call Wilze, who are not ashamed to say that they have a greater right to eat their parents than the worms have."—*Ed.*

were used for one of these, and they were arranged for two
rowers. Such a boat çost about 30 shillings. (Wakefield, Vol.
2, p. 97.) Instead of woven willows—a wooden frame! What
an advance in 1,800 years and after nearly 700 years of the
"civilising" influence of the foremost maritime nation in the
world!

As for the rest, several signs of progress can be seen.
Under King Cormac Ulfadha, who was placed on the throne
in the second half of the third century, his son-in-law, Finn
McCumhal, is said to have reorganised the Irish militia—
the Fianna Eirionn*—probably on the lines of the Roman
legion with differentiation between light troops and troops
of the line; all the later Irish armies on which we have detailed information have the following categories of troops:
the kerne—light troops—and the galloglas—heavy troops
or troops of the line. Finn's heroic deeds are celebrated in
many old songs, some of which still exist; these and perhaps
a few Scottish-Gaelic traditions form the basis of Macpherson's *Ossian* (Irish Oisin, son of Finn), in which Finn appears as Fingal and the scene is transferred to Scotland.[199]
In Irish folk-lore Finn lives on as Finn Mac-Caul, a giant,
to whom some wonderful feat of strength is ascribed in almost
every locality of the island.

Christianity must have penetrated Ireland quite early, at
least the east coast of it. Otherwise the fact that so many
Irishmen played an important part in Church-history even
long before Patrick cannot be explained. Pelagius the Heretic is usually taken to be a Welsh monk from Bangor; but
there was also an ancient Irish monastery, Bangor, or rather
Banchor at Carrickfergus. That he comes from the Irish
monastery is proved by Hieronymus, who describes him as
being "stupid and heavy with Scottish gruel" ("*scotorum
pultibus praegravatus*").[200] This is the first mention of Irish
oatmeal gruel (Irish lite, Anglo-Irish stirabout), which even
then, before the introduction of potatoes, was the staple food
of the Irish people and after that continued to be so along-

---

* Feini, Fenier, is the name given to the Irish nation throughout the
Senchus Mor. Feinechus, Fenchus, Law of the Fenians, often stands for
the *Senchus* or for another lost law-book. Feine, grad feine also de-
signates the plebs, the lowest free class of people.

side with the latter. Pelagius's chief followers were Celestius and Albinus, also Scots, that is, Irishmen. According to Gennadius,[201] Celestius wrote three detailed letters to his parents from the monastery, and from them it can be seen that alphabetical writing was known in Ireland in the fourth century.

The Irish people are called Scots and the land Scotia in all the writings of the early Middle Ages; we find this term used by Claudianus, Isidorus, Beda, the geographer of Ravenna, Eginhard and even by Alfred the Great: "Hibernia, which we call Scotland" ("Igbernia the ve Scotland hatadh").[202] The present Scotland was called Caledonia by foreigners and Alba, Albania by the inhabitants; the transfer of the name Scotia, Scotland, to the northern area of the eastern isle did not occur until the eleventh century. The first substantial emigration of Irish Scots to Alba is taken to have been in the middle of the third century; Ammianus Marcellinus already knows them there in the year 360.[203] The emigrants used the shortest sea-route, from Antrim to the peninsula of Kintyre; Nennius explicitly says that the Britons, who then occupied all the Scottish lowlands up to the Clyde and Forth, were attacked by the Scots *from the west*, by the Picts from the north.[204] Further, the seventh of the ancient Welsh historical *Triads*[205] reports that the gwyddyl ffichti (see below) came to Alba over the Norse Sea (Môr Llychlin) and settled on the coast. Incidentally, the fact that the sea between Scotland and the Hebrides is called the Norse Sea shows that this *Triad* was written after the Norse conquest of the Hebrides. Large numbers of Scots came over again at about the year 500, and they gradually formed a kingdom, independent of both Ireland and the Picts. They finally subdued the Picts in the ninth century under Kenneth MacAlpin and created the state to which the name Scotland, Scotia was transferred, probably first by the Norsemen about 150 years later.

Invasions of Wales by the gwyddyl ffichti or Gaelic Picts are mentioned in ancient Welsh sources (Nennius, the *Triads*) of the fifth and sixth centuries. These are generally accepted as being invasions of Irish Scots. Gwyddyl is the Welsh form of gavidheal, as the Irish call themselves. The origin of the term Picts can be investigated by someone else.

Patricius (Irish Patrick, Patraic, as the Celts always pronounce their c as k in the Ancient Roman way) brought Christianity to dominance in the second quarter of the fifth century without any violent convulsions. Trade with Britain, which had been of long standing, also became livelier at this time; architects and building workers came over and the Irish learned from them to build with mortar, while up to then they had only known *dry-stone* building. As mortar building occurs between the seventh and twelfth centuries, and then only in church buildings, that is proof enough that its introduction is connected with that of Christianity, and further, that from then on the clergy, as the representative of foreign culture, severed itself completely from the people in its intellectual development. Whilst the people made no, or only extremely slow, social advances, there soon developed amongst the clergy a literary learning which was extraordinary for the time and which, in accordance with the custom then, manifested itself mostly in zeal for converting heathens and founding monasteries. Columba converted the British Scots and the Picts; Gallus (founder of St. Gallen) and Fridolin the Allemanni, Kilian the Franks on the Main, Virgilius the city of Salzburg. All five were Irish. The Anglo-Saxons were also converted to Christianity mainly by Irish missionaries. Furthermore, Ireland was known throughout Europe as a nursery of learning, so much so that Charlemagne summoned an Irish monk, Albinus, to teach at Pavia, where another Irishman, Dungal, followed him later. The most important of the many Irish scholars, who were famous at that time but are now mostly forgotten, was the "Father", or as Erdmann calls him, the "Carolus Magnus"* of philosophy in the Middle Ages—*Johannes Scotus Erigena*. Hegel says of him, "Real philosophy began first with him."[206] He alone understood Greek in Western Europe in the ninth century, and by his translation of the writings attributed to Dionysius the Areopagite, he restored the link with the last branch of the old philosophy, the Alexandrian Neoplatonic school.[207] His teaching was very bold for the time. He denied the "eternity of damnation", even for the devil, and brushed close to Pantheism. Contemporary ortho-

---

* Charles the Great.—*Ed.*

doxy, therefore, did not fail to slander him. It took a full two hundred years before the branch of learning founded by Erigena was developed by Anselm of Canterbury.*

Before this development of culture could have an effect on the people, it was interrupted by the raids of the Norsemen. The raids, which form the main staple product of Scandinavian, and particularly Danish, patriotism, occurred too late, and the nations from which they originated were too small for them to result in conquest, colonisation, and the forming of states on a large scale as had been the case with the earlier invasions of the Germans. Their advantage which they bequeathed on historical development is infinitesimal in comparison with the immense and fruitless (even for the Scandinavians themselves) disturbances they caused.

Ireland was far from being inhabited by a single nation at the end of the eighth century. Supreme royal power over the whole island existed only in appearance, and by no means always at that. The provincial kings, whose number and territories were continually changing, fought amongst themselves, and the smaller territorial princes likewise carried on their private feuds. On the whole, however, these internal wars seem to have been governed by certain customs which held the ravages within definite limits, so that the country did not suffer too much. But this was not to last. In 795, a few years after the English had been first raided by the same plundering nation, Norsemen landed on the Isle of Rathlin, off the coast of Antrim, and burnt everything down; in 798, they landed near Dublin, and after this they are mentioned nearly every year in the annals as heathens, foreigners, pirates, and never without additional reports of the losccadh (burning down) of one or more places. Their colonies on the Orkneys, Shetlands and Hebrides (Southern Isles, Sudhreyjar in the old Norse sagas) served them as operational bases against Ireland, and against what was later

---

* More about Erigena's doctrine and works is to be found in Erdmann's *Grundriss der Geschichte der Philosophie*, 2. Aufl., Berlin, 1869, Bd. I, S. 241-47. Erigena, who was not a clergyman, shows real Irish wit. When Charles the Bald, King of France, who was sitting opposite him at table, asked him the difference between a Scot and a sot, Erigena answered: "The width of a table."

known as Scotland, and against England. In the middle
of the ninth century, they were in possession of Dublin,*
which, according to Giraldus, they rebuilt for the first time
into a proper city. He also attributes the building of Limerick and Waterford to them. The name Waterford is only
a nonsensical anglicisation of the ancient Norse Vedhrafiördhr, which means either storm-bay [*Wetterföhrde*] or
ram-bay [*Widderbucht*]. Naturally, as soon as the Norsemen
settled down in the land, their prime necessity was to have
fortified harbour-towns. The population of these long remained Scandinavian, but in the twelfth century it had long
since assimilated Irish speech and customs. The quarrelling
of the Irish princes amongst themselves greatly simplified
pillage and settlement for the Norsemen, and even the temporary conquest of the whole island. The extent to which
the Scandinavians considered Ireland as one of their regular
pillage grounds is shown by the so-called death-song of Ragnar Lodbrók, the *Krákumál*, composed about the year 1000
in the snaketower of King Ella of Northumberland.[209] In
this song all the ancient pagan savagery is massed together,
as if for the last time, and under the pretext of celebrating
King Ragnar's heroic deeds in song, all the Nordic peoples'
raids in their own lands, on coasts from Dünamünde to Flan-
ders, Scotland (here already called Skotland, perhaps for the
first time) and Ireland are briefly pictured. About Ireland
is said:

"We hew'd with our swords, heap'd high the slain,
Glad was the wolf's brother of the furious battle's
                                        feast;
Iron struck brass-shields; Ireland's ruler,
                                        Marsteinn,
Did not starve the murder-wolf or eagle;
In Vedhrafiördhr the raven was given a sacrifice.

We hew'd with our swords, started a game at dawn,
A merry battle against three kings at Lindiseyri;
Not many could boast that they fled unhurt from there.

---

* The assertion of Snorri in the *Haraldsaga*,[208] that Harald Hårfagr's
sons, Thorgils and Frodi, were the first of the Norsemen to occupy
Dublin—that is, at least 50 years later than stated—is in direct con-
tradiction with all Irish accounts which are unimpeachable for this
period. Evidently Snorri is confusing Harald Hårfagr's son Thorgils
with the Thorgils (Turgesius) mentioned later.

Falcon fought wolf for flesh, the wolf's fury devoured
many;
The blood of the Irish flow'd in streams on the beach
in the battle."*

By the first half of the ninth century, a Norse Viking Thorgils, called Turgesius by the Irish, had succeeded in submitting all Ireland to his rule. But, with his death in 844, his kingdom fell apart, and the Norsemen were driven out. The invasions and battles continued with varying success. Finally, at the beginning of the eleventh century, Ireland's national hero, Brian Borumha, originally King of only a part of Munster, gained the kingship of all Ireland and gave the decisive battle to the concentrated force of the invading Norsemen on the 23rd April (Good Friday), 1014, at Clontarf, close to Dublin, as a result of which the power of the invaders was broken forever.

The Norsemen who had settled in Ireland, and on whom Leinster was dependent (the King of Leinster, Maolmordha, had come to the throne in 999 with their help and was maintained there by it), had sent messengers to the Hebrides, the Orkneys, Denmark and Norway asking for reinforcements, in anticipation of the impending decisive battle. Help came to them in large numbers. The *Niâlssaga*[211] recounts how Jarl Sigurd Laudrisson armed himself for the departure on the Orkneys, and how Thorstein Siduhallsson, Hrafn the Red and Erlinger of Straumey went with him, and how he arrived in Dublin (Durflin) with all his army on Palm Sunday.

---

* "Hiuggu ver medh hiörvi, hverr lâthverr of annan;
gladhr vardh gera brôdhir getu vidh sôknar laeti,
lêt ei örn nê ylgi, sâ er Îrlandi styrdhi,
(môt vardh mâlms ok rîtar) Marsteinn konungr fasta;
vardh î Vedhra firdhi valtafn gefit hrafni.

Hiuggu ver medh hiörvi, hâdhum sudhr at morni
leik fyrir Lindiseyri vidh lofdhûnga threnna;
fârr âtti thvî fagna (fêll margr î gyn ûlfi,
haukr sleit hold medh vargi), at hann heill thadhan kaemi;
Yra blôdh î oegi aerit fêll um skaeru."

Vedhrafiördhr is, as we have said, Waterford; I do not know whether Lindiseyri has been discovered anywhere. On no account does it mean Leinster as Johnstone translates it[210]; eyri (sandy neck of land, Danish öre) points to a quite distinct locality. Valtafn can also mean falcon feed and is generally translated as such here, but as the raven is Odin's holy bird, the word obviously has both meanings.

"Brodhir had already arrived with his whole force. Brodhir tried to learn by means of sorcery how the battle would turn out, and the answer was this: if the battle was fought on a Friday, King Brian would win the victory but die; and that if it was fought before that time, then all who were against him would fall. Then Brodhir said that they should not fight before Friday."

There are two versions of the battle itself, that of the Irish annals and the Scandinavian one of the *Niálssaga*. According to the latter:

"King Brian had come up to the fortified town" (Dublin) "with his entire army, and on Friday the army" (of the Norsemen) "issued from the town. Both hosts arranged themselves in battle array. Brodhir headed one wing, King Sigtrygg" (King of the Dublin Norsemen according to the *Annals of Inisfallen*) "the other. We must say that King Brian did not wish to give battle on Good Friday; therefore a shield-burg was set about him and his army stationed in front of that. Ulf Hraeda headed the wing facing Brodhir, and Ospak and his sons headed the wing facing Sigtrygg, but Kerthialfadh stood in the middle and had the flag carried before him."

When the battle began Brodhir was driven into a wood by Ulf Hraeda where he found safety. Jarl Sigurd had a hard struggle against Kerthialfadh, who fought his way to the flag and slew the flag-bearer as well as the next man who seized the flag; then all refused to carry the flag and Jarl Sigurd took the flag from the staff and hid it in his clothing. Soon after he was pierced by a spear, and with this his part of the army appears to have been defeated. Meanwhile Ospak attacked the Norsemen in the rear and defeated Sigtrygg's wing after a hard fought battle.

"Thereupon the entire host took to flight. Thorstein Hallson stopped while the others were fleeing and tied his shoe thong. Then Kerthialfadh asked him why he was not running too.

" 'Because I can't get home this evening anyway,' said Thorstein, 'as I live out in Iceland!' Kerthialfadh spared him."

Brodhir now saw from his hiding-place that Brian's army was pursuing those who fled from the battle and that few people remained at the shield-burg. Then he ran out of the wood, broke through the shield-burg and slew the King. (Brian, who was 88, was obviously not capable of joining in the battle and had remained in the camp.)

"Then Brodhir shouted: 'Let it pass from mouth to mouth that Brodhir felled Brian!' "

But the pursuers returned, surrounded Brodhir and seized him alive.

"Ulf Hraeda slit open his belly, led him round and round an oak-tree, and in this way unwound all his intestines out of his body, and Brodhir did not die before they were all pulled out of him. Brodhir's men were slain to the last man."

According to the *Annals of Inisfallen* the Norse army was divided into three sections. The first consisted of the Dublin Norsemen and 1,000 Norwegian volunteers, who all wore long shirts of mail. The second was made up of the Irish auxiliary forces from Leinster under King Maolmordha. The third consisted of reinforcements from the Islands and Scandinavia under Bruadhair, the commander of the fleet that had brought them, and Lodar, the Jarl of the Orkneys. Against these Brian also placed his troops in three sections; but the names of the leaders given here do not correspond with those given in the *Niâlssaga*, and the account of the battle is insignificant. The following account, given in the *Four Masters*, is shorter and clearer:

"A.D. 1013 [given here as everywhere mistakenly for 1014]. The foreigners of the west of Europe assembled against Brian and Maelseachlainn" (usually called Malachy, King of Meath under Brian's High Kingship); "and they took with them ten hundred men with coats of mail. A spirited, fierce, violent, vengeful, and furious battle was fought between them—the likeness of which was not to be found at that time—at Cluaintarbh" (Meadow of the Bulls, now Clontarf) "on the Friday before Easter precisely. In this battle were slain Brian ... in the eighty-eighth year of his age; Murchadh, his son, in the sixty-third year of his age; Conaing, ... the son of Brian's brother; Toirdhealbhach, son of Murchadh..." (there follow a multitude of names). "The" (enemy) "forces were afterwards routed by dint of battling, bravery, and striking, by Maelseachlainn, from Tulcainn to Athcliath" (Dublin), "against the foreigners and the Leinstermen; and there fell Maolmordha, son of Murchadh, son of Finn, King of Leinster.... There was a countless slaughter of the Leinstermen along with them. There were also slain Dubhgall, son of Amhlanibh" (usually called Anlaf or Olaf), "and Gillaciarain, son of Gluniairn, two tanists of the foreigners, Sichfrith, son of Lodar, Earl of the Orkneys *(iarla Insi h Oirc)*; Brodar, chief of the Danes, who was the person that slew Brian. The ten hundred in armour were cut to pieces, and at the least three thousand of the foreigners were there slain."

The *Niâlssaga* was written in Iceland approximately 100 years after the battle; the Irish annals are based, at least in part, on contemporary information. The two are completely independent of each other. Yet not only do they correspond in all the main points, but they also complete each other. We can only find out who Brodhir and Sigtrygg were from the Irish annals. Sigurd Laudrisson is the name of Sichfrith, Lodar's son. Sichfrith is in fact the correct Anglo-Saxon form of the ancient Norse name, Sigurd. In Ireland, Scandinavian names appear—on coins as well as in the annals—mainly in their Anglo-Saxon forms, not in the ancient Norse. In the *Niâlssaga* the names of Brian's generals are adapted for easier pronunciation by the Scandinavians. One of the names, Ulf Hraeda, is, in fact, ancient Norse, but it would be risky as some do to conclude from this that Brian had Norsemen in his army too. Ospak and Kerthialfadh appear to be Celtic names; the latter might be a distortion of the Toirdhealbhach mentioned in the *Four Masters*. The date of the battle—given as the Friday after Palm Sunday in the one, and as the Friday before Easter in the other—is the same in both, as is also the place of the battle. Although this is given as Kantaraburg (otherwise Canterbury)[212] in the *Niâlssaga*, it is also explicitly said to be close to the gates of Dublin. The course of the battle is reported more precisely in the *Four Masters*: The Norsemen attacked Brian's army on the Plain of Clontarf. From there they were thrown back beyond the Tolka, a little stream near the northern part of Dublin, towards the city. Both report that Brodhir slew King Brian, but more detailed accounts are given only in the Norse source.

It can be seen that our reports on this battle are quite informative and authentic, considering the barbarity of that time. There are not many eleventh-century battles on which such reliable and corroborating accounts are available from both sides. This does not prevent Professor Goldwin Smith from describing it as a "shadowy conflict" (*Ir. His.*, p. 48). Certainly, the most robust facts quite often take on a "shadowy" form in our Professor's head.

After their defeat at Clontarf, the Norse raids became less frequent and less dangerous. The Dublin Norsemen soon came under the domination of the neighbouring Irish princes, and, after one or two generations, were assimilated by the native population . The only compensation the Irish got for the devastation caused by the Scandinavians was

three or four cities and the beginnings of a trading bourgeoisie.

The further back we go into history, the more the characteristics distinguishing different peoples of the same race
disappear. This is partly because of the nature of the sources,
which in the measure in which they are older become thinner
and contain only the most essential information, and partly
because of the development of the peoples themselves. The
less remote the individual branches are from the original
stock, the nearer they are to each other and the more they
resemble each other. Jacob Grimm has always quite correctly treated the information given by Roman historians,
who described the War of the Cimbri,[213] Adam of Bremen and Saxo Grammaticus, all the literary written records
from *Beowulf* and *Hildebrandslied* to the *Eddas*[214] and the
sagas, all the books of law from the *Leges barbarorum*[215] to
the ancient Danish and ancient Swedish laws and the old
Germanic judicial procedures as equally valuable sources of
information on the German national character, customs and
legal conditions. A specific characteristic may be of purely
local significance, but the character reflected in it is common to the whole race; and the older the sources used, the
more local differences disappear.

Just as the Scandinavians and the Germans differed less
in the seventh and eighth centuries than they do today, so also
must the Irish Celts and the Gallic Celts have originally
resembled each other more than present-day Irishmen and
Frenchmen. Therefore we should not be surprised to find in
Caesar's description of the Gauls many features which are
ascribed to the Irish by Giraldus some twelve hundred
years later, and which, furthermore, are discernible in the
Irish national character even today, in spite of the admixture
of Germanic blood. . . .

Published in the book
Marx-Engels Archives,
Vol. X, Russ. ed., Moscow, 1948

Translated from the
German

Frederick Engels

From THE PREPARATORY MATERIAL
FOR THE "HISTORY OF IRELAND"[216]

DRAFT PLAN

1. Natural conditions
2. Ancient Ireland
3. English conquests
   1) First invasion
   2) Pale and Irishry
   3) Subjugation and expropriation. 152...-1691
4. English rule
   1) Penal Laws.[217] 1691-1780
   2) Rebellion and Union. 1780-1801
   3) Ireland in the United Kingdom
   a) The period of the small peasants. 1801-1846
   b) The period of extermination. 1846-1870

Published in the book
Marx-Engels Archives,
Vol. X, Russ. ed., Moscow, 1948

Translated from the
German

NOTES FOR THE "HISTORY OF IRELAND"

Ir[ish] literature?—17th century, poet[ical], histor[ical], jurid[ical], then completely suppressed due to the extirpation of the Ir[ish] *literary* language—exists *only in manuscript*—publication is beginning only now—this is [possible] only with an oppressed people. See Serbs, etc.

———

The English knew how to reconcile people of the most diverse races with their rule. The Welsh, who held so tenaciously to their nationality and language, have fused completely with the British Empire. The Scottish Celts, though rebellious until 1745[218] and since almost completely exterminated first by the government and then by their own aristocracy, do not even think of rebellion. The French of the Channel Isles fought bitterly against France during the Great Revolution. Even the Frisians of Heligoland,[219] which Denmark sold to Britain, are satisfied with their lot; and a long time will probably pass before the laurels of Sadowa and the conquests of the North-German Confederation[220] wrench from their throats a pained wail about unification with the "great fatherland". Only with the Irish the English could not cope. The reason for this is the enormous resilience of the Irish race. After the most savage suppression, after every attempt to exterminate them, the Irish, following a short respite, stood stronger than ever before: it seemed they drew their main strength from the very foreign garrison forced on them in order to oppress them. Within two generations, often within one, the foreigners became more Irish than the Irish, *Hiberniores ipsis Hibernis*. The more the Irish accepted the English language and forgot their own, the more Irish they became.

The bourgeoisie turns everything into a commodity, hence also the writing of history. It is part of its being, of its condition for existence, to falsify all goods: it falsified the writing of history. And the best-paid historiography is that which is best falsified for the purposes of the bourgeoisie. Witness Macaulay, who, for that very reason, is the inept G. Smith's unequalled paragon.

Queen's Evidence.—Rewards for Evidence.
England is the only country where the state openly dares to bribe witnesses, [be it] by an offer of exemption from punishment, be it by ready cash. That prices are fixed for the betrayal of the sojourn of a political persecutee is comprehensible, but that they say: who gives me evidence *on grounds of which* somebody can be sentenced as the contriver of some crime or another—this infamy is something not only the Code, but also Pr[ussian] common law have left to Eng[lish] law. That collateral evidence is required alongside with that given by the informer is useless; generally

there is suspicion of somebody, or else it is fabricated, and
the informer only has to adjust his lies accordingly.

Whether this pretty usage *[saubere Usus]* has its roots
already in Eng[lish] legal proceedings is hard to say, but it
is certain that it has received its *development* on Irish soil
at the time of the Tories[221] and the penal laws.

_____

On March 15, 1870, when the government removed an
Irish sheriff (Coote of Monaghan) on the plea that he had
packed the jury panel, G. H. Moore, M. P. for Mayo, said in
Parliament:

"If Capt. Coote had done all the things of which he had been
accused, he had only followed the practice which, in political cases,
?
had been habitually sanctioned by the Institute Executive."

As one instance out of many that might be cited, he would
mention *that though County Cork had a proportion of 500,000
Catholics against 50,000 Protestants, at the time of the Fenian
trials in 1865,[222] a jury panel was called, composed of 360
Protestants and 40 Catholics!*

_____

The German Legion of 1806-13 was also sent to Ireland.
Thus, the good Hanoverians who refused to put up with
French [bondage] rule, *were used by the English to preserve
the English rule in Ireland!*

_____

The agrarian murders in Ireland cannot be suppressed
because and *as long as* they are the only effective remedy
against the extermination of the people by the landlords.
They *help,* that is why they continue, and will continue, in
spite of all the coercive laws. Their number varies, as it does
with all social phenomena; they can even become epidemic
in certain circumstances, when they occur at quite insignificant occasions. The epidemic can be suppressed, but the sickness itself cannot.

Published in the book   Translated from the
Marx-Engels Archives,   German
Vol. X, Russ. ed., Moscow, 1948

CHRONOLOGY OF IRELAND[223]

?	Immigration of the Scots (Milesians).
200 B.C. ?	King Kimbaoth.
A.D. 2 ?	King Conary the Great?
258 ?	First Scottish settlement in Albany (Scotland).
	King Cormac Ulfadha.—Finn McCumhal.
396	Irish invasion of Great Britain. King Nial of the Nine Hostages.
406	Dathy, last of the Irish heathen kings.
403	St. Patrick brought to Ireland from France as slave. He fled in 410.
422*	Returned as converter and died in 465.
684	Egfrid, King of Northumberland, sailed his navy to Ireland.
795	First Danish invasion, thenceforth regularly renewed (first invasion of England in 787).
818-33	King Concobar.
839-46	Feidlim, King of Munster.
844	Turgesius died and Danes were expelled.
849	New Danish invasion.
853	Olaf, Ivar and Sitrick arrived. Nose-money tribute.
901-08	Cormac McCulinan, King of Munster.
902	Leinster expelled Danes from Dublin.
926	Muirkeartach's first victory over Danes.
937	Battle of Brunanburh. Olaf of Dublin takes part.[224]
939	Muirkeartach—ruler of all Ireland.
943	Muirkeartach died.
944	King Donogh died.
969	Mahon, King of Munster, and his brother Brian Boromhe (King Kennedy's son) defeated Limerick Danes at Sulchoide and, pursuing them, captured Limerick, which they burned.

---

* Slip of the pen. St. Patrick begins as missionary in Ireland in 432.—*Ed.*

976	Mahon assassinated by another chieftain,* Maolmua. Brian Boru, King of all Munster, defeated Maolmua and other chieftains involved in the plot, conquered Iniscathy (Shannon estuary) from the Danes and expelled them from the other Shannon islands.
980	Malachy the Great (of the Hy Nials) became King of Tara (at that time there were only two kingdoms in Ireland—Cashel and Tara); defeated the Danes at Tara, subjugated them and freed all Irish war prisoners (c. 2,000). Leinster and other vassal chieftains [*Unterfürsten*] plotted against Brian, but were foiled.
982	Malachy overran Brian's possessions.
983	Malachy overran Leinster. Brian made war. They signed an agreement consummating the division of Ireland, with Leinster remaining a tributary of the Southern Kingdom.
988	Another war broke out between the two with changing fortune, until
997	the agreement formalising the division was reaffirmed.
998-1000	The two made common cause in war against Danes, achieving notable success.
1000	Again war between the two; Malachy, the weaker, submitted *before* the battle.
1001	Brian Boru became King of Tara and all Ireland.
1008	Defeated the rebellious Southern Hy Nials at Athlone. General peace set in.
1013	Sitrick=Sigtrygg, the Danish King of Dublin, and his allies from Leinster invaded Meath, where Malachy was local king, and defeated him. Brian denied Malachy help, but in summer marched against and ravaged Leinster.

Ms. says *Fürst*, i.e., prince.—*Ed.*

1014   Large-scale invasion of Ireland by the Norsemen. They made Dublin their main base. Brian marched on Dublin. Battle of Clontarf on April 23 (Good Friday). The Danes defeated (described in *Niálssaga*; see Dietrich, [*Altnordisches Lesebuch*] p. 52). Brian was assassinated in his tent by the Norwegian Admiral Brôdar; his son Morrough fell too. After the battle strife broke out anew over succession and supremacy.

1015   *Malachy* again became King of Ireland and repulsed a new Danish invasion. Numerous inland risings and new clashes with the Danes who never recovered after Clontarf.

1022   Malachy abdicated and withdrew to a cloister, where he soon died. *No new supreme king was elected.* Wars of succession followed in Munster until

1064   Turlough, Brian Boru's nephew, became King and

1072   annexed Dublin, Leinster and Meath.

1070   Murchad, the first *Irish* King of the Dublin Danes, who now assimilate rapidly.
Ulster was also finally subjugated by Turlough.

1086   Turlough died. Wars of succession followed.

1090   Treaty of Lough Neagh: Murkertach, son of Turlough, made King of the South, and Domnal O'Lochlin, chief of the Hy Nials, King of the North. But war broke out between them at once, lasting 28 years. In 1103 Murkertach was defeated.

1114   Murkertach, who fell sick, abdicated in favour of Dermot, his brother.

1121   Domnal O'Lochlin died. New wars of succession followed.

1088   Tigernach (pronounced Tiarna), the chronicler, died.

1086   Marianus Scotus died in Mayence.

1136   Tordelvac O'Connor, King of Connaught,

	made King of all Ireland, but continuously attacked by the kings of Munster, until
1151	the Momons were totally defeated at Moinmor and Munster was subjugated. But a rising followed at once
1153	by Murtogh O'Lochlin, King of Tyrone, chief of Ulster and member of the Hy Nials, who, however, was also defeated.
1152	Synod in Kells. Resolutions against simony, usury, priest marriage and concubinage. Later, a prescript by Cardinal Legate Paparo, *introducing payment of tithe in Ireland.*
1156	Tordelwach died. His son Roderic O'Connor—King of Connaught; but Murtogh O'Lochlin made King of all Ireland, meeting but little resistance from Roderic. Otherwise, peace.
1166	Murtogh died. Roderic O'Connor became King of Ireland. Held
1167	counsel with all chiefs and prelates *at Athboy*, where a retinue of 30,000 people gathered. This was exactly four years before the English invasion!
1153	Dermot McMurchad, King of Leinster, abducted Dervorgilla, wife of Tiernan O'Ruark, chief of Breffny in East Connaught.
1154	Tordelwach forced him to return her and protected O'Ruark. However, his successor O'Lochlin sided with Dermot, while Roderic again on O'Ruark's side.
1166	Roderic sent reinforcements to help O'Ruark and drove out Dermot, who fled
1168	to England and appealed for help to Henry II. The latter had soon after 1155 obtained from Pope Adrian IV (an Englishman by name of Breakspear) a bull allowing him in return for recognising extended temporal papal court authority to conquer Ireland in order to reform the Irish

church, with every Irish household paying the Pope 1d. yearly.

1169-71   Conquest of South and East Ireland by the English.[225]

1173   Marauding by the English.

1174   Strongbow and Hervey of Mount Maurice defeated by Donald O'Brian. General uprising. Raymond Le Gros brought 30 knights, 100 men-at-arms and 30 archers from England and restored order. He became Strongbow's son-in-law and enfeoffed *Idrone, Fethard and Glascarrig*; captured Limerick from Donald O'Brian.

1175   O'Brian beleaguered Limerick, but was defeated at Cashel. *Here Irishmen, the princes of Ossory and Kinsale, sided with the English.* Roderic and O'Brian accepted defeat. Roderic was reaffirmed as King of all Ireland under English suzerainty, exclusive of Leinster, Meath and the coast from Waterford to Dungarvan. These were put directly under English rule. Roderic acknowledged that the Kings of England were for all time Lords Paramount in Ireland and the fee of the soil should be in them. Meanwhile, old laws remained and chieftains retained full power in Roderic's possessions, making war on each other as before.

1176   Strongbow died.

1177   English invasion of Ulster under de Courcy failed. Ditto of Connaught under Milo de Cogan without pretext and just as unsuccessful. The Irish *laid waste the land and withdrew to the hills*, attacking the English as the latter withdrew, and defeating them.

1178   De Courcy defeated in Ulster and pressed back to Downpatrick.

1182   De Cogan (Milo) assassinated in Desmond. Uprising in Munster. Strife among Irish, as

	a result of which Roderic abdicated in favour of his son, Connor Manmoy.
1184-85	New reinforcements of the English. Continuous plunder of the country, especially of Ulster, by the English.
1185	John (Lackland), 12 years old, sent to Ireland as Lord. His retinue insulted the Irish chiefs, and a general uprising broke out. Irish clans, long subdued in the Pale, were driven out by the English and their land confiscated. Even Welsh were mistreated by John's men. *Now the Irish began a small war with some success*, destroying isolated forts and detachments. But soon they resumed wars against each other, so that by and large the English held their ground.
1189	Henry II died. Uprisings against the English broke out continuously until the end of the century. Continuous internal wars between the Irish and those Irishmen who fought on the side of the English.
1198	Strife broke out among the English barons. After Roderic's death a war of succession began in Connaught between his sons Carrach, supported by William de Burgh (of the Fitz-Adelms), and Cathal, backed by J. de Courcy and Walter de Lacy.
	Soon thereafter the rivalry between John de Courcy and Hugh de Lacy culminated in
1205	de Courcy's capture by the King and the transfer of his county in Ulster to de Lacy.
1205-16	Ireland mostly quiet until John's death.
1216	*HENRY III.* Ten years old. Earl Pembroke, Strongbow's heir in Leinster, Earl Marshal of England, appointed administrator. Magna Carta[226] extended to Ireland (i.e., for the English).
1219-20	War between William Earl Pembroke (son of the above) and Hugh de Lacy over some

border land, with O'Neill of Tyrone helping de Lacy.

1245   Maurice Fitz-Gerald, Lord Justice of Ireland, supplied an Irish army which included Feidlim, King of Connaught, to aid King Henry in the war against Wales. This campaign was conducted voluntarily by the Irish barons, for they were not obligated to serve outside Ireland; "may this not be considered a precedent".*

1244 and 1254   Henry ordered the indigenous Irish chiefs to provide him with troops in Scotland and Gascogne. Nothing is known of whether they complied.

1255   Irish troops sailed to help Earl of Chester and the Welsh against the English, but were defeated before landing by Prince Edward (later I). Thereupon, Irish troops dispatched to help the King *against* the Welsh.

1259   Uprising of the McCarthys of Desmond, almost all of whose land was given over to the Geraldines.[227] The Geraldines were expelled, but the success was not lasting, because other chiefs denied help.

1264   Feud between the de Burghs and Geraldines, until finally the Irish Parliament (?) in Kilkenny and the new Lord Justice Barry put an end to it.

1270   A new strong uprising of the Irish, but only destruction and a small war resulted; English power remained vigorous.

1272   *EDWARD I.* Early in his reign, the Irish (of the Pale) petitioned that English law be extended to them.

That same year, 1272, the Irish rose again.

Invasion of Ireland by Scots, followed by a

---

* Undertaking given by Henry III to the Anglo-Irish barons in the Act of the 28th year of his reign.—*Ed.*

raid of Scotland by Richard de Burgh and Sir Eustace de Poer with Irish troops employing their favourite method of *smoking* the Scots *out of the caves.*

1276-80   Many wars against the Irish.

1277   Wars of succession between the O'Brians of Thomond; Thomas de Clare, son of Earl of Gloucester, took advantage of this to establish himself in the country. In the meantime, the Irish warred among themselves in Connaught, of which Lord Justice Robert de Ufford wrote the King that it would be fine if the rebels killed each other, because it did not cost the King's treasury anything and would help instil peace in the country (Vol. III, p. 33*).

1280   Edward called on lords spiritual and temporal and all the other Englishmen in Ireland to hold counsel about the petition asking for the Irish to be placed under English law. He was in favour (the Irish promised 8,000 marks for it), because the laws of the Irish

From   were "hateful in the sight of God" and so
Davies**   unjust that they could not be considered as laws, though he did not wish to act without the consent of the lords. However, the barons appear not to have taken any notice, with still only a few Irishmen admitted within the pale of English law.

Feuds between the *de Burghs and the Geraldines,* likewise between other barons, throughout Edward's reign. Similar strife between the Irish chiefs.

At last,

1295   Lord Justice Sir John Wogan convened Parliament to settle the feuds, devising an armistice that lasted two years. This Parliament was, of course, no more than a gather-

---

* Th. Moore, *History of Ireland,* Vol. III, Paris, 1840, p. 33.—*Ed.*
** John Davies, *Historical Tracts,* London, 1786.—*Ed.*

ing of barons and prelates. For its decisions see excerpts [from Moore, *History of Ireland*, Book of Excerpts II,] p. 12.[228]

1299 When Anglo-Irish auxiliary troops set out for the Scottish war[229] an uprising occurred in the Maraghie mountains and in Oriel. Peace ensued for a number of years after the troops returned.

1303 Again, Anglo-Irish troops from Ulster set out for Scotland.

1306 Irish rising in Meath crushed in the Battle of Glenfell.

1307 Irish rising in Offaley and Connaught.

1307 *EDWARD II.*

1309 or 1310 Parliament in Kilkenny: acts against gross exactions and general misconduct of the nobility.

1312 The Byrnes and O'Tooles of Wicklow marched on Dublin, while English bondsmen [*Lehnsleute*] in Oriel rebelled.

1307 Robert Bruce, who had fled to Rachlin Island, Antrim County, where he was in hiding all winter, helped by the Irish, set out for Galloway with 300 Scotsmen and 700 Irish troops, but was intercepted by Duncan M'Dowal, a local chief, at embarkation and defeated.

1315 After Robert Bruce's victory at Bannockburn in 1314,[230] Edward Bruce and 6,000 men landed in Antrim, the Irish joining him en masse, and conquered Ulster; he was crowned King of Ireland in Dundalk, defeated the English under de Burgh on the Banne River, Down County, and waited for reinforcements from Scotland. While Feidlim O'Connor of Connaught marched off with the English, Roderic O'Connor rebelled; Connaught was swept by insurrection; but Feidlim defeated Roderic, who was killed in battle; whereupon Feidlim banded with Bruce. Munster, too, rose against the

English; even several of the great lords (English) and many English people made common cause with Bruce. The latter defeated the English in Meath, marched on Kildare and defeated them once more; an insurrection in Leinster, especially Wicklow (Byrnes, O'Tooles and O'Moores), held in check by the English.

1316   Food shortages compelled Bruce to withdraw to Ulster, where he idled. The English Lord Justice, Butler, suppressed the rising in Wicklow, then the English marched against

1316   Feidlim, defeating him (he fell) at Athenry. Robert Bruce arrived in Ireland with a large force, and Carrickfergus surrendered; at the end of the year, Robert Bruce marched on Dublin, but did not dare to attack; instead he headed for Naas and Kilkenny, ravaging the land up to Limerick and thereby cutting himself off from food supplies, losing many men through hunger, especially due to the lateness of the season.

1317   In May, Bruce brought his half-starved army to Ulster and departed for Scotland, leaving the troops to his brother Edward, probably because he was disappointed in the Irish. The Scots were quiet, but the Irish, like the English barons, were again at each other's throats.

1318   Finally, Edward Bruce was defeated and killed by the English at Faughard in Dundalk.

1327   *EDWARD III.* Feud between Maurice Fitz-Thomas, later Earl of Desmond, and Lord Arnold Poer, consequent on which

1328   the Irish rose in Leinster under Donald M'Morrough of the old Dermot clan.

1329   Pacification of feuding barons by Lord Justice Roger Outlaw. The Irish again petitioned that they might be permitted to use the law of England without being obliged to purchase charters of denization, which the King advised the barons to concede, but which the latter again shelved *ad acta*.

1330 New feuds among the barons and risings of the Irish in the south and east, until finally Fitz-Thomas, Earl of Desmond, helped by the O'Brians (who had rebelled shortly before!) defeated the rebels. Soon thereafter O'Brian rebelled again; a new war ensued, in which the de Burghs indulged in plunder and abuse during their march across Fitz-Thomas's estates, causing another feud; Lord Justice Sir John Darcy had to lock up the chiefs of both houses.

1331 New rebellions in Leinster.

1332 Royal decree issued that the Irish and English should have the same law (English), excluding villeins (betagii, classed with the English villanis). But the decree was stillborn. Likewise, a royal ordinance against *absenteeism*; twenty-two absentees (English lords) were to accompany the King on his voyage to Ireland, but this did not materialise.

1339 Irish risings all over Ireland, with here and there assimilated barons on the Irish side.

1341 Sir John Morris, Knight, Lord Justice of Ireland. On pretext of money shortage due to the war against France, he took back all estates, titles and jurisdiction granted by Edward III and Edward II, and demanded settlement of all due, even void, crown debts.

1342 He ordered all Anglo-Irish or Irish officials and judges, or officials and judges with Anglo-Irish or Irish wives to be replaced by *imported Englishmen* (the power of the Anglo-Irish lords was to be broken). Convened Parliament in Dublin in October. Opposed Parliament of Nobles, especially of the Desmonds, in Kilkenny; a protest petition was sent to the King, who acknow-

	ledged receipt, which was as far as matters seem to have gone. Morris's orders of restitution remained in force.
1343	Sir Ralph Ufford, husband of the Countess Dowager of Ulster, was made Lord Justice, and
1345	convened Parliament in Dublin, while Desmond convened one in Callan; Ufford came to grips with him and compelled him to comply. Ufford died in 1346, and the King's fight against the lords seems to have ended for a time.
1353	The confiscated possessions (1342) were returned.
1361	Lionel, Duke of Clarence, third son of Edward, appointed Lord Lieutenant of Ireland. Marched without the Irish lords, whom he slighted, against O'Brian of Thomond, and was defeated; then he called on them for help, and the latter defeated the Irish.
1364	Lionel returned to England.
1367	Parliament of Kilkenny.[231] At this time, Ireland was so peaceful that the King's writ ran in Ulster and Connaught and the revenues of those provinces were regularly accounted for in the Exchequer.
1369-70	New risings of the O'Tooles and others in Leinster, and of O'Connor and O'Brian in the south-west; they were suppressed.
1364	Dublin University founded.*
1377	*RICHARD II.* Almost every Parliament (English) of his reign demanded supplies and men for war in Ireland.
1394	Richard landed in Waterford with 4,000 horsemen and 30,000 archers to reconquer Ireland. The chiefs of Leinster and Ulster, numbering 75, expressed submission. Those

* The official founding date of Dublin University (Trinity College) is 1591.—*Ed.*

of Ulster were to pay the bonaght[232] to the Earl of Ulster, while those of Leinster relinquished all their land and promised help against all other Irish, for which they would keep land thus conquered.

1395   No sooner Richard and his army returned than raids were renewed into the Pale.

1399   Richard marched against Ireland again, but in his absence

1399   *HENRY IV*, Bolingbroke of Lancaster, usurped the English throne and took Richard prisoner on his return.

1402   The O'Byrnes of Wicklow were defeated by John Drake, Mayor of Dublin.

1407   War against McMorrough of Leinster; yielded no decisive results, though by and large favourable for the English.

1410   Parliament in Dublin. An Act made it treason to exact coynye and livery.[233] During an excursion by Thomas Le Botiller, Prior of Kilmainham and Lord Justice, with 1,500 kerns (Irish infantry) against O'Byrne, half went over to the enemy and the English *had to withdraw*. An act was introduced whereby the *Irish* were prohibited to migrate without special licence to assure enough hands for the fields.

1413   *HENRY V*.

1414   Talbot victorious over Irish borderers.*

1417   200 Irish horsemen and 300 infantry under Thomas Butler, Prior of Kilmainham, went to France as auxiliary troops[234]: the horsemen on ponies, unsaddled, clothed in armour, the infantry with shields, spears and large knives. They fought very well and won much acclaim.

1421   New wars with the Irish, the latter being defeated in Leinster and Oriel.

1422   *HENRY VI*.

---

* The reference is to the borders of the Pale.—*Ed.*

1432	Sir Thomas Stanley, Lord Lieutenant, repulsed unusually strong Irish attacks.
1438	For the second time an Act was passed in English Parliament that all people born in Ireland (except beneficed clergymen, English estate holders and a few others) must at once return to the country of their birth. A similar act was passed in Irish Parliament to curb the exodus to England.
1449	Duke of York, heir of Earl March and as such Earl of Ulster and Cork, Lord of Connaught, Clare, Trim and Meath, hence nominally Lord of $\frac{1}{3}$ of Ireland, was appointed Lord Lieutenant *for ten years.* As usual, wars and feuds continued. Throughout the hundred years, the government contended with financial difficulties. Ireland's annual deficit was about £1,500.
1450	York returned to contest the English throne.
1460	York defeated and killed at Wakefield,[235] where he was accompanied by "the flower of all the English colonies (in Ireland), specially of Ulster and Meath, whereof many noblemen and gentlemen were slain at Wakefield" (Davies).*
1460	*EDWARD IV.*
1463-67	Earl of Desmond became Lord Lieutenant; ascendancy of the Geraldines. Carlow, Ross, Dunbar's Island and Dungarvan bestowed to Desmond; he was also made beneficiary of a large annuity chargeable on the principal seigniories belonging to the Crown in the Pale. But Desmond was too Irish and too popular, and hence.
1467	Lord Worcester became his successor, imprisoning Desmond, indicting him under the Statute of Kilkenny for alliance and intermarriage with the Irish. (It was through this marital connection with the Irish that

---

* John Davies, *Historical Tracts*, London, 1786.—*Ed.*

Desmond was able to uphold the King's authority in Munster; as for the Statute, it was long out of use in the south.) Parliament of Drogheda found Desmond attainted of treason for "alliance, fostering, and alterage with the King's enemies, for furnishing them with horses, harness, and arms, and supporting them against the King's subjects". He was beheaded in Drogheda on February 5, 1468.

1468   Worcester recalled, while Earl Kildare, the Geraldine, though also attainted, was restored and even made Lord Lieutenant.

1476   John, Earl of Ormond (attainted under Edward as follower of Henry VI), restored to all his possessions and in high favour. The Butlers rose, the Geraldines fell, but regained favour in 1478.

1478   Thomas, Earl Kildare, died. His son, Gerald Fitz-Thomas, Earl Kildare, was made Lord Deputy (of the Duke of Clarence, who was Lord Lieutenant).

1483   *EDWARD V and RICHARD III.*

1485   *HENRY VII.* Confirmed the Yorkists (the Geraldines and others) in their Irish offices, and installed no Lancasterites beside them. However, Thomas, Earl Ormond (attainted by Edward IV), was reinstated in his Irish and English estates and made member of the English Privy Council (he was brother of James).

1486   In Dublin, posing as young Earl of Warwick, son of the Duke of Clarence, Lambert Simnel was crowned King Edward VI. Kildare and the Pale, excluding Waterford, the Butlers and a few foreign bishops, swore allegiance, and the Duchess of Burgundy, sister of Edward IV, sent 2,000 German mercenaries under Martin Schwarz, to support him. These and Irish levies were then sent to England, landed in Furness,

	and pushed forward
1487	to *Stoke* (Nottinghamshire) on June 6, where they were annihilated. "The Iryshemen, although they foughte hardely and stuck to it valiantly, yet because they were after the manner of their country almost naked, without harneys or armour, they were stricken down and slain like dull and brute beasts" (Hall).* Simnel was captured and sent to the royal kitchen as scullion [*Spiessdreher*] (Gordon).** *Kildare*, whose power the King feared, was pardoned and remained Lord Deputy. Dubliners, however, were penalised and their ships, goods and merchandise given by the King to the Waterforders.

and these were mostly *degenerate* English![236]

1488	Sir Richard Edgecomb sent to Ireland with 500 men to receive the new oath and proclaim the official pardon for the rebellion.
1489	Henry invited the Irish lords to Greenwich and chastised them; they would have crowned apes if he had stayed away much longer, he said, and made ex-King Simnel serve them at table.
	Continuous wars among the natives.
1492	Kildare suddenly deposed and W. Fitz-Symons, Archbishop of Dublin, made Lord Deputy. Thereupon the border Irish rebelled and raided the Pale. *Perkin Warbeck*, the false Richard of York, landed in Cork; the city took his side, but Warbeck left at once, going to the court of the French King.
1494	Sir Edward Poynings sent to Ireland as Lord Deputy with 1,000 men and diverse English jurists. Parliament of Drogheda.

---

* Ed. Hall, *Chronicle, containing the History of England during the Reign of Henry IV and the Succeeding Monarchs to the End of the Reign of Henry VIII*, London, 1809.—*Ed.*

** J. Gordon, *A History of Ireland, from the Earliest Account to the Accomplishment of the Union with Great Britain in 1801*, vols. I-IY, London, 1806.—*Ed.*

Re Poynings's Act see Butt.*	The Poynings's Act: no parliament in Ireland may convene in council (English Privy Council) without approval of the King. Kildare, too, attainted of treason and sent to England as prisoner,
1496	but regains favour and is appointed Lord Lieutenant of Ireland. From then on Kildare was loyal to the King and waged violent wars against the Irish.
1497	Warbeck, who returned to Ireland (Cork) from Scotland, was joined by Earl Desmond, but, after unsuccessfully besieging Waterford, went to Cornwall. (This is now contested by virtue of a letter by Henry VII, according to which Warbeck landed "in the wylde Irisherie" in difficult circumstances and would have been captured by Kildare and Desmond if he had not made a hasty escape.)
1496-1500	Kildare's wars against the Irishry in Ulster, Connaught and Munster (Davies says [in *Hist. Tracts*, ed. 1786, p. 48] those were his "private quarrels", which is confirmed in detail by Gordon), all of them victorious, until finally Ulick Burke, Lord Clanricarde, called MacWilliam, a son-in-law of Kildare, chief of a mighty troop of "degenerate English", placed himself at the head of a general uprising in the south and west. Kildare set out with his entire Anglo-Irish force and a few Irish and
On August 19, 1504	defeated the rebels in Axtberg (Knoc-tuadh), seven miles off Galway; Galway and Athenry surrendered, and the spirit of the Irish was thereby broken (?!) (in the country where Black Rent[237] was paid until 1528!!). Kildare's arrogance as first Irish lord was ever in evidence in government matters and wars.
1509	*HENRY VIII.*

* J. Butt, *The Irish People and the Irish Land*, Dublin, 1867.—*Ed.*

Kildare continued his campaigns against the Irish. In 1509, he undertook a big campaign against James, eldest son of Earl Desmond, O'Brian, etc.

1513   Kildare died. His son Gerald, Lord Deputy, warred on against the Irish until 1517, was mostly successful, yet as always the victories were not decisive, and he had to begin all over again after a few years. However, like his father, he was very popular among the Irish, who considered him "rather as the chief of a great leading sept than as acknowledged ruler of the whole kingdom" by virtue of his Irish nature and many family ties with the Irish. In 1519, Kildare fell out of favour through Wolsey and was recalled to England.

1520   Thomas Howard, Earl of Surrey, was appointed Lord Lieutenant. An Englishman, he held the Irish in check. He reconciled two old enemies, Earl Desmond, the assimilated Geraldine who often espoused the Irish cause, with Earl Ormond, follower of the English, but not for long. On the whole, he acted skilfully, though this did not prevent continuous wars. He resigned and was

1521   followed by Sir Piers Butler, eighth Earl of Ormond who, though married to the sister of Earl Kildare,

1522-23   destroyed a number of the latter's castles. War between the two. At last, Ormond was dismissed and

1524   *Kildare* made Deputy.

In 1523, Desmond entered into an alliance with Francis I of France, who intended to, but did not, invade Ireland. Desmond was persecuted, concealed himself and remained undiscovered.

1526   Kildare was again recalled to England and thrown into the Tower, then released upon security.

(Ormond relinquished his title of Earl of Ormond in favour of Sir Thomas Boleyn and became Earl of Ossory.)

1528   O'Connor of Offaley treacherously captured a Deputy (of the Lord Lieutenant Richard Nugent, Lord of Delvin). This O'Connor was Kildare's son-in-law. Violent strife followed among the Anglo-Irish.

1530   Kildare returned in the retinue of the new Lord Deputy, Sir William Skeffington. He extended his Irish family ties, giving his daughter away in marriage to Fergananym O'Carrol, and laid waste the estates of his rival, Ormond-Ossory.

1532   Kildare again made Lord Lieutenant. Prosecuted war against all his enemies as enemies of the Crown, and fortified and armed his castles to resist the King if the necessity arose; however, he was again recalled to England, and on his departure

1534   his 21-year-old son Thomas (Lord Thomas Fitz-Gerald) stayed behind as his Deputy. The latter was led to believe that his father had been beheaded in the Tower and that he, too, and all his family, would suffer the same fate. He rode to the Council with 140 horsemen, laid down all his insignia of office and publicly withdrew his allegiance to the King. Then he started a rebellion. The Council took refuge in Dublin Castle, which Fitz-Gerald beleaguered. Fitz-Gerald also plundered Ossory's estates, but without marked success. In the meantime, Dublin townsmen captured the force besieging the Castle and Fitz-Gerald concluded an armistice with Ossory in order to take Dublin, but was defeated. Ossory meanwhile (though threatened in the south by the rebellious Desmond) laid waste Carlow and Kildare. Fitz-Gerald was excommunicated because his troops caused the death of the Archbishop of Dublin.—The war was fought half-heartedly by both sides, though most of the Pale was ravaged, until finally O'Connor (from Offaley) and then Lord Thomas Fitz-Gerald surrendered in 1535 and Fitz-Gerald was shipped to England. He surrendered on a solemn promise of pardon (Gordon [Vol. I], p. 238).

1536 The five uncles of Fitz-Gerald, of whom three *had opposed* the rebellion, and ten other lords were invited to a feast by Lord Grey and there put under guard (Gordon [Vol. I], p. 238) and sent to London. They and Lord Thomas Fitz-Gerald were executed in Tyburn (the elder Kildare died in London earlier). Thereby the power of the Geraldines was providentially terminated. Only a 12-year-old boy escaped abroad.

1536 ff. Lord Leonard Grey, Lord Deputy, made war on the indigenous population, especially the O'Connors.

1538 Peaceful expedition (hosting) by Grey to Galway through Offaley, Ely O'Carrol, Ormond, Arrah and Thomond. MacWilliam deposed as chief of Clanricarde and the captaincy given to Ulick de Burgh, later Earl of Clanricarde. All chiefs whose possessions Grey crossed, were made to swear allegiance, but, as Ormond wrote Cromwell, "neither from them nor any other from all the Irishry" could faith be expected once the troops departed.

1539 **According to O'Conor* the confederation was *directed against the Reformation*.** Large confederation of the northern chiefs and of Desmond and the Fitz-Geralds in the south to reinstate Gerald Fitz-Gerald, son of the executed Earl Kildare, in his rights. Gradually, the confederation expanded. The allies sought the help of the Emperor and of France, reviving the *idea*

---

* Matthew O'Conor, *The History of the Irish Catholics*, Dublin, 1813.—*Ed.*

of Ireland as an independent kingdom under O'Neill. The confederates also contacted the King of Scotland, who was also against the Reformation,[238] now an issue against the King in Irish matters. (The confederation fell apart after the Battle of Ballahoe [O'Conor, p. 10], of which no details are available.)

In the autumn, Lord Grey traversed the south once more at the head of his troop, but without any special success, though compelling Gerald Fitz-Gerald (and his friends) to flee to France and later to Italy. (Queen Mary reinstated him.) Otherwise, there was peace and order in Ireland, and only the bastard Geraldines (a completely assimilated family) were, "by the permission of God, killing one another" (Lord Grey's letter). John Alen, Lord Chancellor, wrote Cromwell: "I never did see, in my time, so great a resort to law as there is this term, which is a good sign of quiet and obedience. This country was in no such quiet these many years."

1540
See Gordon. Lord Grey recalled and soon executed. Some clashes with the Irish, though nothing of significance, for by and large the country was calm. Sir Anthony St. Leger, Lord Deputy, subdued the Cavenaghs of Carlow, the O'Moores of Leix and diverse other minor clans. O'Connor submitted too, and so did O'Donnell. As for O'Neill, the King entered into negotiations with him.

1541 By an Act of Parliament Henry was proclaimed *King* of Ireland.

From now on *the Irish chiefs became vassals [of the King] and came under English law* (probably a consequence of the unsuccessful confederation of 1539).

Turlogh O'Toole of North Wicklow was the first to go to England of his own volition, followed by Earl Desmond, who was at once made member of the King's Council. Irish lords and Irish nobles appeared in 1541 Parliament; they had not done so in many years or had never appeared there before. Ormond translated the English speeches to the Irish.

1542   O'Neill submitted and became Earl of Tyrone, while his son was made Lord Duncannon.

This time the peace was real; Desmond even ordered the arrest of two other Geraldines engaged in a feud, Lord Roche and the White Knight,* both were dispatched to Dublin and slept in the same bed, suffering each other quite well. O'Brian became Earl Thomond and MacWilliam became eighth Earl of Clanricarde. These Irish chiefs were so lacking in money that the government had to provide them with clothes in which to come to Parliament (see *Davies*).

All these lords acknowledged the King's supremacy.

1544   Again, Irish kerns served in the English army in France.

1545   Likewise against the Scots, though actually they did not land in Scotland.

England owed all these successes, the first real subjugation of Ireland, to *St. Leger*.

1547   *EDWARD VI.*

1550   French envoys went to O'Donnell and O'Neill in Ulster.

1550   New liturgy introduced in Ireland. Long debates among the clergy, while English soldiers plundered cloisters and churches, and destroyed sacred pictures. By and large, however, only among the higher classes were there a few converts to the new religion.

---

* A member of the Geraldines also known as Fitzgibbon.—*Ed.*

1552	War of succession between the sons of Earl Tyrone (O'Neill) in Ulster. In the south, feuds between Earl Thomond and his relatives, and in Connaught between Clanricarde and another de Burgh.
1553	*MARY.* St. Leger reappointed Lord Deputy in Ireland until 1558. Gerald Fitz-Gerald reinstated as eleventh Earl Kildare (and Baron of Offaley). Continued feuds between the chiefs.
1556?	After 13 years an Irish Parliament was finally reconvened, repealing all acts against the Pope and others passed since the Act of the 20th year of Henry VIII.
1557	Leix was incorporated in the Pale as Queen's County and Offaley as King's County,[239] the Moores and O'Connors having been banished under Edward VI and now almost all annihilated (see Gordon).
1558	*ELIZABETH.* New oath of supremacy taken from which only two Irish bishops abstained; the entire Irish Parliament took the oath, making the Reformation in the Pale official and formalising it on paper. All acts of 1556 (?) were declared null and void.
1560	Feud between Shane O'Neill ("The O'Neill") and the Dublin government, which would make Calwagh O'Donnell of Donegal Earl Tyrconnel if he agreed to help it, but O'Neill takes him prisoner. Finally,
1561	Shane submits directly to the Queen and goes to her in England, but encounters difficulties in obtaining an audience. When Matthew's son, then Earl of Tyrone, died, he returned to Ireland and in time claimed supremacy (independence) in all Ulster, but
1564	finally made peace and submitted to the Queen.
1565	Open war between Desmond and Ormond,

with Desmond wounded and captured by the latter.

1564 To win the Queen's favour, O'Neill made war on the island Scots settled along the coast of Ulster (Antrim) and defeated them. But Elizabeth and her representatives did not keep their word and endeavoured to trip up O'Neill. Again, a war broke out. Ulster was ravaged by an English army, but O'Neill withdrew to his unapproachable hills. Most of the chiefs of Ulster

1567 submitted, as did O'Neill's subjects, leaving O'Neill no choice but to flee to the Antrim Scots, where he was assassinated on the instigation of Piers, an English officer (see Gordon).

1570 Desmond captured and shipped to England. Rising of the Geraldines under James Fitz-Maurice, who took Kilmallock and turned to Spain for help. But order was soon restored by Sir John Perrot, Lord President of Munster, and Fitz-Maurice was compelled to submit.

Excommunication of Elizabeth[240] is joyfully received in Ireland.

Uprising of Clanricarde's sons.

Thomond (who fled to France) plots to assassinate Sir Edward Fitton, Lord President of Connaught; later, Thomond regained the Queen's favour through the English Ambassador in France.

Act of attainder against Shane O'Neill, whereby more than half of Ulster went to the Crown. The Lord Deputy in Council was also empowered to accept surrenders See Davies, and re-grant under English tenure (see Gorp. 200 ff. don).

Another Act declared the old clan system of chieftainship totally abolished, unless granted by the Crown. This reservation made the Act illusory, for the Crown had to tolerate what it could not hinder. Seven new counties with sheriffs (?) and other officials established (see Davies), but without assizes.

1572   Sir Thomas Smith tried to establish an English plantation in Ulster, but it was too weak and the indigenous population wiped out the colonists.

1579   Landing by James Fitz-Maurice, brother of Earl Desmond, in Smerwick, Kerry County, with three ships and 100 men, Catholics of different nationalities; but he and his Irish followers were killed when requisitioning in Tipperary. Thereupon, the invasion was soon defeated.

Leix and Offaley still rebellious, especially Rory Oge O'Moore, who was killed in 1578. After the invasion of Smerwick was repulsed, a rising by Desmond followed, whose betrayal was now confirmed in captured papers. He was defeated, his castles were seized, but he escaped.

1580   Rising in Wicklow under Lord Baltinglass. Setback for the English infantry, which ventured into the hills and valleys, in the Valley of Glendalough, says Gordon ([Vol. I], p. 271).

Landing of 700 Spaniards in Smerwick with arms for 5,000. However, their fort was captured by Lord Grey de Wilton, the Lord Deputy, and all of them massacred after surrendering and placing themselves at the discretion of the victors.

1583   Desmond, who stalked undiscovered in the south escaping from pursuit, was killed by peasants whose cattle he seized. He was the last of the Fitz-Geralds to be Earl Desmond.

1584   Sir John Perrot was reappointed Lord Deputy. He was instructed, among other things, "to consider how Munster may be repeopled and how the forfeited lands in

Ireland (Desmond and others) may be disposed of to the advantage of Queen and subject".

1587   As son of Matthew of Dungannon, heir of the earldom, Hugh O'Neill petitioned Irish Parliament to name him Earl of Tyrone and allow him possession of the estates. He led a troop of horsemen in the service of the Queen against Desmond, but had secret designs of becoming more than just Earl of Tyrone. He was granted the title and then from the Queen also his possessions on condition that he should claim no authority over the lords bordering on his county.

1588   Sir John Perrot returned to England, saying he found the Irish much more manageable than the Anglo-Irish and even the English Government. Fell into disfavour and died in the Tower.

The government in Dublin—it was still Perrot—arrested Hugh O'Donnell, son of *the* O'Donnell, and two sons of Shane O'Neill by resorting to subterfuge (they were given drink aboard a ship), and brought them to Dublin as hostages to ensure the loyalty of the old O'Donnell; they were held in captivity for three years.

1591   "Red Hugh" (O'Donnell) escaped and at home was (with his father's consent) proclaimed chief of Tyrconnel; he concluded an alliance with O'Neill Tyrone (who had flirted with both sides, until he had reason to fear for his life). O'Neill taught his men war craft (he had a bodyguard of 600 infantry and introduced a system of short-term training [*Krümpersystem*]), and laid in equipment and ammunition.

1597   Sir John Norris sent to Ireland with troops as Lord General to restore the imperilled authority of the Queen, but died the same year.

1592-96
{ Tyrone declared himself *the* O'Neill, which amounted to high treason.* He concluded an alliance with the other O'Neills, the Magennisses, M'Mahons and O'Donnells, and was appointed allied commander; when he heard that 2,000 fresh English troops were en route, he struck out, capturing and demolishing Fort Portmor on Blackwater, but was compelled by Bagenal (his brother-in-law), who was Marshal of Ireland, to lift the siege of Monaghan. However, on getting reinforcements he made Bagenal retreat. When the English advanced with fresh forces, O'Neill set fire to his own town of Dungannon and many villages, withdrawing into his forests. It came to light that he had offered Ireland to the King of Spain in return for 3,000 troops and money subsidies. Meanwhile, the insurgents in the north, whom Sir John Perrot had armed against the Antrim Scots and who had many veteran soldiers among them, were now very strong. Hence,

1596   new negotiations were begun. Tyrone submitted, and the insurgents demanded religious freedoms, which were finally granted by the Queen. But again hopeful news arrived of munition shipments from Spain, prompting Tyrone to blockade

1598   Fort Blackwater; he decisively defeated Marshal Bagenal (whom he killed with his own hands), who had hurried to the rescue. Now, the rest of Ulster rose too.

1599   Devereux, Earl Essex, the Queen's favourite, was sent to Ireland with 20,000 infantry and 2,000 cavalry. He wasted the summer in a march on Munster, his rearguard being defeated by the O'Moores on the return march, and finally, after his army was dec-

---

* After Shane O'Neill's rising adoption of this title implied rebellion against English dominion.—*Ed.*

imated by disease, went to Ulster, where
O'Neill Tyrone inveigled him in parleys,
and he lost more time. (Tyrone demanded
freedom to practise Catholicism, confirmation of the Ulster chiefs in their possessions
of the past 200 years, and all officials and
judges and half the garrison to be Irish.)
In the end Essex returned to England and
Charles Blount, Lord Mountjoy, replaced
him as Lord Deputy, with Sir George Carew
(author of *Pacata Hibernia*) as Lord President of Munster.

In the meantime, Tyrone went to Munster
to incite the local chiefs, especially James
Fitz-Thomas, Earl of Desmond, and Florence
McCarthy. Mountjoy sent strong troops to
the northern border forts of the Pale,
Dundalk, Carlingford, and others, while
marching on Ulster and issuing the order to
cut off Tyrone's retreat at Athlone or
Limerick. But Tyrone escaped by forced
marches, whereupon Mountjoy deployed
strong garrisons to Lough Foyle (Derry?)
and Ballyshannon, which kept the Ulster
people in check.

A campaign against the O'Moores of Leix.
The English totally destroyed the harvest.

1600   Carew planned to assassinate the Sugan Earl
(straw rope earl) of Desmond and McCarthy.
Mountjoy restored order in Kildare and
Carlow, and all Ireland was subjugated
save Tyrone.

Coinage of Ireland embased by Elizabeth.

1601   Two Spanish ships dropped anchor at Kil-
beg, Donegal, bringing arms, equipment and
money for Tyrone.

Twice, a price was set on Tyrone's head:
£2,000 if alive and £1,000 if dead. But this
was futile, as were the prices on the heads
of the insurgent chiefs hiding in Munster.
However, the Sugan Earl was finally cap-

tured. No one could be found for money to show the way through the forests to Tyrone's possessions.

Attempt on Tyrone by an assassin hired by the English Government; it failed.

On September 22, five thousand Spaniards landed at Kinsale and occupied the town. Mountjoy laid siege, with part of the southern Catholics declaring against the Spaniards or neutral, while the bulk sided with them. Tyrone, Tyrrell, O'Donnell, etc., marched against Mountjoy and fortified themselves in a swampy area, cut off his supplies, but were prevailed upon by the Spaniards to give battle on December 23 and were totally defeated. O'Donnell escaped to Spain, Tyrone to his possessions, while the Spaniards surrendered on a promise to be allowed to depart freely.

O'Donnell was active in Spain for Ireland. Mountjoy went north and laid waste all Tyrone.

1602   Fort Dunboy (at Bantry), the last fort of the Spanish (it belonged to Daniel O'Sullivan), was captured and its Irish garrison massacred.

1603   Finally, peace was concluded between Mountjoy and Tyrone, whereupon the latter submitted, but was confirmed in his possessions. Then Elizabeth died. All Ireland was subjugated for the first time.

1603   *JAMES I.* Everybody expected him to restore the Catholic religion. It was at once reintroduced in Waterford, Cashel, Clonmel and Limerick, but these were quickly brought to their senses by Mountjoy. James, however, demanded that all officials, barristers and graduates of universities gave the oath of supremacy and also restored the Act of Uniformity.[241] He at once purged the Dublin Council of Catholics. Although

the penal laws against the Papists were
upheld, they were not applied. But in

1605   all Catholic priests were banished on pain
of death (Sir Arthur Chichester was now
Deputy) and, according to O'Conor, Catholic
church services were banned by proclamation.

Gavelkind and tanistry[242] were again repealed by a judgement of the King's Bench,
the English inheritance law introduced, the
land of Irish smallholders directly confirmed
by the Crown and these placed directly
under Crown protection, whereby clanship
was visibly broken, while *all duties of the
clan people were converted into money rent
to their landlord.* Yet all this was done
gradually. Tyrone and Roderic O'Donnell,
brother of Red Hugh, went to England,
where the former was confirmed in his possessions and the latter made Earl of Tyrconnel. Both of them were so closely watched
by spies that Tyrone complained he could
not drink a full carouse of sack, but the state
was advertised thereof within a few hours
after.

1607   Land litigation between O'Neill Tyrone and
Sir Donogh (Donald Ballagh) O'Shane
(O'Cahan), a neighbouring chief, before the
Lord Deputy and an *English court*; this
convinced Tyrone that he must either submit
completely, or rebel again. But now there
were English forts and garrisons in his possessions, and the clanship was weakened.
Ireland herself was too weak, and salvation
could come only from abroad. Hence a plot
by Tyrone, Tyrconnel and Richard Nugent,

The existence Baron Delvin, to rebel with Spanish help.
of this plot   The plot was betrayed by Earl Howth, who
strongly   had just become Protestant. Tyrone and
doubted   Tyrconnel were summoned before the Dub-
even by Smith lin Council, escaped to France and from

[*Irish History*...],
p. 100. See
[Excerpts] IX,
[p.] 13.[243]

there to Brussels. Introduction of English law and the many court charges instantly lodged against him brought home to Tyrone that it was all over *now* with chieftainship. Finally, he went to Rome, where he died in 1616. The main branch of the Hy Nials ended shortly with the assassination of his son in Brussels.

James, meanwhile, found it necessary to declare publicly that the two earls did not flee religious persecutions, because never persecuted on religious grounds. *But who would believe that?*

1608   Uprising by Sir Cahir O'Doherty, Chief of Inish-Owen, who captured Culmore Fort by a trick, attacked Derry, and held out for five months, until finally killed.

Plantation of Ulster, where the Crown acquired 800,000 acres (English) or almost all Donegal, Tyrone, Coleraine, Fermanagh, Cavan and Armagh (supremacy converted into land holdings!) through the forfeiture of Tyrone, Tyrconnel, O'Doherty, etc. Each holding was divided into lots of three classes: 1) 2,000 English acres for servitors of the Crown, either the great officers of state or rich adventurers from England; 2) 1,500 acres for servitors of the Crown in Ireland with permission to take either English or Irish tenants; 3) 1,000 acres for the natives. The City of London received large grants in Derry on the condition of spending £20,000 for building the towns of Derry and Coleraine. A standing army was formed to guard the Colony. Thus, six out of 32 counties were expropriated and thoroughly plundered.

The Brehon Laws[244] were simultaneously completely abolished and replaced by English law, but, as if to render the state of outlawry of the Irish complete, while thus forbidden the use of their own country's law, they were still shut out as aliens and enemies from the law of their masters.

1613 The first Parliament in 27 years, and the first to represent more than just the Pale, opened in Dublin. Since the previous Parliament 17 new counties were constituted and 40 boroughs incorporated, of which most were mere villages consisting of a few houses built by Ulster undertakers.[245] Though the lords of the Pale remonstrated, new boroughs were constantly fabricated to assure a Protestant majority, the manoeuvre proving eminently successful. This caused recusant members to secede, but the matter was later settled. No anti-Catholic bills were tabled, but in recompense the Catholics voted for bills of attainder against Tyrone, etc.—This was a despicable thing to do, because nothing had been proved, but it justified the confiscations in Ulster.—Further, a bill was passed whereby all laws against Irish enemies were abolished and all put under the jurisdiction of English law.

1623 Royal proclamation that all Catholic priests secular and regular had to leave the Kingdom in 40 days, after which all persons were prohibited to converse with them.

1613 Commission instituted to inquire into defective titles to land in Ireland and escheated lands. It declared all land between Arklow and Slane rivers and many estates in Leitrim, Longford, Westmeath, King's and Queen's counties, totalling 82,500 acres, as Crown property. All was confiscated and granted to English and Irish colonists as in Ulster.

See O'Conor, 18, 2.[246]

A feeling of general insecurity arises among landholders, because resumption by the Crown under Henry VII of all land granted since Edward I, as well as the land of absentees, and various other similar juridical discoveries were now used to contest everything. Besides, many titles to land had either been lost or defective. A whole class of "discoverers" (of flaws in titles) appeared, consisting of "needy adventurers from England"; whenever the jurymen decided against the King, they were locked up. The Attorney-General declared that, with all Irish having been expelled when possession was first taken of the Pale, no Irish could have even an acre of freehold[247] in the five counties.

Wholesale resettlement of clans followed. Seven clans moved from Queen's County to Kerry; 25 landowners, mostly O'Ferrels, were expropriated without compensation. Instructive was the case of Byrnes of Wicklow (from Carte's *Life of Ormonde* in Matthew O'Conor's *History of the Irish Catholics*).[248]

1625   *CHARLES I.* Very short of money, he lost no time in coming to terms with the Catholic lords and gentry in Ireland. For three years they paid him £40,000 annually, in return for which he granted the following "graces": "Recusants[249] to be allowed to practise in courts of law, and sue the livery of their lands out of the Courts of Wards, on taking an oath of civil allegiance instead of the oath of supremacy; that the claims of the Crown (to defective titled lands) should be limited to the last 60 years*; that the inhabitants of Connaught be permitted to make a new enrolment of their estates", i.e., that their estates should be assured for them (etc., etc., 51 points in all), "and that a Parliament should be held to confirm these graces and establish every man in the undisturbed pos-

---

* In other words, the King undertook not to claim land held in hereditary possession for over 60 years.—*Ed.*

session of his own land". Further, reforms of
all kinds, extortions through courts of law
and soldiers, monopolies and penal laws
against religion, and promise of an "act of
oblivion and general pardon" (see O'Conor).
Lord Falkland convened Parliament to confirm these graces, but not under the Great
Seal of England (as required by the Acts of
Henry VIII and Elizabeth); the English
Council protested and Parliament did not
take place.

The Lords Justices indulged in flagrant
persecutions, confiscating 16 monasteries
because the Carmelites had held public services.

1633   *Sir Thomas Wentworth*, later Earl of Strafford, Lord Deputy. At that time the Irish
Channel teemed with pirates, and he could
not cross without being escorted by a
warship. He quickly alienated everybody.
Only a few members of the Privy Council
were admitted to sittings. Ireland was ruled
in accordance with the theory of the absolute
royal prerogative. Catholics and Protestants
alike were compelled by threats and cajolery
jointly to pay £20,000 more in voluntary
taxes. An order was issued that no one of
any rank could leave Ireland without the
permission of the Lord Deputy, and that no
complaint could be lodged against him
before the English royal court unless first
submitted to him.

Finally, however, a Parliament was
necessary to obtain money, however much
Wentworth dreaded it due to the question of graces, and particularly the restriction of Crown claims to 60 years, which
made a difference of £20,000 annually.

Wentworth saw to it that many army officers were chosen, which placed him in a
position to tilt the scales between the Catho-

lics and Protestants and thereby squeeze money out of both by threats.

1634 Parliament opened. Wentworth insisted on subsidies at once for a number of years and the Commons foolishly conceded six subsidies, whereupon a convocation of the clergy also conceded eight subsidies of £3,000 each. The lords, however, demanded redress of grievances and confirmation of graces, to which Wentworth replied brazenly that he had never even sent them to the King (which was untrue).

The same Parliament passed the two Statutes of Wills and Uses, whereby the Crown was allowed to interfere in the upbringing of the "heirs apparent" of big landowners, hoping thus gradually to convert them to Protestantism.

1635
Wentworth intended to drive out *all* Connaught landowners and recultivate the whole province. Leland, Vol. III,* quoted by O'Conor.

Violation of graces begun in Connaught. Wentworth came before the Grand Jury of Roscommon, where all landowners were gathered ("being anxious," he said, "to have persons of such means as might answer the King a round fine in the castle chamber in case they should prevaricate"), and told them that the best means of enriching the county was a *plantation* like Ulster; hence, they should investigate the King's title to the estates concerned. A proclamation was issued "that by an easy composition they should be allowed to buy indefeasible titles". The Justices of the Peace all being bribed ("more or less in the pound of the first year's rent were bestowed by the King upon the Lord Chief Justice and Lord Chief Baron of Ireland") while the juries were either packed or intimidated, the verdicts always favoured the King, as in the case of Sligo and Mayo. In Galway, however, there was resistance

---

* Th. Leland, *The History of Ireland from the Invasion of Henry II*, vols. I-III, London, 1773.—*Ed.*

and the juries decided against the King, but Wentworth importuned and harassed the landowners so that they finally transferred title to their estates to the King and pleaded for mercy. But Wentworth now wished the jury to announce it had judged falsely and admit perjury. This was rejected, whereupon the Sheriff was fined £1,000 and the members of the jury £4,000 each and were to be held in Dublin Castle until payment and remorse.

Furthermore, people were imprisoned right and left for harmless speeches and brought before *military courts*, which naturally found them guilty.

1636

To protect the English wool trade Wentworth banned wool exports even to England, except against licenses sold by himself, pocketing much money in this way; he introduced *cultivation and weaving of flax* successfully in Ireland (but with profit for himself).

Wentworth's principle was to rule Ireland so that she could not exist without the Crown. Hence, a government *salt monopoly* was introduced.

1640

When the Scottish war broke out,[250] Wentworth was made Earl of Strafford and Lord Lieutenant of Ireland, a title no one had held since Essex. A new Irish Parliament voted in four new subsidies. Strafford recruited 8,000 infantry and 1,000 cavalry to reinforce the troops in Ireland. However, these 9,000 were nearly all Catholics.

Each subsidy of about £40,000.

In June, Parliament reconvened and since most officers were away, the Catholics were in the majority. It was now agreed 1) to reduce incomes of the priesthood, 2) to redistribute the subsidies for this reason, because the Lord Lieutenant's distribution was unlawful and unjust. Charles ordered

Meanwhile (end of 1640) Long Parliament convened, whose

opposition began.

the page on which these decisions were recorded to be torn out of the Journal of the Commons and Lords.

1641 February

But Parliament decided to send to Charles a deputation with a Remonstrance of Grievances. Despite Strafford's objections, the deputation arrived in England. Apart from the delay in confirming the graces, the grievances listed arbitrary interventions and decisions by the Lord Lieutenant; chicanery of the courts of law, heavy penalties to suppress freedom of speech and press; unlawful powers of special tribunals; insecurity of person and property, and monopoly; total of 16 items.

See O'Conor.[251]

Strafford indicted by Long Parliament and executed. His various tyrannies in Ireland were held up against him, including the charge that he had established a tobacco monopoly for his own profit. As to the charge that he had collected taxes with military help and applied martial law, he maintained that this had always been so in Ireland and that the Provost Marshal had always hung people "who were going up and down the country and could not give a good accord of themselves" (what good was it, therefore, to introduce English law if it worked against the nation and could only be applied per martial law?).

All that could be said for Strafford was that he had applied the Penal Code against Catholics solely to extort money (for the Crown).

A new conspiracy in the north: Roger O'Moore, whose ancestors had been driven out of Offaley (in Edward's and Mary's reigns), Lord Maguire, Baron of Iniskillen, who still had remnants of his clan in Fermanagh, Hugh McMahon, Tyrone's grandson, Colonel Byrn and Sir Phelim O'Neill,

1641

The whole
story sounds
apocryphal,
resting on the
hearsay
evidence of

strongly supported by Irish driven out by the plantation. Also supported by many Connaught chiefs recently expelled by Strafford. Earl Antrim plotted with them in the name of the King, who would, since the Irish Government gravitated towards Parliament, deal with them and the Lords of the Pale, and would depose that government. Dublin Castle was to be captured first, October 23, but the conspiracy was betrayed and Sir William Parsons, one of the Lords Justices,* had everyone within reach arrested (McMahon, Maguire, etc.), while O'Moore and others escaped.

Meanwhile, fighting broke out in Ulster and Phelim O'Neill, *ass and pig* (see O'Conor[252]), captured Charlemount by treachery; all other castles in the eight northern counties were attacked and captured, or quickly starved out. In eight days everything was captured and Phelim had gathered 30,000 men.

(The Lords Justices and generally the now dominant party in Ireland planned to exterminate all Irish and Anglo-Irish Catholics and replace them with English and Scottish Protestants—see Cromwell's plan.) After outbreak of the revolt in Ulster, a company was formed in London in February [1642], petitioning Parliament to sell the ten million acres to be confiscated in Ireland, using the proceeds to prosecute a war of annihilation; the company offered to be middleman.

After outbreak of the rebellion in October, a large congregation of Catholics in Multifarnam Abbey, Westmeath County, debated the policy of whether to kill or simply drive out the Protestants. Phelim settled the issue by having Lord Charlemount and his other

---

* The other was Lord Justice John Borlase.—*Ed.*

Dr. H. Jones; the congregation seems never to have taken place, or to have been of a different nature.

Evidently, the rising was due to the refusal to convene Parliament. See O'Conor.[253]

prisoners killed, and by letting all Englishmen and Scots be massacred in three parishes; furious over the fall of Newry he also ordered the burning of the town and cathedral of Armagh despite its surrender, and had 100 people killed. It is possible, however, that the killing of the Catholics of Island-Magee at Carrickfergus by government troops occurred earlier and provoked the Catholics.

Leitrim (the O'Rourkes), the O'Ferrells of Longford (where plantations were also laid out) and the O'Byrnes of Wicklow rebelled on October 12; Wexford and Carlow, the Tooles and Cavanaghs, that is, all the Irish clans driven out by James, joined the rising and advanced to the walls of Dublin.

All quiet in Munster until December, but Lord President Sir William St. Leger provoked the gentlemen to rise under Philip O'Dwyer by his arrogance and by calling them all rebels. They captured Cashel.

In Connaught, where Lord Ranelagh was Lord President, the rising was also general, compelling Ranelagh to resign. Galway alone was saved for the government by Lord Clanricarde (the same Clanricarde whose property Wentworth and his tribunals had ravaged), but he, too, was put under restraint by the Lords Justices. The rising was just what the latter wanted; they wished no submission save in battle, for that entailed *forfeiture of lands*. Except Galway and a few castles in Roscommon all Connaught was engulfed by the insurrection.

Phelim O'Neill now beleaguered Drogheda; at Julian's Town Bridge, three miles from Drogheda, he drove a small force sent to relieve the besieged back to Dublin, causing much fear there; regiments went over to the rebels and Sir Charles Coote, then besieging Wicklow, was hastily recalled.

The lords and gents of the Pale, whom the government had supplied with some arms but who were at once required to return them as Catholics and told to leave Dublin and go to their estates, where they could do nothing unarmed but submit to the insurgents and thereby become traitors, could not hold out any longer. Sir Charles Coote, Governor of Dublin, roamed up and down the Pale and did nothing but "kill, burn, and destroy" in accordance with his instructions. Men of estate were taken along as prisoners to assure the King's escheats upon attainders, while the rest of the population were executed under martial law, including a Catholic priest, Father Higgins, who was under Earl Ormond's protection and had a safe-conduct.

The Lords Justices ordered the prisoners, McMahon and others, to be tortured to determine whether the King was behind the rebellion, but in vain.

Drogheda was bravely defended by Sir Henry Tichbourne, a soldier of the Cromwell school. He repulsed an escalade. Whereupon the town was merely blockaded, its food stores running low. Finally in February [1642], after a three months' siege, Marquis Ormond with 3,000 infantry and 500 horsemen arrived to relieve the beleaguered town and the Irish withdrew at once.

In view of the ravages inflicted by government troops in the Pale, even by Ormond, the Catholic Lords of the Pale arranged a meeting with Roger O'Moore, Byrn and McMahon, whereupon, following the Irish plea that they had risen for the King's rights and that his Irish subjects should be just as free as those in England, an alliance was

It appears that from March to October the clergy and then the clergy and gentry were dominant, and from October on the Commons were also represented. See *O'Conor*.

The people of the Pale were still craving for peace with the government and made frequent approaches. The Irish also demanded a reversal of attainders.

concluded—the first between Irish and Anglo-Irish of the Pale—and the Pale revolted. This was followed by the desertion of those few Catholics outside the Pale who had hesitated.

Catholic priests reappeared from hiding, holding synods in Kells on March 22, 1642, and particularly in Kilkenny in May 1642, deciding to send envoys to the Emperor, the King of France, and the Pope. Soon thereafter money, arms, equipment and officers (mostly Irish who had served in foreign armies) arrived from all parts of Europe to help the Irish. A General Assembly was then instituted in Kilkenny in October with two chambers: a Council of 12 persons to govern the judiciary, the judges, etc., and a Supreme Council, serving as the provisional government. Supreme Commanders were appointed for the provinces: Owen O'Neill, the Spanish colonel, in Ulster, Preston in Leinster, Garret Barry in Munster and Colonel John Burke in Connaught. An address was sent to the King, setting forth the grounds for the movement and the wishes of the Irish Catholics, in which they called themselves the National Assembly.

Owen O'Neill had been commander of Arras during the French siege in 1640 and in contrast to Sir Phelim O'Neill was closely enough related to the royal family to be declared The O'Neill. Besides, he was a good officer.

Thomas Preston, brother of Lord Gormanstown, Colonel in Imperial* and Spanish service, had distinguished himself during the Dutch siege of Löwen. He brought three ships, cannon, small arms and equipment, with four colonels, several engineers and 500 other Irish officers.

---

* Austrian.—*Ed.*

At this time, Ormond defeated an Irish detachment under Lord Mountgarret in Kildare (at Kilrush). Thereafter, Preston was defeated at Tymahoo and some other (?) detachment at Raconell. In spite of this, the insurgents were doing well. Finally, Charles, who needed support against the English Parliament, authorised Ormond to negotiate a year's armistice. The negotiations began, and an armistice followed. Meanwhile, the Lords Justices continued to act in the spirit of the Parliament. "The parliament pamphlets were by them received as oracles, their commands obeyed as laws, and the extirpation preached as a gospel." And to leave the rebels no avenue of escape, submissions by individuals were turned down. Even the quietest Catholics of the Pale, Lord Dunsany, Sir J. Netterville, and others, were imprisoned, tortured and arraigned wholesale for high treason on the strength of thus obtained confessions. Estates were seized en masse and their owners flung into gaol. More than "1,000 indictments were found by a Grand Jury against such men in two days", and another about 2,000 were "in reserve on the record".

Scarampi, the Pope's legate, arrived in Kilkenny with troops and military supplies. He reinforced the old Irish party, which primarily proposed to restore the Catholic religion to its full splendour, refused to trust the King, denounced the armistice, paid none of the subsidies demanded by the King and meant to fight the King and the English Parliament. The King was not to be trusted for had he not betrayed Strafford after promising that not a hair on his head would be touched.

1643   The Anglo-Irish moderates were opposed to this, finally bringing about a year's armistice on the basis of previously negotiated articles (their content?). When billets had been arranged for the respective armies and the armistice ratified by the Lords Justices and the Council on September 19, 1643, the Irish agreed to pay the King £30,000, half in money and half in cattle.

At once, five regiments were dispatched from Ireland to reinforce the King's army in England.

Indignation ran high in Ireland, as in England, over this armistice (that is, among the Catholics in Ireland and the Parliament party in England). The Lords Justices and the Council in Dublin, likewise opposed, obstructed it in every way they could. English Parliament pronounced Marquis Ormond "traitor against the three kingdoms". The Cavaliers,[254] too, were discontented. The 20,000 English and Scots in Ulster "vowed to live and die in opposition to the cessation".

Meanwhile, a new Remonstrance to the King was drawn up by the Catholics in Trim, enumerating their grievances, demanding redress and then placing 10,000 troops at the King's disposal.

That was the famous Remonstrance of *Trim.*

However, simultaneously, Ormond marched on Rossa and defeated General Preston (what about the armistice?).

Four parties in Ireland: 1) Irish Catholic, 2) Anglo-Irish Catholic (the bulk of the Confederates was recruited from these two parties), 3) the King's party, and 4) the Puritans.

For all this see O'Conor. While Ormond negotiated with the Confederates in Kilkenny to extort money for the King and, if possible, hoodwink them over the agreed points, the King invited Confederate delegates to Oxford. The delegates arrived with brusque demands: complete freedom of religion and repeal of the penal laws against Catholics; a free Parliament with suspension of Poynings's Law of 1494 while it sat (because it said nothing could be done without the English Council); repeal of all Irish Acts and Ordinances since August 1641; also a general amnesty and an Act of Limitation for Security of Estates; offices should be impartially granted to Catholics; passage of an Act establishing the independence of the Irish state and Parliament from the English; investigation of the massacres (committed by both sides during the war). The delegates of the Irish Protestants (who also came to Oxford) demanded, on the other hand, that all penal laws be preserved, the Catholic priests banned and Catholics excluded from all offices. The Solemn League and Covenant[255] was established at this time; Monroe and his Scots in the north accepted it at once, and so did most officers and men of the King's army under Ormond. English Parliament put Monroe in command of all troops in Ulster and he captured Belfast, where there were many Royalists, in a surprise attack.

Ormond, in the meantime, obtained the King's permission to amnesty "as to life and lands" all rebels returning into the King's service, as the chief means of breaking up the Confederation, which succeeded in many respects. O'Neill was now so badly off in the north that he had to plead for arms and equipment in Kilkenny, which he received; he was also appointed commander in Connaught, while Lord Castlehaven was made Supreme Commander.

Rinuccini, Archbishop of Fermo, was now

the Pope's nuncio, arriving with considerable arms and equipment.

1645 Charles commanded Ormond to conclude peace with the Irish at any cost, in order to release the army for England. He was quite willing to suspend Poynings's Act "for such bills as might be agreed upon" and to abolish the penal laws. But Ormond baulked, possibly because he was too much a Protestant, but probably because he knew that it was farthest from Charles's mind to keep his word. (?) Hence,

1646 Lord Herbert, Earl of Glamorgan, was sent to Kilkenny, concluding a treaty with the Confederates whereby the latter remained in possession of all churches and church revenues that had not in fact passed into Protestant possession and were allowed to hold public church services; the Catholic clergy was not to be punished for exercising their jurisdiction over their parishes. In return, 10,000 men under Glamorgan were placed at the King's disposal and two-thirds of the church revenues for three years were allotted for the upkeep of this army. For this Glamorgan was empowered by Charles above his signature and private seal. The treaty consisted of two parts, one public and the other secret (which contained the stipulation on religion). It was farthest from Charles's mind ever to ratify the treaty. As Hallam said, "his want of faith was not to the Protestant but to the Catholic".

But the secret was soon out. Sir Charles Coote, a Puritan, was sent to Connaught to capture Sligo, in which he succeeded, but M. O'Kelly, Catholic Archbishop of Tuam, tried to recapture it, falling in battle. A copy of the secret treaty was found in his belongings and made public at once.

The situation became extremely confused

Limerick, for example, stood neutral, because preoccupied with internal conflicts. In Connaught, three Presidents: one for the King, another for English Parliament (Coote) and one more for the Supreme Council of the Confederation.

The King disavowed Glamorgan, the treaty therefore became null and void, and the peace earlier concluded by Ormond was ratified by the Irish Commissioners on March 28.

Naturally, this did not suit the Covenanters, and Monroe had 60 men and 18 women massacred in Newry. O'Neill with 5,000 infantry and 500 cavalry marched against Armagh towards the end of May and stationed himself at Benburb, where on June 5 he was attacked by Monroe, whom he totally defeated, whereupon Monroe, who had lost all his artillery, abandoned Portedown, Downpatrick and other places.[256]

Published in the book   Translated from the
Marx-Engels Archives,   German
Vol. X, Russ. ed., Moscow, 1948

VARIA ON THE HISTORY OF THE IRISH CONFISCATIONS[257]

16th Century. Henry VIII

1536. Parliament in Dublin introduces the Oath of Supremacy and the King is given the privilege of taking the pick of all ecclesiastical livings. Quite different in the doing, however, for the subsequent insurrections were directed, among other things, against the Oath. Yet refusal to take the Oath of Supremacy was high treason in Ireland just as in England (Murphy,* p. 249).

---

* J. Murphy, *Ireland, Industrial, Political and Social,* London, 1870.—*Ed.*

16th Century. Edward VI and Mary

Confiscations in Queen's and King's Counties. During the reign of Edward VI, as was usual in Ireland, the O'Moores of Leix and O'Connors of Offaley[258] carried on a feud with some lords of the Pale. The government qualified this as rebellion. General Bellingham, later Lord Deputy, was sent against them and forced them to submit. Advised to see the King and submit to him in person as O'Neill had done successfully in 1542. O'Moore and O'Connor, unlike O'Neill, were imprisoned and their estates were confiscated. But that was not the last of it. The inhabitants declared that the land belonged to the clans, not to the chiefs, who therefore could not forfeit it, and were, at most, liable to forfeiting their private domains. They declined to move out. The government sent troops, and had the land cleared after unintermittent fighting and extermination of the population (Murphy, p. 255).

This was the pattern [der ganze Grundriss] for all subsequent confiscations under Elizabeth and James. The Irish were denied all rights against the Anglo-Irish of the Pale, with resistance treated as rebellion. That sort of thing became usual.

By Acts in the 3rd and 4th years of the reign of Philip and Mary, c. 1 and 2, the Lord Deputy, the Earl of Sussex, was endowed with "full power and authority ... to give and to grant to all and every Their Majesties' subjects, English or Irish ... at his election and pleasure, such estates in fee simple, fee tail,[259] leases for term of years, life or lives" in these two counties "as for the more sure planting or strength of the countries with good subjects shall be thought unto his wisdom and discretion meet and convenient" (Murphy, p. 256).

16th Century. Elizabeth

English policy under Elizabeth: to keep Ireland in a state of division and strife. "Should we exert ourselves," the English government averred, "in reducing Ireland to order and civility, it must soon acquire power, consequence and riches. *The inhabitants will be thus alienated from England*; they will cast themselves into the arms of some foreign power, or erect themselves into an independent and separate state. Let us rather connive at their disorders, for a weak and disordered people can never attempt to detach themselves from the Crown of England." Thus Sir Henry Sidney and Sir John Perrot, successive Lord Deputies (the last-named the best that they ever had, in 1584-87), about the "horrid policy" against which they protest (Leland, Vol. II, p. 292[*] and Murphy, p. 246). Perrot's intention of granting the Irish equal rights with the Anglo-Irish and obviating confiscations was blocked by the English party in Dublin. (*Yet he it was who had O'Donnell's son brought aboard a ship, filled with drink and borne away.*[**])

Tyrone's rebellion, among other things, against religious persecution: "he and other lords of Ulster entered into a secret combination, about this time, that they would defend the Roman Catholic religion ... that they would suffer no sheriffs nor garrisons to be within the compass of their territories, and that they would ... jointly resist all invasions of the English" (Camden).[***] The conduct of Deputy Mountjoy in this war is described by Camden: "He made incursions on all sides, spoiled the corn, burnt all the houses and villages that could be found, and did so gall the rebels, that, pent in with garrisons and streightened more and more every day, they were reduced to live like wild beasts, skulking up and down the woods and deserts" (Murphy, p. 251).

See *Holinshed Chronicles* (p. 460) on how Ireland is laid waste in this war. Half the population is said to have been done in.

According to the returns for 1602 by John Tyrrell, the Mayor of Dublin, prices there climbed: wheat from 36/- to 180/- the quarter, barley malt from 10/- to 43/- and oat malt from 5/- to 22/- the barrel, peas from 5/- to 40/- the peck, oats from 3/4 to 20/- the barrel, beef from 26/8 to 160/- the carcass, mutton ditto from 3/- to 26/-, veal ditto from 10/- to 29/-, lamb from 1/- to 6/-, and a pig from 8/- to 30/- (Leland, Vol. II, p. 422).

---

[*] Th. Leland, *The History of Ireland from the Invasion of Henry II*, Vol. II, London, 1773.—*Ed.*

[**] See p. 238.—*Ed.*

[***] W. Camden, *Annals, or the History of Elizabeth, late Queen of England*, London, 1635. Quoted from Murphy.—*Ed.*

Desmond had estates confiscated in all counties of Munster except Clare, and also in Dublin. They were worth £7,000 per annum. Irish Parliament of 1586 expropriated 140 land-owners by confiscation in Munster alone under the Act of the 28th year of Elizabeth's reign, c. 7 and 8. McGeoghegan lists the names of the grantees of Desmond's estates,* with some families still nearly all in possession until 1847 (*? probably cum grano salis*).

The annual Crown rent on these estates was 2d to 3d per acre, with no indigenous Irish admitted as tenants and the government undertaking to keep adequate garrisons.

Neither provision was observed. Some estates were abandoned by the grantees and reoccupied by the Irish. Many of the undertakers stayed in England and appointed agents, who were "ignorant, negligent, and corrupt" (Leland, Vol. III).

17th Century. James I

Penal Laws against Catholics (Elizabeth, in 2nd year of reign, 1560, c. 1) are applied more and more since the beginning of the reign of James I, it becoming dangerous to practise Catholicism. Under Elizabeth 2 cl. 1, the fine of 12d was imposed for every non-attendance of a Protestant Church service and, in 1605, under James, imprisonment was added by Royal Proclamation and, hence, unlawfully. This did not help. Besides, in 1605 all Catholic priests were ordered out of Ireland in 40 days on pain of death.

Surrenders of Estates and Regrants (see Davies, 7[b260]). These followed the pronouncement of tanistry and gavelkind as unlawful by the Court of King's Bench in the Hilary Term in the third year of the reign of James I.[261] A Royal Proclamation stipulated surrender of estates and regrant under new valid titles. Most Irish chiefs came forward to receive incontestable title at last, but this was made conditional on their giving up the clan relationship in favour of the English landlord-tenant relationship (Murphy, p. 261). This in 1605 (see "Chronology").**

---

* Engels borrows the reference to McGeoghegan's *Histoire d'Irlande* from Murphy's book, p. 258.—*Ed.*

** See p. 242.—*Ed.*

Plantation of Ulster. According to Leland, Irish under-tenants and servants were tacitly exempted from the Oath of Supremacy, whereas all the other planters were compelled to take it. Carte says that all Irish settlers, especially natives, who were allowed part of their land, were exempted, *but this was irrelevant because trial for refusing to take the Oath was impracticable.*

The Scottish Presbyterians in Ulster also resisted taking the Oath of Supremacy, and *this was suffered* by the authorities (Murphy, p. 266). That may have been useful for the Irish as well.—Carte estimates the number of English settlers in Ulster in 1641 at 20,000 and of Scottish settlers at 100,000 (*Life of Ormonde*, Vol. I, p. 177).*

Sir Arthur Chichester, Lord Deputy, was rewarded for his services in this plantation with the territory of Innoshowen(?) "and all the lands possessed by O'Dogherty, a tract of country far exceeding the allotments generally made to northern undertakers" (Leland, Vol. II, p. 438). As early as 1633 these estates were valued at £10,000 per annum (*Strafford's State Letters*, Vol. II, p. 294). Chichester was the ancestor of Marquis of Donegal, who would have £300,000 per annum for his Belfast estate alone, if another of his ancestors had not surrendered it to others under long leases (Murphy, p. 265).

The plantation of Ulster culminated the first period, with a new means discovered for confiscation: *defective titles.* This is effective under James and Charles, until Cromwell renews the invasion. See extracts from Carte, 2[a, b].262

Another effective pretext for confiscation was that old Crown rents, long forgotten by Crown and landowners, were still due from many estates. These were now pulled out and, wherever unpaid, the estate was forfeited. No receipts existed, and that was enough (Murphy, p. 269).

Concerning the attempt to confiscate Connaught (see "Chronology",** and O'Conor, *The History of the Irish Catholics*263), *recall James's dirty trick [schöne Schweinerei]:*

When the people of Connaught surrendered their titles to a specially appointed Royal Commission in 1616 and had these reconveyed by new patents, they paying £3,000 for

---

* Th. Carte, *A History of the Life of James Duke of Ormonde from his Birth in 1610 to his Death in 1688*, Vol. VI, London, 1736.—*Ed.*
** See p. 244.—*Ed.*

their enrolment in Chancery, the titles were not registered. A new commission was named *on this pretext* in 1623 to declare them null and void by reason of deliberate default, an oversight that depended not on the landowners but the government. (See Carte, *Life of Ormonde*, Vol. I, pp. 47 and 48.) In the meantime, James died.

A Court of Wards for Ireland was established in 1614. Carte avers in *The Life of Ormonde*, Vol. I, p. 517, that no lawful basis existed for it as in England, being meant to bring up Catholic heirs in the Protestant religion and English customs. Its president was the good Sir William Parsons, who had helped plan it.

17th Century. Charles I

That the Irish insisted in the graces* that "three score years' possession (of an estate) should conclude His Majesty's title" was understandable, for this was the *law of England* (*Strafford's State Letters*, Vol. I, p. 279), enacted by the Act of the 21st year of James's reign (Murphy, p. 274). *Yet English law applied to the Irish only in so far as it suited the English government.*

Strafford wrote the English Secretary of State on December 16, 1634, that in his Irish Parliament "the Protestants are the majority, and this may be of great use to confirm and settle His Majesty's title to the Plantations of Connaught and Ormond; for this you may be sure of, *all the Protestants are for Plantations, all the others are against them*; so as these, being the great number, you can want no help they can give you therein. Nay, in case there be no title to be made good to these countries in the Crown, yet should not I despair, forth of reasons of state, and for the strength and security of the Kingdom, to have them passed to the King by an immediate Act of Parliament (*State Letters*, Vol. I, p. 353).

Outside Connaught, too, money was extorted continuously on pain of inquiry into titles. The O'Byrnes of Wicklow, for example, twice paid £15,000 to preserve a portion of their estates, while the City of London paid £70,000 to prevent confiscation of its plantations in Colrain and Derry for alleged breach of covenant (Leland, Vol. III, p. 40).

---

* See pp. 245-46.—*Ed.*

The Court of High Commission[264] [the Irish Star Chamber]
established by Wentworth in the year 1633, after the English
model, "with the same formality and the same tremendous
powers" (Leland, Vol. III, p. 29), and this naturally without
Parliament's consent in order "to bring the people here to a
conformity in religion, and, in the way to that, raise, perhaps,
a good revenue to the Crown" (January 31, 1633, *State
Letters*, Vol. I, p. 188). The Court saw to it that all newly-
appointed officials, doctors, barristers, etc., and all those who
"sued out livery of their estates" should take the Oath of
Supremacy, which, as McAuley observed, was a religious in-
quisition where that of the Star Chamber was political.

Then the Castle Chamber, called Star Chamber[265] as in
England, which, Lord Deputy Chichester said, was "the
proper court to punish jurors who will not find a verdict for
the King upon good evidence" (oft-quoted passage from
Desiderata Curiosa Hibernicae, Vol. I, p. 262*).

It is said therein [(in the Remonstrance of Trim) the agents
complain] that the penalties there employed consisted in
"imprisonment and loss of ears" and "fines, pillory, boring
through the tongue, marking on the forehead with an iron
and other infamous punishments", as this is also indicated
in the indictment of Strafford (Murphy, p. 279).

When Strafford went to Connaught in 1635, he took with
him *4,000 horse* "as good lookers on, while the plantations
were settling" (Strafford, *State Letters*, Vol. I, p. 454). In
Galway he imposed fines not only on the jury that would
not find a verdict for the King, but also the sheriff "for returning so insufficient, indeed, we conceive, so packed a jury,
in £1,000 to His Majesty" (August 1635, Vol. I, p. 451).

As, by the 28th Act of Henry VIII, c. 5, 6 and 13, all
recourse to the Pope's jurisdiction was prohibited and all
Irish came under the Protestant ecclesiastical courts, whose
verdict could be appealed against to the King alone. They
took cognizance to all marriages, baptisms, burials, wills,
and administrations, and punished recusants for not going to
church under the 2nd Act of Elizabeth, c. 2, and also collected
the tithes. Bishop Burnet (*Life of Dr. Bedel, Bishop of Kil-
more*, p. 89) said these courts were "often managed by a

---

* Quoted according to Murphy, p. 279.—*Ed.*

chancellor *that bought his place* and so thought he had a right to all the profits he could make out of it. And their whole business seemed to be nothing but oppression and extortion. . . . The *officers* of the court thought they had a sort of right to oppress the natives and that all was well got that was wrung from them . . . they made it their business to draw people into trouble by vexatious suits, and to hold them so long in that, for 3d. worth of the tithe of turf, they would be put to a £5 charge". In the graces, which never materialised, Protestant clergymen were to have been forbidden "to keep private prisons of their own" for spiritual offences, so that offenders should be committed to the King's public gaols (Murphy, p. 281).

See Spenser, excerpt 5[a] about the Protestant clergy.[266]

Borlase and Parsons encouraged the rebellion everywhere. According to Lord Castlehaven's *Memoirs*, they said: "The more rebels, the more confiscations." Leland (Vol. III, p. 166), too, observes that, as before, "extensive forfeitures were the favourite object of the chief governors and their friends".

By that time, the Irish Royalist army was to have been 50,000 strong through reinforcement from England and Scotland.

See Carte, *The Life of Ormonde*, Vol. III, p. 61, for the instructions to the army.[267]

The motto of the Kilkenny Confederates was: *Pro deo, pro rege, et patria Hibernia unanimes* (for God, King and Ireland unanimous); *so that is where the Prussians lifted it from!* (Borlase, *Irish Rebellion*, p. 128).*

17th Century. Cromwell

Drogheda Massacre.[268] After a successful assault "quarter had been promised to all who should lay down their arms—a promise observed until all resistance was at an end. But at the moment that the city was completely reduced, Cromwell . . . issued his fatal orders that the garrison should be put to the sword. His soldiers, many of them with reluctance, butchered the prisoners. The governor and all his gallant offi-

---

* Ed. Borlase, *The History of Execrable Irish Rebellion*, London, 1680.—*Ed.*

cers, betrayed to slaughter by the cowardice of some of their troops, were massacred without mercy. For five days this hideous execution was continued with every circumstance of horror" (Leland, Vol. III, p. 361). A number of Catholic ecclesiastics found within the walls were bayoneted. "Thirty persons only remained unslaughtered ... and these were instantly transported as slaves to Barbadoes" (Leland, Vol. III, p. 362).

Petty (*Political Anatomy of Ireland*, Dublin edition of Petty's tracts, 1769, pp. 312-15) estimates that 112,000 British and 504,000 Irish inhabitants of Ireland died in the war of 1641-52. In 1653, soldiers' debentures[269] were sold at 4/- to 5/- in the pound, so that with 20/- being the price [nominal] of two acres of land, and there being 8 million acres of good land in Ireland, all Ireland was purchasable for £1 million, though in 1641 it was worth £8 million. Petty estimates the value of livestock in Ireland in 1641 at £4 million, and in 1652 at less than £500,000, so that Dublin had to get meat from Wales. Corn was 12/- per barrel in 1641 and 50/- in 1652. Houses in Ireland worth £2 million in 1641, were worth less than £500,000 in 1653.

Leland, too, admits in Vol. III, p. 171, that "the favourite idea of both the Irish Government and the English Parliament (from 1642 onwards) was the utter extermination of all the Catholics of Ireland".

See Lingard (*History of England*, Vol. VII, 4th ed., p. 102, Note) on the *transportation of Irish* as slaves to the West Indies (figures vary from 6,000 to 100,000). Of the 1,000 boys and 1,000 girls to be sent to Jamaica, the commissioners wrote in 1655: "Although we must use force in taking them up, yet it *is so much for their own good and likely* to be of such great advantage to the public, that you may have such number of them as you shall think fit" (*Thurloe's Papers*, Vol. IV, p. 23).*

"By the first Act of Settlement, the forfeiture of two-thirds of their estates had been pronounced against those who had borne arms against the Parliament and one-third of their estates against those who had resided in Ireland any time from October 1, 1649, to March 1, 1650, and had

---

* Quoted from Murphy, p. 298.—*Ed.*

not manifested their constant good affection to Parliament. The Parliament had power to give them, in lieu thereof, other lands to the proportion of value thereof." The second Act concerned resettlement (see Prendergast, *Cromwellian Settlement of Ireland*, Book of Excerpts VII, 1ª).[270]

Distribution of land to soldiers was limited to those who had served under Cromwell from 1649 (Murphy, p. 302).

See Carte, *Life of Ormonde*, Vol. II, p. 301, about some cases of land surveying, especially by adventurers.[271]

According to Leland (Vol. III, p. 410), the Commissioners in Dublin and Athlone kept considerable domains for themselves.

A plantation acre is equal to 1 acre, 2 roods, 19 perches, 5 yards, and 2 $^1/_4$ feet imperial statute measure, or 121 plantation acres may be taken as equal to 196 statute acres (Murphy, p. 302).

17th Century. Charles II

As a result of confiscations under Cromwell and Charles II, the 7,708,238 statute acres confiscated by Cromwell were distributed *finally, by 1675,* as follows:

Statute acres

1) *To Englishmen*

Adventurers	787,326
Soldiers	2,385,915
"Forty-Nine" Officers	450,380
Duke of York	169,431
Provisors	477,873
Duke of Ormond and Colonel Butler	257,516
Bishops' Augmentations	31,596
. . . . Total	4,560,037

2) *To Irishmen*

Decrees of Innocence	1,176,520
Provisors	491,001
King's Letters of Restitution	46,398
Nominees in Possession	68,360
Transplantation	541,530
. . . . Total	2,323,809

Remaining still unappropriated in 1675, being part of towns or land possessed by English or Irish without title or doubtful	824,392
Total in statute acres	*7,708,238*

On "Forty-Nine" officers see O'Conor and Notes.[272] The Duke of York received a grant of all the lands held by the regicides who had been attainted. Provisors were persons in whose favour provisoes had been made by the Acts of Settlement and of Explanation.[273] Nominees were the Catholics named by the King restored to their mansions and 2,000 acres contiguous.

At that time the profitable lands of Ireland were estimated at two-thirds of all land, or 12,500,000 statute acres. Of the rest, considerable tracts were occupied without title by soldiers and adventurers. In 1675, the *twelve and a half million acres of arable* were distributed as follows:

Granted to English Protestants of profitable land forfeited under the Commonwealth	4,560,037
Previously possessed by English Protestant Colonists and by the Church	3,900,000
Granted to the Irish . . . . · ·	2,323,809
Previously possessed by "good affectioned" Irish	600,000
Unappropriated as above	824,391
Statute acres	*12,208,237*

This table was compiled by Murphy; the figure of 3,900,000 acres was taken from the Account published by the Cromwellian proprietors and the rest on the basis of the *Grace Manuscript quoted by Lingard* and the Report of the Commissioners to the English House of Commons, December 15, 1699. It accords with Petty (*Political Anatomy*), who wrote: "Of the whole 7,500,000 plantation acres of good land (in Ireland) the English and Protestants and the Church have this Christmas (1672) 5,140,000 (= 8,352,500 statute acres) and the Irish have near half as much" (Murphy, pp. 314 and 315).

17th Century. William III[274]

By the Acts of Settlement and Explanation, 2,323,809
  statute acres were granted to the Irish, they having
  600,000 previously in their possession

Totalling	2,923,809	statute acres	
Of these lands, 1,060,792 plantation acres were escheated under William worth £211,623 6s. 3d. per annum (Report of the Commissioners of the House of Commons, 1699) . .	1,723,787	"	"
The rest	1,200,022	"	"
or as Murphy calculated (he probably erred when subtracting)	1,240,022	"	"
In addition, restituted by special favour of the King on pardoning (65 persons)	125,000	"	"
The Court of Claims restored (792 persons)	388,500	"	"
Total	513,500		

Making the total possessed by
  the Irish *1,753,522**

Compiled by Murphy on the basis of the Report of the
Commissioners of the House of Commons (English) in
December 1699.

Published in the book
Marx-Engels Archives,
Vol. X, Russ. ed., Moscow, 1948

Printed according to the
manuscript in German
and English. Part of
the manuscript translated
from the German

---

* Engels points out that Murphy may have erred in his calculation
by 40,000 acres, in which case the total would have been 1,713,522.
—*Ed.*

Frederick Engels

NOTES FOR THE PREFACE TO A COLLECTION OF IRISH SONGS[275]

Some Irish folk-music is very ancient, some has arisen in the last three to four hundred years, and some only in the last century. Especially much was written at that time by one of the last Irish bards, Carolan. In the past these bards or harpists—poets, composers and singers in one person—were quite numerous. Every Irish chieftain had his own bard in his castle. Many travelled the country as wandering singers, persecuted by the English, who correctly saw in them the main bearers of the national, anti-English tradition. Ancient songs about the victories of Finn Mac Cumhal (whom Macpherson stole from the Irish and turned into a Scot under the name Fingal in his *Ossian*, which is entirely based on Irish songs), about the magnificence of the ancient royal palace of Tara, the heroic deeds of King Brian Borumha, and later songs about the battles of Irish chieftains against the Sassenach (Englishmen) were all preserved in the living memory of the nation by the bards. And they also celebrated the exploits of contemporary Irish chieftains in their fight for independence. When in the 17th century, however, the Irish people were completely crushed by Elizabeth, James I, Oliver Cromwell and William of Orange, their landholdings robbed and given to English invaders, the Irish people outlawed in their own land and transformed into a nation of outcasts, the wandering singers were hounded in the same way as the Catholic priests, and had gradually died out by the beginning of this century. Their names are lost, of their poetry only fragments have survived, the most beautiful legacy they have left their enslaved, but unconquered people is their music.

Irish poems are all written in four-line verses. For this reason a four-line rhythm always lies at the basis of most, especially the ancient, Irish melodies, though sometimes it may be a little hidden, and frequently a refrain or conclusion on the harp follows it. Some of these ancient songs are even now, when in the largest part of Ireland Irish is understood only by the old people or even not at all, known only by their Irish names or first words. But the greater, more recent part, has English names or texts.

The melancholy dominating most of these songs is still the expression of the national disposition today. How could it be otherwise amongst a people whose conquerors are always inventing new, up-to-date methods of oppression? The latest method, which was introduced forty years ago and pushed to the extreme in the last twenty years, consists in the mass eviction of Irishmen from their homes and farms— which, in Ireland, is the same as eviction from the country. Since 1841 the population has dropped by two and a half million, and over three million Irishmen have emigrated. All this has been done for the profit of the big landowners of English descent, and on their instigation. If it goes on like this for another thirty years, there will be Irishmen only in America.

Published in the journal
Movimento Operaio No. 2,
Milano, 1955

Translated from the
German

Karl Marx and Frederick Engels

EXCERPTS FROM LETTERS ON IRELAND WRITTEN BETWEEN 1869 AND 1872

MARX TO ENGELS

March 1, 1869

Also received Foster on Saturday evening.* The book is indeed significant for its time. First, because in it Ricardo's theory is fully developed and better than in Ricardo—on money, rate of exchange, etc. Secondly, because one sees here how those asses, the Bank of England, Commission of Inquiry,[276] and the theoreticians, racked their brains over the problem: England debtor to Ireland. Despite this, the rate of exchange is always against Ireland and money is exported from Ireland to England. Foster solves the puzzle for them, viz., the depreciation of Irish paper money. It is true that two years before him (1802) Blake had fully elucidated this difference between the *nominal* and the *real* rate of exchange,[277] about which, by the way, Petty had already said all that was necessary, only after him all this had been forgotten again.

The Irish amnesty is the lousiest of its kind ever. *D'abord*, most of the amnestied had almost served the term after which all penal servitude men are given tickets of leave. And secondly, the chief ringleaders were kept in gaol "because" Fenianism is of "American" origin, and hence the more criminal. That is why such Yankee-Irishmen as Costello are released while the Anglo-Irish are kept under lock and key.

If ever a mountain gave birth to a mouse, it is this ministry of all talents,[278] and indeed in every respect.

---

* J. L. Foster, *An Essay on the Principle of Commercial Exchanges, and more particularly of the Exchange between Great Britain and Ireland*, London, 1804.—*Ed.*

I sent you earlier the report of Pollock and Knox[279] (the same lousy London police magistrate, formerly a *Times* man, who distinguished himself so greatly in the Hyde Park row) on the treatment of Irish "convicts" in England. One of these "convicts" has exposed John Bull's unheard-of infamies and the lies of that blockhead Knox in *The Irishman*.

ENGELS TO MARX

September 27, 1869

We returned safely from Ireland on Thursday, a week ago; were in Dublin, the Wicklow Mountains, Killarney and Cork. Had quite a good time but both women* came back even *hiberniores* than they had been before they left. Weather fine on the whole. According to the papers you are having even worse weather there than we are here.

Learned from Trench's *Realities of Irish Life* why Ireland is so "overpopulated". That worthy gentleman proves by examples that on the average the land is cultivated so well by the Irish peasants that an outlay of £10-15 per acre, which is *completely recouped* in 1-4 years, *raises* its rental value from 1 to 20 and from 4 to 25-30 shillings per acre. *This* profit is to be pocketed by the landlords.

Mr. Trench is in turn nicely checked by his own statements to Senior, which the latter has had published. Trench tells the liberal Senior that if he were an Irish peasant he would be a Ribbonman[280] too!...

Ireland's trade has grown enormously in the past 14 years. The port of Dublin was unrecognisable. On Queenstown Quay I heard a lot of Italian, also Serbian, French and Danish or Norwegian spoken. There are indeed a good many "Italians" in Cork, as the comedy has it. The country itself, however, seems downright depopulated, and one is immediately led to think that there are far too few people. The state of war is also noticeable everywhere. There are squads of Royal Irish all over the place, with sheath-knives, and occasionally a revolver at their side and a police baton in

---

* Engels's wife Lydia (Lizzy) Burns and Marx's daughter Eleanor.
—*Ed.*

their hand; in Dublin a horse-drawn battery drove right through the centre of town, a thing I have never seen in England, and there are soldiers literally everywhere.

The worst about the Irish is that they become corruptible as soon as they stop being peasants and turn bourgeois. True, this is the case with most peasant nations. But in Ireland it is particularly bad. That is also why the press is so terribly lousy.

ENGELS TO MARX

October 24, 1869

Irish history shows one what a misfortune it is for a nation to have subjugated another nation. All the abominations of the English have their origin in the Irish Pale. I have still to plough my way through the Cromwellian period, but this much seems certain to me, that things would have taken another turn in England, too, but for the necessity for military rule in Ireland and the creation of a new aristocracy there.

ENGELS TO MARX

November 1, 1869

It is really lucky that the *Bee-Hive* now flaunts its bourgeois colouring as impudently as stupidly. I've never seen such a filthy issue as yesterday's.[281] This cringing before Gladstone and the whole bourgeois-patronising-philanthropic tone should soon break the neck of that paper and make it necessary to have a real workers' paper. It is very good that the only workers' paper is becoming more and more bourgeois precisely at a time when the workers are sobering up from their liberal intoxication. But Sam Morley should have had more brains than to place such stupid fools on it and allow them to lay on the bourgeois varnish so thickly and obviously.

The Fenian demonstration in London[282] is merely another proof of what the official publicity of the press is worth. A couple of hundred thousand people assemble and stage the most imposing demonstration London has seen for years, and as the interest of respectability requires it, the entire London press without exception can describe it as a shabby failure.

MARX TO ENGELS

November 6, 1869

Within the next few days I'll send you a volume which I happened to pick up, containing all sorts of pamphlets about Ireland. Those by Ensor (whom I quoted in *Capital*[283]) have all sorts of piquant things. Ensor was a political economist of English origin (his father was still living there at the time of Ensor's birth), a Protestant, and, in spite of all that, one of the most resolute Repealers before 1830. Being himself indifferent to religious matters, he can be witty in defending Catholicism against the Protestants. The first pamphlet in the book is by Arthur O'Connor. I expected more of it, since this O'Connor played a considerable role in 1798[284] and I found good essays by him on Castlereagh's administration in Cobbett's *Political Register*. Tussy must look through Cobbett some time to see what he has on Ireland.

ENGELS TO MARX

November 17, 1869

The best joke of the Irish is to propose O'Donovan Rossa as candidate for Tipperary. If this succeeds, Gladstone will find himself in a fine fix. And now another amnesty in Italy!

I hope to read the details about the debates, etc., in the International* next Sunday in the *Bee-Hive*. Should there be any documents, please send them on to me. Last Sunday the *Bee-Hive* had nothing about the International although it did report on the wedding of the Duke of Abercorn's daughters.

Prendergast's *Cromwellian Settlement*** is *out of print*. You

---

* See pp. 152-56.—*Ed.*
** J. P. Prendergast, *The Cromwellian Settlement of Ireland*, London, 1865.—*Ed.*

would therefore greatly oblige me by *ordering* it immediately *at a second-hand bookseller's*. Butt's *Irish People*: none in London. Other Irish pamphlets, for example, those of Lords Rosse and Lifford: cannot find. Such are the answers my bookseller received from his London agent, and he told me at the same time that in general the English book trade cannot take it upon itself to obtain publications appearing in Ireland, since it is not the custom to have a correspondent in Dublin, but only in London. I'll write directly to Duffy in Dublin.

I've found some very useful things about Ireland here: Wolfe Tone's *Memoirs*, etc., in the *catalogue*. Whenever I ask for these things in the library, they are not to be found, like Wakefield.* Some old fellow must have had all the stuff together and returned it en masse, so that the whole lot is hidden away somewhere. But in any case these things must be found.

Goldwin Smith of *Irish History and Irish Character* is a wise bourgeois thinker. Ireland was intended by providence as a grazing land, the prophet Léonce de Lavergne foretold it, *ergo pereat*** the Irish people!

MARX TO ENGELS

November 18, 1869

The *Bee-Hive suppressed* the report (by Eccarius) of the latest meeting*** on the *pretext* that it had arrived too late. The real reason was that

1) it *did not wish it to be known* that the General Council would take up the Irish question at its next meeting;

2) the report contained references objectionable to it (i.e., to Mr. Potter) about the Land and Labour League.[285] In fact, Mr. Potter *failed* ignominiously as nominee to the League's Committee.

Last Tuesday I opened the discussion on point 1: *the atti-*

---

* E. Wakefield, *An Account of Ireland, Statistical and Political*, vols. I-II, London, 1812.—*Ed.*

** So perish.—*Ed.*

*** Of the General Council, International Working Men's Association. —*Ed.*

tude of the British Government to the Irish Amnesty Question.* I spoke for about an hour and a quarter, much cheered, and then proposed the following resolutions on Point 1:

"Resolved,

"that in his reply to the Irish demands for the release of the imprisoned Irish patriots—a reply contained in his letter to Mr. O'Shea, etc., etc.—Mr. Gladstone deliberately insults the Irish Nation;

"that he clogs political amnesty with conditions alike degrading to the victims of misgovernment and the people they belong to;

"that having, in the teeth of his responsible position, publicly and enthusiastically cheered on the American slaveholders' Rebellion, he now steps in to preach to the Irish people the doctrine of passive obedience;

"that his whole proceedings with reference to the Irish Amnesty question are the true and genuine offspring of that '*policy of conquest*', by the fiery denunciation of which Mr. Gladstone ousted his Tory rivals from office;

"that the *General Council* of the *International Working Men's Association* express their admiration of the spirited, firm and highsouled manner in which the Irish people carry on their Amnesty movement;

"that these resolutions be communicated to all branches of, and working men's bodies connected with, the *International Working Men's Association* in Europe and America."

Harris (an O'Brien man) *seconded* my proposal. However, the President (Lucraft) pointed to the clock (we could stay until 11 only); the matter was therefore left over to next Tuesday. All the same, Lucraft, Weston, Hales, etc., in fact the whole Council, tentatively declared for the proposal in informal way.

Milner, another O'Brienite, said the language of the resolution was too weak (i.e., not declamatory enough); furthermore, he demands that everything I said to substantiate the case should be inserted in the resolutions. (A fine kettle of fish!)

Thus, with the debate continuing on Tuesday, now the

* See pp. 152-56.—*Ed.*

time for you to tell or write me what you may wish to *amend* or *add*. In the latter case, if, for example, you wish to add a paragraph about amnesties elsewhere in Europe, say in Italy, write it at once in the form of a resolution.

ENGELS TO MARX

November 19, 1869

I think an appendix on amnesties in the rest of Europe would only weaken the resolution, since besides Russia (which would be very good by *itself*) Prussia would have to be mentioned too, because of those condemned for the Guelf conspiracy.[286] On the other hand, I should polish up the language a bit: Paragraph 2. I should insert "imprisoned" or something of the sort before "victims", so that it is evident at first sight who is meant....

...Lizzy immediately conveyed a vote of thanks to you for the resolution and is vexed that she cannot be there on Tuesday.

MARX TO ENGELS

November 26, 1869

Last Tuesday's sitting* was full of fiery, heated, vehement speech. Mr. Muddlehead** or the devil knows what that fellow's name is—a Chartist and an old friend of Harney's—had providently brought Odger and Applegarth along. On the other hand Weston and Lucraft were absent because they were attending an Irish ball. Reynolds had published my resolutions in his Saturday issue and also an abstract of my speech*** (as well as Eccarius could do that; he's no stenographer), and Reynolds had printed it right on the front page of the paper, after his leading article. This seems to have scared those flirting with Gladstone. Hence the appearance of Odger and a long rambling speech by Muddershead, who got it in the neck badly from Milner (himself an

---

* Marx is referring to the General Council of November 23, 1869. See pp. 157-158.—*Ed.*
** Marx is referring to Mottershead.—*Ed.*
*** See pp. 152-56.—*Ed.*

Irishman). Applegarth sat next to me and therefore did not dare to speak *against* them; on the contrary, he spoke *for* them, evidently with an uneasy conscience. *Odger* said that if the resolutions were rushed to a vote he would have to say aye. But unanimity was surely better and could be attained by means of a few minor amendments, etc. I thereupon declared—as it was precisely *he* that I wanted to get into a mess—that *he* should submit his amendments at the next session. At the last session, although many of our most reliable members were absent, we would thus have carried the resolution against *one single* opposing vote. Tuesday we shall be there in full force.

ENGELS TO MARX

November 29, 1869

I have discovered here in the Free Library and the Chetham Library (which you know)[287] a large number of very valuable sources (besides the books with second-hand information), but unfortunately neither Young* nor Prendergast, nor the English issue of the Brehon Law** published by the English Government. However, I have found Wakefield again and various things by old Petty. Last week I studied the tracts of old Sir John Davies (Attorney-General for Ireland under James).*** I don't know whether you've read them, they are the main source, but you must have found quotations from them hundreds of times. It is a downright shame that the original sources are not available everywhere, one gets infinitely more from them than from elaborations on them, which make everything that is clear and simple in the original confused and complicated. The tracts show clearly that communal ownership of land was *Anno 1600* still *in full force* in Ireland and was adduced by Mr. Davies in his counsel's speech on the confiscation of the forfeited land in Ulster as a proof that the land did not belong to individual owners (peasants) and hence belonged either to

---

  * A. Young, *A Tour in Ireland*, vols. I-II, London, 1780.—*Ed.*
 ** See pp. 193-94.—*Ed.*
***John Davies, *Historical Tracts*, London, 1786.—*Ed.*

the Lord, who had forfeited it, or else from the very start to the Crown. I've never read anything more beautiful than this speech. Reallotments were made every two or three years. In another pamphlet he describes in detail the incomes, etc., of the head of the clan. I've *never* seen these things quoted, and if they are of any use to you, I'll send you details of them. At the same time I've caught Monsieur Goldwin Smith beautifully. That man never read Davies and that is why he makes the most absurd assertions to exonerate the English. But I shall get that fellow.

MARX TO LUDWIG KUGELMANN

November 29, 1869

You will probably have seen in the *Volksstaat*[288] the resolutions against Gladstone proposed by me on the question of the Irish amnesty.* I have now attacked Gladstone—and it has attracted attention here—just as I had formerly attacked Palmerston.** The demagogic refugees here love to fall upon the continental despots from a safe distance. That sort of thing attracts me only when it is done *vultu instantis tyranni.****

Nevertheless, both my utterance on this Irish amnesty question and my further proposal to the General Council to discuss the attitude of the English working class to Ireland and to pass resolutions on it have of course other objects besides that of speaking out loudly and decidedly for the oppressed Irish against their oppressors.

I have become more and more convinced—and the only question is to drive this conviction home to the English working class—that it can never do anything decisive here in England until it separates its policy with regard to Ireland most definitely from the policy of the ruling classes, until it not only makes common cause with the Irish but actually takes the initiative in ·dissolving the Union established in 1801 and replacing it by a free federal relationship. And

---

* See pp. 155-56.—*Ed.*
** See pp. 70-71.—*Ed.*
*** Right to the face of the tyrant.—*Ed.*

this must be done, not as a matter of sympathy with Ireland but as a demand made in the interests of the English proletariat. If not, the English people will remain tied to the leading-strings of the ruling classes, because *it* will have to join with them in a common front against Ireland. Every one of its movements in England herself is crippled by the strife with the Irish, who form a very important section of the working class in England. *The prime condition* of emancipation here—the overthrow of the English landed oligarchy—remains impossible because its position here cannot be stormed so long as it maintains its strongly entrenched outposts in Ireland. But there, once affairs are in the hands of the Irish people itself, once it is made its own legislator and ruler, once it becomes autonomous, the abolition of the landed aristocracy (to a large extent the *same persons* as the English landlords) will be infinitely easier than here, because in Ireland it is not merely a simple economic question but at the same time a *national* question, since the landlords there are not, like those in England, the traditional dignitaries and representatives of the nation, but its mortally hated oppressors. And not only does England's internal social development remain crippled by her present relations with Ireland; her foreign policy, and particularly her policy with regard to Russia and the United States of America, suffers the same fate.

But since the English working class undoubtedly throws the decisive weight into the scale of social emancipation generally, the lever has to be applied here. As a matter of fact, the English republic under Cromwell met shipwreck in—Ireland. *Non bis in idem!** But the Irish have played a capital joke on the English government by electing the "convict felon" O'Donovan Rossa to Parliament. The government papers are already threatening a renewed suspension of the Habeas Corpus Act,[289] a renewed system of terror. In fact, England never has and never *can*—so long as the present relations last—rule Ireland otherwise than by the most abominable reign of terror and the most reprehensible corruption.

---

* Not twice the same thing!—*Ed.*

MARX TO ENGELS

December 4, 1869

Dear Fred,

The resolutions were carried unanimously, despite Odger's constant *verbal* amendments. I let him have his way on one point only, agreeing to omit the word "deliberate" before "insults" in paragraph 1.* I did that on pretence that everything a Prime Minister publicly did must be presumed *eo ipso to be deliberate.* The true reason was that I knew that, as soon as the first paragraph was accepted in substance, all further resistance would be useless. I'm sending you two *National Reformers* containing a report on the first two meetings,** but nothing yet about the last. This report is also badly written and lots of things are definitely wrong (due to misunderstanding), yet it is better than Eccarius's reports in *Reynolds's.* They are by Harris, whose currency panacea you'll also find in the latest issue of the *National Reformer.*

With the exception of Mottershead, who acted like John Bull, and Odger, as always, like a diplomat, the English delegates behaved excellently. The general debate on the attitude of the English working class to the Irish question begins on Tuesday.[290]

Here one has to fight not only prejudices, but also the stupidity and wretchedness of the *Irish* leaders in Dublin. *The Irishman* (Pigott) knew about the proceedings and resolutions not only from *Reynolds's,* to which he subscribes and which he often quotes. They (the resolutions) were sent him directly by an Irishman*** as early as November 17. Up to now, *deliberately not a word.* The ass acted in a similar way during our debates and the petition for the three Manchester men.**** The "Irish" question must be treated as something quite separate, apart from the rest of the world, namely, it must be *concealed*, that *English* workers sympathise with the Irish! What a stupid beast! And this in respect of the *Inter-*

---

* See p. 158.—*Ed.*
** Reference is to the meetings of the General Council on November 16 and 23, 1869.—*Ed.*
*** Probably by G. Milner.—*Ed.*
**** See pp. 368-72.—*Ed.*

national which has press organs all over Europe and the
United States! This week he received the resolutions officially, signed by the Foreign Secretaries. They've also been
sent to the *People.** *Nous verrons.* Mottershead subscribes
to *The Irishman* and will not fail to use this opportunity to
poke fun at the *"highsouled"* Irishmen.

But I'll play a trick on Pigott. I'll write to Eccarius today
and ask him to send the resolutions with the signatures, etc.,
to Isaac Butt, who is President of the Irish Working Men's
Association. Butt is not Pigott.

ENGELS TO MARX

December 9, 1869

I half expected that about *The Irishman.* Ireland still
remains the *sacra insula,* whose aspirations must on no account be mixed up with the profane class struggles of the
rest of the sinful world. Partially, this is certainly honest
madness on the part of these people, but it is equally certain
that it is partially also a calculated policy of the leaders
in order to maintain their domination over the peasant.
Added to this, a nation of peasants always has to take its
literary representatives from the bourgeoisie of the towns
and their ideologists, and in this respect Dublin (I mean
Catholic Dublin) is to Ireland much what Copenhagen is to
Denmark. But to these gentry the whole labour movement is
pure heresy and the Irish peasant must not on any account
be allowed to know that the socialist workers are his sole
allies in Europe.

In other respects, too, *The Irishman* is extremely lousy
this week. If it is ready to retreat *in this way*, the minute it
is threatened with a suspension of the Habeas Corpus Act,
the former sabre-rattling was all the more out of place. And
now even the fear that some more political prisoners may
be elected! On the one hand, the Irish are warned, and
quite rightly, not to let themselves be inveigled into unlawful
action; on the other, they are to be prevented from doing
the only lawful thing that is pertinent and revolutionary

---

* Probably to *The New-York Irish People.—Ed.*

and alone able to break successfully with the established practice of electing place-hunting lawyers and to impress the English Liberals. It is obvious that Pigott is afraid that others might outstrip him.

You will remember, by the way, that O'Connell always incited the Irish against the Chartists although or, to be more exact, because they too had inscribed Repeal on their banner.

MARX TO ENGELS

December 10, 1869

The way I shall put forward the matter next Tuesday is this: that quite apart from all phrases about "international" and "humane" *justice for Ireland*—which are taken for granted in the *International Council*—*it is in the direct and absolute interest of the English working class to get rid of their present connection with Ireland.* And this is my fullest conviction, and for reasons which in part I *cannot* tell the English workers themselves. For a long time I believed that it would be possible to overthrow the Irish regime by English working-class ascendancy. I always expressed this point of view in the *New-York Tribune.** Deeper study has now convinced me of the opposite. The English working class will *never accomplish anything* before it has got rid of Ireland. The lever must be applied in Ireland. That is why the Irish question is so important for the social movement in general.

I have read a lot of *Davies* in extracts. The book itself I had only glanced through superficially in the Museum.** So you would do me a great favour if you would copy out for me the passages relating to *common property*. You *must* get *"Curran's Speeches" edited by Davies (London: James Duffy, 22, Paternoster Row).* I meant to give it to you when you were in London. It **is** now circulating among the English members of the Central Council and God knows when I shall see it again. For the period *1779-1800* (Union) it is of decisive importance, not only because of *Curran's speeches* (especially those held *in courts*; I consider *Curran the only great*

---

* See pp. 54-58.—*Ed.*
** The British Museum Library.—*Ed.*

lawyer (people's advocate) of the eighteenth century and the *noblest personality*, while *Grattan* was a parliamentary rogue), but because you will find quoted there *all the sources* for the *United Irishmen*.[291] This period is of the highest interest, scientifically and dramatically. Firstly, the deeds of the English in 1588-89 repeated (and perhaps even intensified) in 1788-89. Secondly, a class movement can easily be traced in the Irish movement itself. *Thirdly*, the infamous policy of Pitt. *Fourthly*, and that will be very irksome to the English gentlemen, the proof that Ireland came to grief because, in fact, from a revolutionary standpoint, *the Irish were too far advanced for the English King and Church mob*, while on the other hand the English reaction in England had its roots (as in Cromwell's time) in the subjugation of Ireland. *This period* must be described in at least one chapter.[292] Put John Bull in the pillory!...

As to the present *Irish movement*, there are three important factors: 1) opposition to lawyers and trading politicians and blarney; 2) opposition to the dictates of the priests, who (the *superior ones*) are traitors, as in O'Connell's time as well as in 1798-1800; 3) the coming out of the *agricultural labouring class* against the farming class at the last meetings. (Similar happenings in 1795-1800.)

The rise of *The Irishman* was due only to the suppression of the *Fenian* press. For a long time it had been in opposition to Fenianism. Luby, etc., of the *Irish People*, etc., were educated men who treated religion as a bagatelle. The government put them in prison and then came the Pigotts & Co. *The Irishman* will amount to anything only until those people come out of prison. It is aware of this although it is now making *political capital* by declaiming in behalf of the "felon convicts".

MARX TO ENGELS

December 17, 1869

Our Irish resolutions* have been sent to all trade unions that maintain ties with us. Only one has protested, a small branch of the curriers, saying they are political and not

---

* See pp. 155-56.—*Ed.*

within the Council's sphere of action. We are sending a deputation to enlighten them. Mr. Odger now understands how useful it was for him that he voted *for* the resolutions despite all sorts of diplomatic objections. As a result the 3,000-4,000 Irish electors in Southwark have promised him their votes.

ENGELS TO MARX

January 19, 1870

I have at last discovered a copy of Prendergast in a local library and hope that I shall be able to obtain it. To my good or bad fortune, the old Irish laws are also to appear soon, and I shall thus have to wade through those as well. The more I study the subject, the clearer it is to me that Ireland has been stunted in her development by the English invasion and thrown centuries back. And this ever since the 12th century; furthermore, it should be borne in mind, of course, that three centuries of Danish invasions and plunder had by then substantially drained the country. But these latter had ceased over a hundred years earlier.

In recent years, research on Ireland has become somewhat more critical, particularly as far as Petrie's* studies of antiquity are concerned; he impelled me also to read some Celtic-Irish (naturally with a parallel translation). It does not seem all that difficult, but I shall not delve deeper into the stuff, I have had enough philological nonsense. In the next few days, when I get the book, I'll see how the old laws have been dealt with.

ENGELS TO MARX

January 25, 1870

I've at last received Prendergast and—as it always happens—two copies at once, namely, W. H. Smith and Sons have also got hold of one. I shall have finished with it

---

* G. Petrie, *The Ecclesiastical Architecture of Ireland, anterior to the Anglo-Norman Invasion*, Dublin, 1845.—*Ed.*

tonight. The book is important because it contains many excerpts from unprinted Bills. No wonder it is out of print. Longman and Co. must have been furious at having to put their name on *such* a book, and since there certainly was little demand for it in England (Mudies have *not a single copy*) they shall sell the edition for pulping as soon as they can or, possibly, to a company of Irish landlords (for the same purpose) and certainly will not print a second. What Prendergast says about the Anglo-Norman period is correct inasmuch as the Irish and Anglo-Irish, who lived at some distance from the Pale, continued during that period the same lazy life as before the invasion, and inasmuch as the wars of that period too were more "easy-going" (with few exceptions), and did not have the distinctly devastating character they assumed in the 16th century and which afterwards became the rule. But his theory that the enormous amiability of the Irishmen, and especially the Irish women, *immediately* disarms even the most hostile immigrant, is just thoroughly Irish, since the Irish way of thinking lacks all sense of proportion.

A new edition of Giraldus Cambrensis has appeared: *Giraldi Cambrensis Opera*, ed. J. S. Brewer, London, Longman and Co., 1863, *at least 3* volumes; could you find out the price for me and whether it would be possible to get cheaply, second hand, the whole work or at least the volume containing "Topographia Hibernica" and perhaps also "Hibernia expugnata"?

In order not to make a fool of myself over Cromwell, I'll have to put in a lot more work on the English history of the period. That will do no harm, but it will take up a lot of time.

ENGELS TO MARX

February 17, 1870

And thus the mountain Gladstone has successfully given birth to his Irish mouse.[293] I really don't know what the Tories could have against this Bill, which is so indulgent with the Irish landlords and finally places their interests in the tested hands of the Irish lawyers. Nevertheless, even this slight restriction of the eviction right will put an end to excessive emigration and the conversion of arable land into pastures. But it is very amusing if the brave Gladstone thinks he has settled the Irish question by means of this new prospect of endless lawsuits.

Is it possible to get a copy of the Bill? It would be important for me to follow the debates on the individual clauses.

MARX TO ENGELS

February 19, 1870

The best part of Gladstone's speech is the long introduction, in which he says that even the "beneficent" laws of the English have always the reverse effect in practice. What better proof does that fellow need that England is not called upon to be the lawgiver and ruler of Ireland!

His measures are a pretty piece of patchwork. The main thing in them is to lure the lawyers with the prospect of lawsuits and the landlords with the prospect of "state assistance".

Odger's election scandal was doubly useful: the pig-Whigs saw for the first time that they must let the workers into Parliament, or else the Tories will get in. Secondly, it is a lesson to Mr. Odger and his accomplices. He would have got in despite Waterlow[294] *if some of the Irish workers had not abstained from voting*, because he had behaved so trimming during the debate in the General Council,* which they knew of from *Reynolds's*.

You'll receive the Irish Bill next week.

MARX TO ENGELS

March 5, 1870

All sorts of things have happened in Fenian affairs in the meantime. A letter I wrote to the *Internationale* in Brussels,** and in which I censured the French Republicans for

---

* See pp. 157-58.—*Ed.*
** See pp. 164-69.—*Ed.*

their narrow national aims, has been printed, and the editors have announced that they will publish their remarks *this* week. You must know that in the letter of the Central Council to the *Genevans*—which was conveyed also to the Brussels people and the main centres of the International in France—I developed in detail the importance of the Irish question for the working-class movement in general (owing to its repercussion in England).*

Soon after, Jennychen was driven to anger by that disgusting article in the *Daily News*, the officious paper of the Gladstone Ministry, in which this bitchy publication turns to the "liberal" brothers in France and cautions them not to confuse the cases of Rochefort and O'Donovan Rossa.²⁹⁵ The *Marseillaise* has really fallen into the trap, it believes the *Daily News* and in addition has published a wretched article by that gossip-monger Talandier, in which this *ex-procureur de la République*,** now a teacher of French at the military school in Woolwich (also ex-private tutor with Herzen, on whom he wrote a passionate obituary), attacks the Irish for their Catholic faith and accuses them of having brought about Odger's failure—because of his participation in the Garibaldi committee. Besides, he adds, they support Mitchel despite his taking side with the slaveholders, as though Odger himself did not stick to Gladstone despite his even greater support for the slaveholders.

So Jennychen—*ira facit poetam****—besides a private letter, wrote an article to the *Marseillaise* which was *printed*. In addition, she received a letter from the *rédacteur de la rédaction*, a copy of which I am enclosing. Today she sends another letter to the *Marseillaise*, which, in connection with Gladstone's reply (this week) to the interpellation about the treatment of the prisoners, contains excerpts from O'Donovan Rossa's letter (see *Irishman, Feb. 5, 70).**** In it Gladstone is presented to the French not only as a monster by Rossa's letter (inasmuch as Gladstone is in fact responsible for the entire treatment of the prisoners under the Tories too), but at the same time as a ridiculous hypocrite, being the author

---

* See pp. 160-63.—*Ed.*
** Ex-public prosecutor in a first-instance court.—*Ed.*
*** Ire makes a poet (paraphrasing from Juvenal's first satire).—*Ed.*
**** See pp. 379-84.—*Ed.*

of the *Prayers, The Propagation of the Gospel, The Functions of Laymen in the Church* and *Ecce Homo*.

With these two papers—the *Internationale* and the *Marseillaise*—we shall now unmask the English to the Continent. If you should happen, one day or the other, to find something suitable for one of these papers, you too should participate in our good work.

MARX TO PAUL AND LAURA LAFARGUE

March 5, 1870

Here, at home, as you are fully aware, the Fenians' sway is paramount. Tussy is one of their head centres.[296] Jenny writes on their behalf in the *"Marseillaise"* under the pseudonym of J. Williams. I have not only treated the same theme in the Brussels *"Internationale"*, and caused resolutions of the Central Council to be passed against their gaolers. In a circular, addressed by the Council to our corresponding committees, I have explained the merits of the Irish Question.*

You understand at once that I am not only acted upon by feelings of humanity. There is something besides. To accelerate the social development in Europe, you must push on the catastrophe of official England. To do so, you must attack her in Ireland. That's her weakest point. Ireland lost, the British "Empire" is gone, and the class war in England, till now somnolent and chronic, will assume acute forms. But England is the metropolis of landlordism and capitalism all over the world.

ENGELS TO MARX

March 7, 1870

When I read that story about the *Marseillaise* in the "Irishman in Paris" on Saturday afternoon, I knew immediately in what part of the world this Mr. Williams could be found, but, silly as it may be of me, I couldn't account for

---

* See pp. 160-63.—*Ed.*

the first name.[297] It is a very good story, and the naïve letter with Rochefort's naïve demand that O'Donovan Rossa be asked for a contribution to the *Marseillaise* gives Jenny an excellent opportunity to raise the question of the treatment of prisoners and to open the eyes of the *bons hommes* over there.

Why don't you have the letter of the General Council to the Genevans published? The central sections in Geneva, Brussels, etc., read these things, but so long as they are not published they do not penetrate into the masses. They should also appear in German in the relevant organs. *You are publishing far too little.*

Please send me the relevant issues of the *Marseillaise* and *Internationale* for a few days. Jennychen's success has been met with a universal hurrah here and the health of Mr. J. Williams has been drunk with all due honours. I am very eager to hear how that story develops. The stupid correspondent of the "Irishman in Paris" should try some time if he can get such things into the newspapers of his friend Ollivier.

A couple of days ago, my bookseller suddenly sent me the *Senchus Mor*, the old Irish laws, and what's more, not the new edition but the *first*. So, with a lot of pushing I have succeeded in *that*. And such difficulties with a book having *Longmans* as its London firm on the title page and published by the government! I haven't been able to look at the stuff yet, as I have in the meantime taken up various modern things (about the 19th century) and must finish with them first.

MARX TO ENGELS

March 19, 1870

Enclosed is a *Marseillaise*, which should, however, be returned with the preceding one. I haven't read it myself yet. The article was written jointly by Jennychen and myself* because she didn't have sufficient time. That is also why she hasn't answered your letter and sends Mrs. Lizzy her thanks for the shamrock[298] provisionally through me.

---

* See pp. 384-88.—*Ed.*

From the enclosed letter from Pigott to Jenny you'll see that Mrs. O'Donovan, to whom Jenny sent a private letter together with 1 *Marseillaise*, took her for a gentleman, even though she signed it Jenny Marx. I answered Pigott today on behalf of Jennychen and took the opportunity to explain to him in short my views on the Irish question.

...The sensation caused by Jennychen's second letter (which contained the condensed translation of O'Donovan's letter) in Paris and London has robbed the loathsome and importunate (but very fluent with gab and pen) Talandier of his sleep. He had denounced the Irish as Catholic idiots in the *Marseillaise*. Now he espouses their cause no less full-mouthed in a review of what has been said in the *Times*, *Daily Telegraph* and *Daily News* about O'Donovan's letter. Since Jennychen's second letter was unsigned (by accident) he apparently flattered himself with the idea that he would be considered the secret sender. This has been frustrated by Jennychen's third letter. This fellow is *du reste* a teacher of French at the military school of Sandhurst.

MARX TO SIGFRID MEYER AND AUGUST VOGT

April 9, 1870

On January 1, 1870,[*] the General Council issued a confidential circular[**] drawn up by me in French (for the reaction upon England only the French, not the German, papers are important) on the relation of the Irish national struggle to the emancipation of the working class, and therefore on the attitude which the International Association should take in regard to the Irish question.

I shall give you here only quite briefly the decisive points. Ireland is the bulwark of the *English landed aristocracy*. The exploitation of that country is not only one of the main sources of this aristocracy's material welfare; it is its greatest *moral* strength. It, in fact, represents the *domination of England over Ireland*. Ireland is therefore the great means by which the English aristocracy maintains *its domination in England herself*.

---

[*] In the manuscript "December 1, 1869", which is a misprint.—*Ed.*
[**] See pp. 160-63.—*Ed.*

If, on the other hand, the English army and police we to withdraw from Ireland tomorrow, you would at once have an agrarian revolution there. But the overthrow of the English aristocracy in Ireland involves as a necessary consequence its overthrow in England. And this would fulfil the preliminary condition for the proletarian revolution in England. The destruction of the English landed aristocracy in Ireland is an infinitely easier operation than in England herself, because in Ireland *the land question* has hitherto been the *exclusive form* of the social question, because it is a question of existence, of *life and death*, for the immense majority of the Irish people, and because it is at the same time inseparable from the *national* question. This quite apart from the Irish being more passionate and revolutionary in character than the English.

As for the English *bourgeoisie*, it has in the first place a common interest with the English aristocracy in turning Ireland into mere pasture land which provides the English market with meat and wool at the cheapest possible prices. It is equally interested in reducing, by eviction and forcible emigration, the Irish population to such a small number that *English capital* (capital invested in land leased for farming) can function there with "security". It has the same interest in clearing the estate of Ireland as it had in the clearing of the agricultural districts of England and Scotland. The £6,000-10,000 absentee-landlord and other Irish revenues which at present flow annually to London have also to be taken into account.

But the English bourgeoisie has, besides, much more important interests in Ireland's present-day economy. Owing to the constantly increasing concentration of tenant farming, Ireland steadily supplies her own surplus to the English labour-market, and thus forces down wages and lowers the moral and material condition of the English working class.

And most important of all! Every industrial and commercial centre in England now possesses a working class *divided* into two *hostile* camps, English proletarians and Irish proletarians. The ordinary English worker hates the Irish worker as a competitor who lowers his standard of life. In relation to the Irish worker he feels himself a member of the *ruling nation* and so turns himself into a tool of the aristocrats and capitalists of his country *against Ireland,* thus strengthening their domination *over himself.* He cherishes religious, social, and national prejudices against the Irish worker. His attitude towards him is much the same as that of the "poor whites" to the "niggers" in the former slave states of the U.S.A. The Irishman pays him back with interest in his own money. He sees in the English worker at once the accomplice and the stupid tool of the *English rule in Ireland.*

This antagonism is artificially kept alive and intensified by the press, the pulpit, the comic papers, in short, by all the means at the disposal of the ruling classes. *This antagonism is the secret of the impotence of the English working class,* despite its organisation. It is the secret by which the capitalist class maintains its power. And that class is fully aware of it.

But the evil does not stop here. It continues across the ocean. The antagonism between English and Irish is the hidden basis of the conflict between the United States and England. It makes any honest and serious co-operation between the working classes of the two countries impossible. It enables the governments of both countries, whenever they think fit, to break the edge off the social conflict by their mutual bullying, and, in case of need, by war with one another.

England, being the metropolis of capital, the power which has hitherto ruled the world market, is for the present the most important country for the workers' revolution, and moreover the *only* country in which the material conditions for this revolution have developed up to a certain degree of maturity. Therefore to hasten the social revolution in England is the most important object of the International Working Men's Association. The sole means of hastening it is to make Ireland independent. Hence it is the task of the International everywhere to put the conflict between England and Ireland in the foreground, and everywhere to side openly with Ireland. And it is the special task of the Central Council in London to awaken a consciousness in the English workers that *for them* the *national emancipation of Ireland* is no question of abstract justice or humanitarian sentiment, but *the first condition of their own social emancipation.*

These roughly are the main points of the circular letter, which thereby at the same time gave the *raisons d'être* of the resolutions of the Central Council on the Irish amnesty.* Shortly afterwards I sent a strong anonymous article on the treatment of the Fenians by the English, etc., against Gladstone, etc., to the *Internationale*** (organ of our Belgian Central Committee*** in Brussels). In this article I at the same time made the charge against the French Republicans (the *Marseillaise* had printed some nonsense on Ireland written here by the wretched Talandier) that in their national egoism they were saving all their wrath for the Empire.

That worked. My daughter Jenny wrote a series of articles to the *Marseillaise* signing them J. Williams (she had called herself Jenny Williams in her private letter to the editorial board), and published, among other things, O'Donovan Rossa's letter.**** Hence immense noise. After many years of cynical refusal *Gladstone* was *thus* finally compelled to agree to a *parliamentary enquiry* into the treatment of the Fenian prisoners. Jenny is now the regular correspondent on Irish affairs for the *Marseillaise*. (*This is naturally to be a secret between us.*) The British Government and press are fiercely annoyed by the fact that the Irish question has thus now come *to the forefront* in France and that these rogues are now being watched and exposed via Paris on the whole Continent.

We hit another bird with the same stone, having forced the Irish leaders, journalists, etc., in Dublin, to get into contact with us, which the *General Council* so far had been unable to achieve!

You have now a great field in America for working along the same lines. *Coalition of the German workers with the Irish workers* (and of course also with the English and American workers who will agree to join) is the greatest job you could start on nowadays. This must be done in the name of the International. The social significance of the Irish question must be made clear.

---

* See pp. 155-56.—*Ed.*
** See pp. 164-69.—*Ed.*
*** Marx is referring to the Belgian Federal Council.—*Ed.*
**** See pp. 379-403.—*Ed.*

MARX TO ENGELS

April 14, 1870

You will receive in the course of this week or at the beginning of next *Landlord and Tenant Right in Ireland. Reports by Poor Law Inspectors. 1870*, also *Agricultural Holdings in Ireland. Returns. 1870.*

The reports by Poor Law Inspectors are interesting. Like their *Reports on Agricultural Wages*, which you have already received, these show, *inter alia*, that since the famine a conflict has broken out between the *labourers*, on the one hand, and *farmers and tenants*, on the other. As regards the *Reports on Wages*—assuming the present figures on wages are correct, and that is probable from other sources—either the *former wage rates* are given *too low* or the earlier Parliamentary Returns on them, which I'll find for you in my Parliamentary Papers, were *too high*. On the whole, it is confirmed that, as I said in the section on Ireland, the rise in wages was more than outweighed by the rise in food prices and that, except in autumn, etc., the relative surplus of the labourers is established correctly despite emigration.* Important in the *Landlord and Tenant Right Reports* is also the fact that the progress in machinery has turned a lot of hand-loom weavers into paupers....

It is clear from the two reports of the Poor Law Commissioners that 1) since the famine the clearing of the estates of labourers' dwellings has begun here *as in England* (not to be confused with the suppression of the 40-sh. free-holders after 1829),

2) that the Encumbered Estates proceedings have put a mass of *small usurers* in place of the turned out flotten land-lords. (The charge of landlords $1/6$ according to the same reports.)

ENGELS TO MARX

April 15, 1870

Your conclusions from the Parliamentary Reports agree with my results. It should, however, be remembered that

---

* See pp. 108-13.—*Ed.*

after 1846 the process of clearing 40-sh. freeholders was at first interspersed with clearing of labourers the reason being that, up to 1829, in order to produce freeholders, leases had to be made for 21 or 31 years *and a life* (if not longer), because a person became a freeholder only if he *could not be turned out* during his lifetime. These leases hardly ever excluded subdividing. These leases were partly still valid in 1846, resp. the consequences, that is, the peasants were still on the estate. The same was the case on the estates which were then in the hands of middlemen (who mostly held leases for 64 years and three lives or even for 99 years) and frequently their leases were revertible only between 1846 and 1860. Thus these processes were more or less interspersed so that the Irish landlord was never or seldom in a situation where he had to decide whether labourers in particular rather than other traditional small tenants should be ejected. Essentially it comes to the same thing in England and in Ireland: the land must be tilled by workers who live in *other Poor Law Unions*, so that the landlord and his tenant can remain exempted from the poor tax. This is also said by Senior or rather by his brother Edward, Poor Law Commissioner in Ireland: The great instrument which is clearing Ireland is the *Poor Law.*

Land sold since the Encumbered Estate Court amounts according to my notes to as much as $1/5$ of the total, the buyers were indeed largely usurers, speculators, etc., *mainly Irish Catholics.* Partly also enriched stock-breeders. Yet even now there are *only about 8,000-9,000 landowners* in Ireland.

ENGELS TO MARX

May 15, 1870

In what Parliamentary Paper could one find how much money is wasted every year on the Commissioners for the Publication of the Ancient Laws and Institutes of Ireland? This is a colossal job (in a small matter). It would also be important to know how much of that money is spent 1) as remuneration for idling commissioners, 2) as salaries for really working understrappers, printing costs, etc. This must

surely be somewhere in a Parliamentary Paper. Those fellows
have been drawing wages *since 1852* and up to now only
two volumes have been published! Three lords, three judges,
three priests, one general, and *one* who professionally special-
ises on Ireland who died long ago.*

KARL MARX TO JENNY MARX
(HIS DAUGHTER)

May 31, 1870

Here things are going on pretty much in the old track.
Fred is quite jolly since he has got rid of *"den verfluchten
Commerce"*. His book on Ireland**—which by the by costs
him a little more time than he had at first supposed—will be
highly interesting. The illustrious Doppelju*** who is so much
up in the most recent Irish history and plays so prominent a
part in it, will there find his archeological material ready cut.

MARX TO FRIEDRICH ADOLF SORGE

November 29, 1871

I come now to the question of MacDonnell.[299]
Before admitting him, the Council instituted a most
searching inquiry as to his integrity, he, like *all* other Irish
politicians, being much calumniated by his own countrymen.
The Council—after most incontrovertible evidence on his
private character—chose him because the *mass of the Irish
workmen in England* have more confidence in him than in
any other person. He is a man quite superior to religious
prejudices and as to his general views, it is absurd to say that
he has any "bourgeois" predilections. He is a proletarian, by
his circumstances of life and by his ideas.
If any accusation is to be brought forward against him,
let it be done in exact terms, and not by vague insinuation.

---

* See pp. 193-94.—*Ed.*
** See pp. 171-209.—*Ed.*
*** The German for the letter "w". Marx is referring to "Williams",
the pseudonym used by Jenny Marx.—*Ed.*

My opinion is that the Irishmen, removed for a long time by imprisonment, are not competent judges. The best proof is—their relations with *The Irishman* whose editor, Pigott, is a mere speculator, and whose manager, Murphy, is a ruffian. That paper—despite the exertions of the General Council for the Irish cause—has always intrigued against us. MacDonnell was constantly attacked in that paper by an Irishman (O'Donnell) connected with Campbell (an officer of the London *Police*) and a habitual drunkard who for a glass of gin will tell the first constable all the secrets he may have to dispose of.

After the nomination of MacDonnell, Murphy attacked and calumniated the *International* (not only MacDonnell) in *The Irishman*, and, *at the same time*, secretly, asked us to nominate him secretary for Ireland.

As to O'Donovan Rossa, I wonder that you quote him still as an authority after what you have written me about him. If any man was obliged, personally, to the *International* and the French Communards, it was he, and you have seen what thanks we have received at his hands.[300]

Let the Irish members of the New York Committee not forget that to be useful to them, we want above all *influence on the Irish in England*, and that for that purpose there exists, as far as we have been able to ascertain, no better man than MacDonnell.

ENGELS TO SIGISMUND BORKHEIM

Beginning of March 1872

Sorge is very naïve to demand a book on Ireland written from *our* standpoint. For the last two years I have been intending to write one, but the war, the Commune and the International have brought everything to a standstill. Meanwhile I recommend the following books to Sorge:

1. *The Cromwellian Settlement of Ireland* by Prendergast. London, Longmans, Sec. Ed. 1870-71.

2. *Memoir on Ireland* by O'Connell. London—Duffy, 1869. For the main historical events

3. *The Irish People and the Irish Land* by Isaac Butt. London—Ridgway.

This is all for the present.

However simple the Irish problem may be, it is nevertheless the result of a long historical struggle and hence has to be studied. A manual explaining it all in about two hours does not exist.

Karl Marx

[POSITION OF THE INTERNATIONAL WORKING MEN'S ASSOCIATION IN GERMANY AND ENGLAND

(From the Speech of September 22, 1871, at the London Conference)][301]

You will be aware of the great antagonism which has existed for a long time between the English and Irish workers, the causes of which are easy to enumerate. This antagonism is rooted in differences of language and religion,* and in the competition which Irish workers created in the labour market. It constitutes an obstacle to revolution in England and is, consequently, skilfully exploited by the government and the upper classes, who are convinced that no bonds are capable of uniting the English workers with the Irish. It is true that no union would be possible in the sphere of politics, but this is not the case in the economic sphere and the two sides are forming International sections which, as such, will have to advance simultaneously towards the same goal. The Irish sections will soon be very numerous.

Published in the book
The London Conference of the First International,
Moscow, 1936, in Russian

Translated from the French

---

* In Martin's draft the words "prolonged oppression of Ireland" are inserted after the word "religion".—*Ed.*

[RELATIONS BETWEEN THE IRISH SECTIONS AND THE BRITISH FEDERAL COUNCIL[302]

(Engels's Record of His Report at the General Council Meeting of May 14, 1872)*]

Citizen Engels said the real purport of this motion was to bring the Irish sections under the jurisdiction of the British Federal Council, a thing to which the Irish sections would never consent, and which the Council had neither the right nor the power to impose upon them. According to the Rules and Regulations, this Council had no power to compel any section or branch to acknowledge the supremacy of any Federal Council whatsoever. It was certainly bound, before admitting or rejecting any new branch, within the jurisdiction of any Federal Council, to consult that Council. But he maintained that the Irish sections in England were no more under the jurisdiction of the British Federal Council than the French, German or Italian** sections in this country. The Irish formed, to all intents and purposes, a distinct nationality of their own, and the fact that they used the English language could not deprive them of the right, common to all, to have an independent national organisation within the International.

Citizen Hales had spoken of the relations between England and Ireland as if they were of the most idyllic nature, something like those between England and France at the time of the Crimean war, when the ruling classes of the two countries never tired of praising each other, and everything breathed the most complete harmony. But the case was quite different. There was the fact of seven centuries of English Conquest

---

* See pp. 407-13.—*Ed.*

** In the Minute book of the General Council there follow the words: "and Polish sections".—*Ed.*

and oppression of Ireland, and so long as that oppression existed, it was an insult to Irish working men to ask them to submit to a British Federal Council. The position of Ireland with regard to England was not that of an equal, it was that of Poland with regard to Russia. What would be said if this Council called upon Polish sections to acknowledge the supremacy of a Russian Federal Council in Petersburg, or upon Prussian Polish, North Schleswig, and Alsatian sections to submit to a Federal Council in Berlin? Yet what it was asked to do with regard to Irish sections was substantially the same thing. If members of a conquering nation called upon the nation they had conquered and continued to hold down to forget their specific nationality and position, to "sink national differences" and so forth, that was not Internationalism, it was nothing else but preaching to them submission to the yoke, and attempting to justify and to perpetuate the dominion of the conqueror under the cloak of Internationalism. It was sanctioning the belief, only too common among the English working men, that they were superior beings compared to the Irish, and as much an aristocracy as the mean whites of the Slave States considered themselves to be with regard to the Negroes.

In a case like that of the Irish, true Internationalism must necessarily be based upon a distinctly national organisation; the Irish, as well as other oppressed nationalities, could enter the Association only as equals with the members of the conquering nation, and under protest against the conquest. The Irish sections, therefore, not only were justified, but even under the necessity to state in the preamble to their rules that their first and most pressing duty, as Irishmen, was to establish their own national independence. The antagonism between Irish and English working men in England had always been one of the most powerful means by which class rule was upheld in England. He recollected the time when he saw Feargus O'Connor and the English Chartists turned out of the Hall of Science in Manchester by the Irish.[303] Now, for the first time, there was a chance of making English and Irish working men act together in harmony for their common emancipation, a result attained by no previous movement in their country. And no sooner had this been effected, than they were called upon to dictate to the Irish, and to tell them they must not carry on the movement in their own way, but submit to be ruled by an English Council! Why, that was introducing into the International the subjugation of the Irish by the English.

If the promoters of this motion were so brimful of the truly International spirits, let them prove it by removing the seat of the British Federal Council to Dublin, and submit to a Council of Irishmen.

As to the pretended collisions between Irish and English branches, they had been provoked by attempts of members of the British Federal Council to meddle with the Irish sections, to get them to give up their specific national character and to come under the rule of the British Council.

Then the Irish sections in England could not be separated from the Irish sections in Ireland; it would not do to have some Irishmen dependent upon a London Federal Council and others upon a Dublin Federal Council. The Irish sections in England were our base of operations with regard to the Irish working men in Ireland; they were more advanced, being placed in more favourable circumstances, and the movement in Ireland could be propagated and organised only through their instrumentality. And were they to wilfully destroy their own base of operations and cut off the only means by which Ireland could be effectually won for the International? For it must not be forgotten that the Irish sections, and rightly so, would never consent to give up their distinct national organisation and submit to the British Council. The question, then, amounted to this: were they to leave the Irish alone, or were they to turn them out of the Association? If the motion was adopted by the Council, the Council would inform the Irish working men, in so many words, that, after the dominion of the English aristocracy over Ireland, after the dominion of the English middle class over Ireland, they must now look forth to the advent of the dominion of the English working class over Ireland.

Published in: Marx and Engels, *Collected Works*, second Russian ed., Vol. 18, Moscow, 1961

Printed according to the text of the book *The General Council of the First International. 1871-1872. Minutes,* Moscow

Karl Marx

From THE REPORT OF THE GENERAL COUNCIL TO THE FIFTH ANNUAL CONGRESS OF THE INTERNATIONAL WORKING MEN'S ASSOCIATION HELD AT THE HAGUE[304]

Finally, the government of Mr. Gladstone, unable to act in Great Britain, at least set forth its good intentions by the police terrorism exercised in Ireland against our sections then in course of formation, and by ordering its representatives abroad to collect information with respect to the International Working Men's Association*....

In its former annual reports, the General Council used to give a review of the progress of the Association since the meeting of the preceding Congress. You will appreciate, Citizens,** the motives which induce us to abstain from that course upon this occasion. Moreover, the reports of the delegates from the various countries, who know best how far their discretion may extend, will in a measure make up for this deficiency. We confine ourselves to the statement that since the Congress at Basle, and chiefly since the London Conference of September, 1871, the International has been extended to the Irish in England and to Ireland itself, to Holland, Denmark, and Portugal, that it has been firmly organised in the United States, and that it has established ramifications in Búenos Aires, Australia, and New Zealand.

Published in
September-October 1872
as a leaflet in German and
in some newspapers published
by the International, including
The International Herald
Nos. 27-29 for October 5, 12
and 19, 1872

Printed according to the
text of *The International Herald*

---

* See pp. 404-06.—*Ed.*
** In the leaflet and the newspaper *Volksstaat* the word "Citizens" is replaced by "Workers".—*Ed.*

III

[Meeting in Hyde Park]

London, November 14, 1872

The *Liberal* English Government has at the moment no less than 42 Irish political prisoners in its prisons and treats them with quite exceptional cruelty, far worse than thieves and murderers. In the good old days of King Bomba, the head of the present *Liberal* cabinet, Mr. Gladstone, travelled to Italy and visited political prisoners in Naples; on his return to England he published a pamphlet which disgraced the Neapolitan Government before Europe for its unworthy treatment of political prisoners.[306]

This does not prevent this selfsame Mr. Gladstone from treating in the very same way the Irish political prisoners, whom he continues to keep under lock and key.

The Irish members of the International in London decided to organise a *giant* demonstration in Hyde Park (the largest public park in London, where all the big popular meetings take place during political campaigns) to demand a general amnesty. They contacted all London's democratic organisations and formed a committee which included MacDonnell (an Irishman), Murray (an Englishman) and Lessner (a German)—all members of the last General Council of the International.

A difficulty arose: at the last session of Parliament the government passed a law which gave it the right to regulate public meetings in London's parks. It made use of this and had the regulation posted up to warn those who wanted to hold such a public meeting that they must give a written notification to the police two days prior to calling it, indicating the names of the speakers. This regulation carefully kept hidden from the London press destroyed with one stroke of the pen one of the most precious rights of London's working people—the right to hold meetings in parks when and how they please. To submit to this regulation would be to sacrifice one of the people's rights.

The Irish, who represent the most revolutionary element of the population, were not men to display such weakness. The committee unanimously decided to act as if it did not know of the existence of this regulation and to hold their meeting in defiance of the government's decree.

Last Sunday* at about three o'clock in the afternoon two enormous processions with bands and banners marched towards Hyde Park. The bands played Irish songs and the *Marseillaise*; almost all the banners were Irish (green with a gold harp in the middle) or red. There were only a few police agents at the entrances to the park and the columns of demonstrators marched in without meeting with any resistance. They assembled at the appointed place and the speeches began.

The spectators numbered at least thirty thousand and at least half had a green ribbon or a green leaf in their buttonhole to show they were Irish; the rest were English, German and French. The crowd was too large for all to be able to hear the speeches, and so a second meeting was organised nearby with other orators speaking on the same theme. Forceful resolutions were adopted demanding a general amnesty and the repeal of the coercion laws which keep Ireland under a permanent state of siege. At about five o'clock the demonstrators formed up into files again and left the park, thus having flouted the regulation of Gladstone's Government.

This is the first time an Irish demonstration has been held in Hyde Park; it was very successful and even the London bourgeois press cannot deny this. It is also the first time the English and Irish sections of our population have united in friendship. These two elements of the working class, whose enmity towards each other was so much in the interests of the government and wealthy classes, are now offering one another the hand of friendship; this gratifying fact is due

---

* November 3, 1872.—*Ed.*

principally to the influence of the last General Council of the International,[307] which has always directed all its efforts to unite the workers of both peoples on a basis of complete equality. This meeting, of the 3rd November, will usher in a new era in the history of London's working-class movement.

You might ask: "What is the government doing? Can it be that it is willing to reconcile itself to this slight? Will it allow its regulation to be flouted with impunity?"

Well, this is what it has done: it placed two police inspectors and two agents by the platforms in Hyde Park and they took down the names of the speakers. On the following day, these two inspectors brought a suit against the speakers before the *Justice of the Peace*. The Justice sent them a summons and they have to appear before him next Saturday. This course of action makes it quite clear that they don't intend to undertake extensive proceedings against them. The government seems to have admitted that the Irish or, as they say here, the Fenians have beaten it and will be satisfied with a small fine. The debate in court will certainly be interesting and I shall inform you of it in my next letter.[308] Of one thing there can be no doubt: the Irish, thanks to their energetic efforts, have saved the right of the people of London to hold meetings in parks when and how they please.

Published in
La Plebe No. 117,
November 17, 1872

Translated from the
Italian

Frederick Engels

FROM THE INTERNATIONAL

[EXCERPT]

On the other hand, the British Section of the International
held a Congress at Manchester on June 1 and 2, which was
undoubtedly an epoch-making event in the English labour
movement. It was attended by 26 delegates who represented
the main centres of English industry as well as several smaller
towns. The report of the Federal Council differed from all
previous documents of this kind by the fact that—in a country
with a tradition of legality—it asserted the right of the working class to *use force in order to realise* its demands.

Congress approved the report and decided that the red
flag is to be the flag of the British Section of the International; the working class demands not only the return of
all landed property to the working people but also of all
means of production; it calls for the eight-hour working day
as a preliminary measure; it sends congratulations to the
Spanish workers who have succeeded in establishing a republic and in electing ten workers to the Cortes; and requests
the English Government immediately to release all Irish
Fenians still imprisoned. Anyone familiar with the history
of the English labour movement will admit that no English
workers' congress has ever advanced such far-reaching
demands. In any case, this Congress and the miserable end of
the separatist, self-appointed Federal Council[309] has determined the attitude of the British Section of the International.

Published in
Der Volksstaat No. 53,
July 2, 1873

Translated from the
German

From THE ENGLISH ELECTIONS

Four weeks ago Gladstone suddenly dissolved Parliament. The inevitable "labour leaders" began to breathe again: either they would get themselves elected or they would again become well-paid itinerant preachers of the cause of the "great Liberal Party". But alas! the day appointed for the elections was so close that they were cheated out of both chances. True enough, a few did stand for Parliament; but since in England every candidate, before he can be voted upon, must contribute two hundred pounds (1,240 thaler) towards the election expenses and the workers had almost nowhere been organised for this purpose, only such of them could stand as candidates seriously as obtained this sum from the bourgeoisie, i.e., as acted *with its gracious permission*. With this the bourgeoisie had done its duty and in the elections themselves allowed them all to suffer a complete fiasco.

Only two workers got in, both miners from coal pits. This trade is very strongly organised in three big unions, has considerable means at its disposal, controls an undisputed majority of the voters in some constituencies and has worked systematically for direct representation in Parliament ever since the Reform Acts were passed. The candidates put up were the secretaries of the three Trade Unions. The one, Halliday, lost out in Wales; the other two came out on top: *MacDonald* in *Stafford* and *Burt* in *Morpeth*. Burt is little known outside of his constituency. MacDonald, however, betrayed the workers of his trade when, during the negotiations on the last mining law,[310] which he attended as the representative of his trade, he sanctioned an amendment which was so grossly in the interests of the capitalists that even the government had not dared to include it in the draft.

At any rate, the ice has been broken and two workers now have seats in the most fashionable debating club of Europe, among those who have declared themselves the first gentlemen of Europe.

Alongside of them sit at least fifty Irish Home Rulers. When the Fenian (Irish-republican) Rebellion of 1867[311] had been quelled and the military leaders of the Fenians had either gradually been caught or driven to emigrate to America, the remnants of the Fenian conspiracy soon lost all importance. Violent insurrection had no prospect of success for many years, at least until such time as England would again be involved in serious difficulties abroad. Hence a legal movement remained the only possibility, and such a movement was undertaken under the banner of the Home Rulers, who wanted the Irish to be "masters in their own house". They made the definite demand that the Imperial Parliament in London should cede to a special Irish Parliament in Dublin the right to legislate on all purely Irish questions; very wisely nothing was said meanwhile about what was to be understood as a purely Irish question. This movement, at first scoffed at by the English press, has become so powerful that Irish M.P.s of the most diverse party complexions—Conservatives and Liberals, Protestants and Catholics (Butt, who leads the movement, is himself a Protestant) and even a native-born Englishman sitting for Galway—have had to join it. For the first time since the days of O'Connell, whose repeal movement collapsed in the general reaction about the same time as the Chartist movement, as a result of the events of 1848—he had died in 1847—a well-knit Irish party once again has entered Parliament, but under circumstances that hardly permit it constantly to compromise à la O'Connell with the Liberals or to have individual members of it sell themselves retail to Liberal governments, as after him has become the fashion.

Thus both motive forces of English political development have now entered Parliament: on the one side the workers, on the other the Irish as a compact national party. And even if they may hardly be expected to play a big role in this

Parliament—the workers will certainly not—the elections of 1874 have indisputably ushered in a new phase in English political development.

Published in
Der Volksstaat No. 26,
March 4, 1874

Translated from the
German

Frederick   Engels

From DIALECTICS OF NATURE[312]

We mentioned the potato and the resulting spread of
scrofula. But what is scrofula compared to the effect which
the reduction of the workers to a potato diet had on the
living conditions of the masses of the people in whole
countries, or compared to the famine the potato blight
brought to Ireland in 1847, which consigned to the grave a
million Irishmen, nourished solely or almost exclusively on
potatoes, and forced the emigration overseas of two million
more?

Published in   Translated from the
Die Neue Zeit Bd. 2, No. 44,   German
Stuttgart, 1895-96

Frederick Engels

From ANTI-DÜHRING[313]

If we confine ourselves to the cultivation of landed property consisting of tracts of considerable size, the question arises: whose landed property is it? And then we find in the early history of all civilised peoples, not the "large landed proprietors" whom Herr Dühring interpolates here with his customary sleight of hand, which he calls "natural dialectics", but tribal and village communities with common ownership of the land. From India to Ireland the cultivation of landed property in tracts of considerable size was originally carried on by such tribal and village communities; sometimes the arable land was tilled jointly for account of the community, and sometimes in separate parcels of land temporarily allotted to families by the community, while woodland and pasture land continued to be used in common. It is once again characteristic of "the most exhaustive specialised studies" made by Herr Dühring "in the domain of politics and law" that he knows nothing of all this; that all his works breathe total ignorance of Maurer's epoch-making writings on the primitive constitution of the German mark,[314] the basis of all German law, and of the ever-increasing mass of literature, chiefly stimulated by Maurer, which is devoted to proving the primitive common ownership of the land among all civilised peoples of Europe and Asia, and to showing the various forms of its existence and dissolution.

Published in *Vorwärts* in 1877 and in the book: F. Engels, *Herrn Eugen Dühring's Umwälzung der Wissenschaft*, Leipzig, 1878

Printed according to the third edition which appeared in Stuttgart in 1894
Translated from the German

Frederick Engels

From THE PREPARATORY NOTES
TO "ANTI-DÜHRING"

When the Indo-Germanic people immigrated into Europe they ousted the original inhabitants by *force* and tilled the land which they held as communal property. The latter can be shown to have existed historically among Celts, Germans and Slavs, and it is still in existence—even in the form of direct bondage (Russia) or indirect bondage (Ireland)—among Slavs, Germans and even Celts [rundale]. After the Lapps and Basques were driven out force was no longer used. Equality, or alternatively, voluntarily conceded preferential treatment obtained within the community. Where communal ownership gave rise to private ownership of land by individual peasants, the division among the members of the community took place purely spontaneously up to the sixteenth century; it was mostly a very gradual process and remnants of communal property generally continued to exist. There was no question of using *force*, force was employed only against the remnants of communal property (in England in the eighteenth and nineteenth centuries, in Germany chiefly in the nineteenth century). Ireland is a special case.

Published in
Marx-Engels Archiv,
Zeitschrift des Marx-Engels
Instituts in Moskau, Band 2,
Frankfurt a. M. 1927

Translated from the
German

From AMERICAN FOOD AND THE LAND QUESTION[315]

This American revolution in farming, together with the revolutionised means of transport as invented by the Americans, sends over to Europe wheat at such low prices that no European farmer can compete with it—at least not while he is expected to pay rent. Look at the year 1879, when this was first felt. The crop was bad in all Western Europe; it was a failure in England. Yet, thanks to American corn, prices remained almost stationary. For the first time the British farmer had a bad crop and low prices of wheat at the same time. Then the farmers began to stir, the landlords felt alarmed. Next year, with a better crop, prices went lower still. The price of corn is now determined by the cost of production in America, plus the cost of transport. And this will be the case more and more every year, in proportion as new prairie-land is put under the plough. The agricultural armies required for that operation—we find them ourselves in Europe by sending over emigrants.

Now, formerly there was this consolation for the farmer and the landlord, that if corn did not pay meat would. The plough-land was turned into grass-land, and everything was pleasant again. But now that resource is cut off too. American meat and American cattle are sent over in ever-increasing quantities. And not only that. There are at least two great cattle-producing countries which are on the alert for methods permitting them to send over to Europe, and especially to England, their immense excess of meat, now wasted. With the present state of science and the rapid progress made in its application, we may be sure that in a very few years— at the very latest—Australian and South American beef and mutton will be brought over in a perfect state of preservation and in enormous quantities. What is then to become of the prosperity of the British farmer, of the long rent-roll of the British landlord? It is all very well to grow gooseberries, strawberries, and so forth—that market is well enough supplied as it is. No doubt the British workman could consume a deal more of these delicacies—but then first raise his wages.

It is scarcely needful to say that the effect of this new American agricultural competition is felt on the Continent too. The small peasant proprietor—mostly mortgaged over head and ears—and paying interest and law expenses where the English and Irish farmer pays rent, he feels it quite as much. It is a peculiar effect of this American competition that it renders not only large landed property, but also small landed property useless, by rendering both unprofitable.

It may be said that this system of land exhaustion, as now practised in the Far West, cannot go on for ever, and things must come right again. Of course, it cannot last for ever; but there is plenty of unexhausted land yet to carry on the process for another century. Moreover, there are other countries offering similar advantages. There is the whole South Russian steppe, where, indeed, commercial men have bought land and done the same thing. There are the vast pampas of the Argentine Republic, there are others still; all lands equally fit for this modern system of giant farming and cheap production. So that before this thing is exhausted it will have lived long enough to kill all the landlords of Europe, great and small, at least twice over.

Well, and the upshot of all this? The upshot will and must be that it will force upon us the nationalisation of the land and its cultivation by co-operative societies under national control. Then, and then alone, it will again pay both the cultivators and the nation to work it, whatever the price of American or any other corn and meat may be. And if the landlords in the meantime, as they seem half inclined to do, actually do go to America, we wish them a pleasant journey.

Published in
The Labour Standard No. 9,
July 2, 1881

Printed according to the
text of the newspaper

Frederick Engels

From BISMARCK AND THE GERMAN WORKING MEN'S PARTY

Then Bismarck succeeded in passing an Act by which Social-Democracy was outlawed.[316] The working men's newspapers, more than fifty, were suppressed, their societies and clubs broken up, their funds seized, their meetings dissolved by the police, and, to crown all, it was enacted that whole towns and districts might be "proclaimed", just as in Ireland. But what even English Coercion Bills have never ventured upon in Ireland Bismarck did in Germany. In every "proclaimed" district the police received the right to expel any man whom it might "reasonably suspect" of socialistic propaganda. Berlin was, of course, at once proclaimed, and hundreds (with their families, thousands) of people were expelled....

In the year from October, 1879, to October, 1880, there were in Prussia alone imprisoned for high treason, treason felony, insulting the Emperor, &c., not less than 1,108 persons; and for political libels, insulting Bismarck, or defiling the Government, &c., not less than 10,094 persons. Eleven thousand two hundred and two political prisoners, that beats even Mr. Forster's Irish exploits!

And what has Bismarck attained with all his coercion? Just as much as Mr. Forster in Ireland. The Social-Democratic Party is in as blooming a condition, and possesses as firm an organisation, as the Irish Land League.[317] A few days ago there were elections for the Town Council of Mannheim. The working-class party nominated sixteen candidates, and carried them all by a majority of nearly three to one. Again, Bebel, member of the German Parliament for Dresden, stood for the representation of the Leipzig district in the Saxon

Parliament. Bebel is himself a working man (a turner), and one of the best, if not the best speaker in Germany. To frustrate his being elected, the Government expelled all his committee. What was the result? That even with a limited suffrage, Bebel was carried by a strong majority. Thus, Bismarck's coercion avails him nothing; on the contrary, it exasperates the people. Those to whom all legal means of asserting themselves are cut off, will one fine morning take to illegal ones, and no one can blame them. How often have Mr. Gladstone and Mr. Forster proclaimed that doctrine? And how do they act now in Ireland?

Published in
The Labour Standard No. 12,
July 23, 1881

Printed according to the
text of the newspaper

From SYNOPSIS OF J. R. GREEN'S

"HISTORY OF THE ENGLISH PEOPLE"[318]

1169-1171: *Leinster* (Ireland) in the hands of English
"adventurers"; *Richard of Clare, Earl of Pembroke,* does
homage for Leinster as an English lordship to *Henry II,
who,* accompanied by Pembroke, visited his "new dominion
which the adventurers had won". [Fourteen years earlier,
Pope *Adrian IV* had made him a present of Ireland. He
(Henry) wanted to use the *trade in English slaves (with
Bristol)* as a pretext for invasion, but nothing came of it at
the time, because of the resistance of the English
baronage.] . . .

After *Henry II* left Ireland, nothing indeed but the
feuds and weakness of the Irish tribes enabled the adventurers to hold the *districts of Drogheda, Dublin, Wexford,
Waterford, and Cork,* which now formed the so-called
English Pale. For their part, the adventurers were compelled
to preserve *"their fealty to the English Crown".* John (Lackland) came with an army, stormed its strongholds and drove
its leading barons into exile, divided the Pale into counties,
ordered the observance of the English law; but the departure
of John and his army to England was a signal for a return
of disorder within *the Pale.* . . . Within the Pale itself, the
English settlers were harried and oppressed by their own
baronage as much as by the Irish marauders. . . . After their
victory at *Bannockburn, Robert Bruce* sent a Scotch force to
Ireland with his brother* at its head; general rising of Ireland
welcomed him; but the danger united *pro nunc** the *barons*

---

* Edward Bruce.—*Ed.*
** For a time.—*Ed.*

of the Pale, and in *1316 they emerged victors on the bloody field of Athenree* by the slaughter of 11,000 of their foes and almost complete annihilation of *the sept of the O'Connors.* Thereafter, the *barons of the Pale* sank more and more into Irish chieftains; the *Fitz-Maurices,* who became *Earls of Desmond* and whose vast territory in Munster was erected into a County Palatine, adopted the dress and manners of the natives around them.

Kilkenny Statute of Edward III[319]: *this Statute forbade the adoption of the Irish language or name or dress by any man of English blood;* it enforced within the Pale the exclusive use of the English law, and made the *use of the native or Brehon law,* which was gaining ground, *an act of treason;* it made treasonable any marriage of the Englishry with persons of *Irish race,* or any adoption of English children by Irish foster-fathers.... However, this did not prevent the fusion of the two races, with the lords of the Pale almost completely denying obedience to English government.... In *1394* Richard II landed with an army at Waterford and received the general submission of the native chiefs. But the lords of the Pale held aloof: no sooner Richard quitted the island, than the Irish in turn refused to carry out their promise of quitting Leinster, and engaged in a fresh contest with the *Earl of March, whom the King had proclaimed as his heir and left behind him as his lieutenant in Ireland.* In the summer of *1398* March was beaten and slain in battle; now Richard II was eager to avenge his cousin's death, and complete the work he had begun by a first invasion (with him as *hostage* was *Henry of Lancaster's son,* later Henry V). The *Percies (Earl of Northumberland and his son Henry Percy or Hotspur)* refused to serve in his army. He banished the Percies, who withdrew into Scotland.

MAY 1399: Richard II [went] to Ireland and left his uncle, *Duke of York,* as regent in his stead.

JUNE 1399: Henry of Lancaster entered the Humber and landed at Ravenspur.

IN THE BEGINNING OF AUGUST 1399 Henry of Lancaster master of the realm when Richard II at last sailed from Waterford and landed at *Milford Haven.* By the treacherous pledges of the Earl of Northumberland the ass Richard was lured *to Flint* for a meeting with Henry of

Lancaster, who took him to London as prisoner, where he was coffered in the Tower.

Published in
Marx-Engels Archives,
Vol. VIII, Russ. ed, Moscow, 1946

Printed according to the manuscript in English and German. Part of the manuscript translated from the German

Frederick Engels

JENNY LONGUET, NÉE MARX[320]

Jenny, the eldest daughter of Karl Marx, died at Argenteuil near Paris on January 11. About eight years ago she married *Charles Longuet*, a former member of the Paris Commune and at present an editor of *Justice*.

Jenny Marx was born on May 1, 1844, grew up in the midst of the international working-class movement and was closely attached to it. Despite a reticence that could almost be taken for shyness, she displayed when necessary a presence of mind and energy which could be envied by many a man.

When the Irish press disclosed the infamous treatment the Fenians sentenced in 1866 and later had to suffer in jail, while the English papers stubbornly ignored the atrocities, and when the Gladstone Government, despite the promises it made during the election campaign, refused to amnesty them or even to ameliorate their conditions, Jenny Marx found a means that caused the pious Mr. Gladstone to take immediate steps. She wrote two articles for Rochefort's *Marseillaise* vividly describing how political prisoners are treated in freedom-loving England.* That was very effective. Disclosures in a big Paris newspaper could not be endured. A few weeks later O'Donovan Rossa and most of the others were free and on their way to America.

In the summer of 1871 Jenny, together with her youngest sister,** visited their brother-in-law Lafargue at Bordeaux. Lafargue, his wife, their sick child and the two girls went from there to Bagneres-de-Luchon, a spa in the Pyrenees.

---

* See pp. 379-403.—*Ed.*
** Eleanor Marx.—*Ed.*

Early one morning a gentleman came to Lafargue and said: "I am a police officer, but a Republican; an order for your arrest has been received. It is known that you were in charge of communications between Bordeaux and the Paris Commune. You have one hour to cross the border."

Lafargue with his wife and child succeeded in getting over the pass into Spain, thereupon the police took revenge by arresting the two girls. Jenny had a letter in her pocket from Gustave Flourens, the leader of the Commune who was killed near Paris. Had the letter been discovered, a journey to New Caledonia was sure to follow for the two sisters. When she was left alone in the office for a moment Jenny opened a dusty old account book, put the letter inside and closed the book again. Perhaps the letter is still there. The two girls were then brought before the prefect, the noble Comte de Kératry, well remembered as a Bonapartist, who closely questioned them. But the cunning of the old diplomat and the brutality of the old cavalry officer were of no avail when faced with Jenny's calm circumspection. He left the room in a fit of rage about "the energy that the women of this family seem to possess". After the dispatch of numerous cables to and from Paris, he finally had to release the two girls, who had been treated in a truly Prussian way during their detention.

These two incidents are very characteristic of Jenny Longuet. The proletariat has lost a valiant fighter. But her grief-stricken father has at least the consolation that hundreds of thousands of workers in Europe and America share his sorrow.

London, January 13, 1883

Published in   Translated from the
Der Sozialdemokrat No. 4,   German
January 18, 1883

Karl Marx and Frederick Engels

EXCERPTS FROM LETTERS ON THE IRISH QUESTION WRITTEN BETWEEN 1877 AND 1882

MARX TO ENGELS

August 1, 1877

The Irish skirmishes in the House of Commons are very amusing. Parnell, etc., told Barry that the worst was the attitude of Butt, who hopes to be appointed judge and has threatened to resign his leadership; and that he could do them great harm in Ireland. Barry mentioned Butt's letter to the General Council of the International. They would like to have this document to prove that his stand-offishness in relation to the intransigents is mere pretence. But how am I to find the thing now?[321]

MARX TO JOHN SWINTON

November 4, 1880

Apart Mr. Gladstone's "sensational" failures abroad—political interest centres here at present on the Irish "Land Question". And why? Mainly because it is the harbinger of the *English "Land Question"*.

Not only that the great landlords of England are also the largest landholders of Ireland, but having once broken down in what is ironically called the "Sister" island, the English landed system will no longer be tenable at home. There are arrayed against it the British farmers, wincing under high rents, and—thanks to the American competition—low prices; the British agricultural labourers, at last impatient of their traditional position of ill-used beasts of burden, and—that British party which styles itself *"Radical"*. The latter consists of two sets of men; first the *ideologues* of the party, eager to overthrow the political power of the aristocracy by mining its material basis, the semi-feudal landed property. But behind these principle-spouters, and hunting them on, lurks another set of men—sharp, close-fisted, calculating *capitalists*, fully aware that the abolition of the old land laws, in the way proposed by the ideologues, cannot but convert land into a commercial article that must ultimately concentrate in the hands of capital.

On the other side, considered as a national entity, John Bull has ugly misgivings lest the aristocratic English landed garrison in Ireland once gone—England's political sway over Ireland will go too!

ENGELS TO JENNY LONGUET

February 24, 1881

My dear Jenny,

Well may the illustrious Regnard recommend his factum to your "charity".[322] This Jacobin defending English respectable Protestantism and English vulgar Liberalism with the historical appareil of that same vulgar Liberalism is indeed an object of deepest charity. But to his "facts".

1) The 30,000 Protestants' massacre of 1641. The Irish Catholics are here in the same position as the Commune de Paris. The Versaillais massacred 30,000 Communards and called that the horrors of the Commune. The English Protestants under Cromwell massacred at least 30,000 Irish and to cover their brutality, *invented* the tale that this was to avenge 30,000 Protestants murdered by the Irish Catholics. The facts are these.

Ulster having been taken from its Irish owners who at that time 1600-1610 held the *land in common*, and handed over to Scotch Protestant military colonists, these colonists did not feel safe in their possessions in the troublous times after 1640. The Puritan English government officials in Dublin spread the rumour that a Scotch Army of Covenant-ers[323] was to land in Ulster and exterminate all Irish and Catholics. Sir W. Parsons, one of the two Chief Justices of Ireland, said that in a 12-month there would not be a Catholic left in Ireland. It was under these menaces, repeated in the English Parliament, that the Irish of Ulster rose on

23rd Oct. 1641. But no massacre took place. All contempo-
raneous sources ascribe to the Irish merely the intention of
a general massacre, and even the two Protestant Chief
Justices* (proclam. 8th Febr. 1642) declare that "the chief
part of their plot, and amongst them a general massacre, had
been *disappointed.* The English and Scotch, however, 4th May
1642, threw Irish women naked into the river (Newry) and
massacred Irishmen. (Prendergast, *Cromwellian Settlement
of Ireland*, 1865.)

2) L'Irlande la Vendée de l'Angleterre.³²⁴ Ireland was
Catholic, Protestant England Republican, therefore Ireland—
English Vendée. There is however this little difference that
the French Revolution intended to *give* the land to the
people, the English Commonwealth intended, in Ireland, to
take the land from the people.

The whole Protestant reformation, as is well known to
most students of history save Regnard, apart from its dogma-
tical squabbles and quibbles, was a vast plan for a confiscation of land. First the land was taken from the Church. Then
the Catholics, in countries where Protestantism was in power,
were declared rebels and their land confiscated.

Now in Ireland the case was peculiar.

"For the English," says Prendergast, "seem to have thought that
god made a mistake in giving such a fine country as Ireland to the
Irish; and for near 700 years they have been trying to remedy it."

The whole agrarian history of Ireland is a series of confiscations of Irish land to be handed over to English settlers.
These settlers, in a very few generations, under the charm
of Celtic society, turned more Irish than the aborigines.
Then a new confiscation and new colonisation took place, and
so *in infinitum.*

In the 17th century, the whole of Ireland except the
newly Scotchified North, was ripe for a fresh confiscation.
So much so, that when the British (Puritan) Parliament ac-
corded to Charles I an army for the reduction of Ireland, it
resolved that the money for this armament should be raised
*upon the security of 2,500,000 acres to be confiscated in
Ireland.* And the "adventurers" who advanced the money

---

* The second Chief Justice of Ireland was Borlase.—*Ed.*

should also appoint the officers of that army. The land was to be divided amongst those adventurers: so that 1,000 acres should be given them, if in Ulster for £200—advanced, in Connaught for £300, in Munster for £450, in Leinster for £600. And if the people rose against this beneficent plan they are Vendéens! If Regnard should ever sit in a National Convention, he may take a leaf out of the proceedings of the Long Parliament, and combat a possible Vendée with these means.

The abolition of the penal laws![325] Why the greater part of them were repealed, not in 1793 but in 1778, when England was threatened by the rise of the American Republic, and the second repeal, 1793, was when the French Republic arose threatening and England required all the soldiers she could get to fight it!

The Grant to Maynooth by Pitt.[326] This pittance was soon repealed by the Tories and only renewed by Sir R. Peel in 1845. But not a word about the other *cadeau que faisait à l'Irlande ce grand homme (c'est la première fois qu'il trouve grâce devant les yeux d'un Jacobin*)*, that other *"dotation"* not only *"considérable"* but actually lavish—the 3 Million £ by which the Union of Ireland with England was bought. The parliamentary documents will show that the one item of the purchase money of rotten and nomination boroughs[327] alone cost no less a sum than *£1,245,000*. (O'Connell memoir on Ireland addressed to the Queen.)

Lord Derby instituted *le système des écoles nationales*.[328] *Very true* but why did he? Consult Fitzgibbon, Ireland in 1868,** the work of a staunch Protestant and Tory, or else the official Report of Commissioners on Education in Ireland 1826. The Irish, neglected by the English government, had taken the education of their children into their own hands. At the time when English fathers and mothers insisted upon their right to send their children to the factory to earn money instead of to the school to learn, at that time in Ireland the peasants vied with each other in forming schools of their own. The schoolmaster was an ambulant teacher,

---

* Present made to Ireland by that great man (this is the first time that he found grace in the eyes of a Jacobin).—*Ed.*

** G. Fitzgibbon, *Ireland in 1868, the Battle-field for English Party Strife*, London, 1868.—*Ed.*

spending a couple of months at each village. A cottage was found for him, each child paid him 2^d· a week and a few sods of turf in winter. The schools were kept, on fine days in summer, in the fields, near a hedge, and then known by the name of hedge-schools. There were also ambulant scholars, who with their books under the arm, wandered from school to school, receiving lodging and food from the peasants without difficulty. In 1812 there were 4,600 such hedge-schools in Ireland and that year's report of the Commissioners says that such education was

"leading to evil rather than good", "that such education *the people are actually obtaining for themselves,* and though we consider it practicable to correct it, *to check its progress appears impossible*: it may be improved *but it cannot be impeded*".

So then, these truly *national* schools did not suit English purposes. To suppress them, the *sham* national schools were established. They are *so little secular* that the reading-book consists of extracts both from the Cath. and Prot. Bibles, agreed upon by the Cath. and Prot. Archbishops of Dublin. Compare with these Irish peasants the English who howl at compulsory school-attendance to this day!

ENGELS TO EDUARD BERNSTEIN

March 12, 1881

On Ireland I shall only say the fòllowing: the people are much too clever not to know that a revolt would spell their ruin; it could have a chance only in the event of a war between England and America. In the meantime, the Irish have forced Gladstone to introduce continental regulations[329] in Parliament and thereby to undermine the whole British parliamentary system. They have also forced Gladstone to disavow all his phrases and to become more Tory than even the worst Tories. The coercion bills have been passed, the Land Bill will be either rejected or castrated by the House of Lords,[330] and then the fun will start, that is, the concealed disintegration of the parties will become public. Since Gladstone's appointment, the Whigs and moderate Tories, that is, the big landowners as a whole, are uniting on the quiet into a big landowners' party. As soon as this matures and family and personal interests are settled, or as soon as, perhaps as a result of the Land Bill, the new party is forced to appear in public, the Ministry and the present majority will immediately fall to pieces. The new conservative party will then be faced by the new bourgeois radical party, but without any backing other than the workers and Irish peasants. And so as to avoid any humbug and trickery from taking place here again, a proletarian radical party is now forming under the leadership of Joseph Cowen (M.P. for Newcastle), who is an old Chartist, half, if not entirely, Communist and a very worthy chap. Ireland is bringing all this about, Ireland is the driving force of the Empire. This is for your private information. More about this soon.

MARX TO JENNY LONGUET

April 11, 1881

Let Longuet read *Parnell*'s speech in Cork in *today's Times*; he will find in it the gist of *what should be said about Gladstone's new Land Act*; and one must not overlook the fact that by his disgraceful preliminary measures (including abolition of freedom of speech for members of the Lower House) Gladstone prepared the conditions under which *mass evictions are taking place in Ireland*, while the *Act* is only pure humbug, since the Lords, who get everything they want from Gladstone and no longer have to tremble before the Land League, will doubtless reject it or castrate it so that the Irish themselves will finally vote *against* it.

ENGELS TO EDUARD BERNSTEIN

April 14, 1881

Argyll's retirement from the Ministry because the Irish Land Bill gives the tenants a certain co-ownership of the land is a bad omen for the fate of the Bill in the Upper House. In the meantime Parnell has successfully begun his agitation tour of *England* in Manchester. The position of the big liberal coalition is becoming more and more critical.

Everything here seems to go slowly, but it is so much more thorough.

MARX TO JENNY LONGUET

April 29, 1881

It is a very fine trick of Gladstone—only the "stupid party" does not understand it—to offer at a moment when landed property in Ireland (as in England) will be depreciated by the import of corn and cattle from the U. St.—to offer them at that very moment the public Exchequer where they can sell that property at a price it does no longer possess!

The real intricacies of the Irish land problem—which indeed are not especially Irish—are so great that the only true way to solve it would be to give the Irish Home Rule and thus force them to solve it themselves. But John Bull is too stupid to understand this.

MARX TO JENNY LONGUET

December 7, 1881

The ever faithful Engels has sent you a number of the *Irish World* at my request, containing a *declaration against landownership* (private) by an Irish bishop. This was *the latest news* that I passed on to *your mamma* and she thought you could perhaps insert it in a French paper to frighten the *French clericals*. In any case, it shows that these gentlemen can pipe any tune.

MARX TO ENGELS

January 5, 1882

A different picture is presented by the 3,000 landlords meeting at Dublin, duce* Abercorn,[331] whose only purpose is "to maintain ... contracts and *the freedom between man*

---

* Under the leadership of.—*Ed.*

and man in this realm". Those fellows' rage over the Assistant Commissioners is funny. By the way, they are quite justified in their polemics against Gladstone, but it is only the coercitive measures of the latter and his 50,000 soldiers, apart from the police, that enable these gentlemen to oppose him in such a critical and threatening manner. The whole uproar naturally is meant only to prepare John Bull for the payment of "compensation costs". Serves him right.

ENGELS TO KARL KAUTSKY

February 7, 1882

One of the real tasks of the 1848 Revolution (and the *real*, not illusory tasks of a revolution are always solved as a result of that revolution) was the restoration of the oppressed and dispersed nationalities of Central Europe, insofar as these were at all viable and, especially, ripe for independence. This task was solved for Italy, Hungary and Germany, according to the then prevailing conditions, by the executors of the revolution's will, Bonaparte, Cavour and Bismarck. Ireland and Poland remained. Ireland can be disregarded here, she affects the conditions of the Continent only very indirectly. But Poland lies in the middle of the Continent and the conservation of her division is precisely the link that has constantly held the Holy Alliance together, and therefore, Poland is of great interest to us....

I therefore hold the view that *two* nations in Europe have not only the right but even the duty to be nationalistic before they become internationalistic: the Irish and the Poles. They are most internationalistic when they are genuinely nationalistic. The Poles understood this during all crises and have proved it on all the battlefields of the revolution. Deprive them of the prospect of restoring Poland or convince them that the new Poland will soon drop into their lap by herself, and it is all over with their interest in the European revolution.

ENGELS TO EDUARD BERNSTEIN

May 3, 1882

Don't let the Association[332] here deceive you about the Democratic Federation.[333] So far it is of no account whatever. It is headed by an ambitious candidate for Parliament by the name of Hyndman, an ex-Conservative, who can get together a big meeting only with the help of the Irish and for specifically Irish purposes. Even then he plays only a third-rate part, otherwise the Irish would give it to him.

Gladstone has discredited himself terribly. His whole Irish policy has suffered shipwreck. He has to drop Forster and the Lord Lieutenant of Ireland, Cowper-Temple (whose stepfather is Palmerston), and must say a *pater peccavi**: The Irish M.P.s**** have been set free, the Coercion Bill has not been extended, the back rents of the farmers are to be partly cancelled and partly taken over by the state against fair amortisation.[334] On the other hand the Tories have already reached the stage where they want to save whatever can still be saved: before the farmers *take* the land they should redeem the rents with the aid of the state, according to the Prussian model, so that the landowners may get at least *something*! The Irish are teaching our leisurely John Bull to get a move on. That's what comes from shooting![335]

ENGELS TO EDUARD BERNSTEIN[336]

June 26, 1882

In Ireland there are two trends in the movement. The first, the earlier, is the *agrarian* trend, which stems from the organised brigandage practised with support of the peasants by the clan chiefs, dispossessed by the English, and also by the big Catholic landowners (in the 17th century these brigands were called *Tories*, and the Tories of today have inherited their name directly from them). This trend gradually developed into natural resistance of the peasants to

---

* Father, I have sinned. An error seems to have crept in since the Lord Lieutenant of Ireland at the time was not William Cowper-Temple but his nephew Francis Cowper.—*Ed.*

** Parnell, Dillon, O'Kelly.—*Ed.*

the intruding English landlords, organised according to localities and provinces. The names Ribbonmen, Whiteboys, Captain Rock, Captain Moonlight, etc., have changed, but the form of resistance—the shooting not only of hated landlords and agents (rent collectors of the landlords) but also of peasants who take over a farm from which another has been forcibly evicted, boycotting, threatening letters, night raids and intimidation, etc.—all this is as old as the present English landownership in Ireland, that is, dates back to the end of the 17th century at the latest. This form of resistance cannot be suppressed, force is useless against it, and it will disappear only with the causes responsible for it. But, as regards its nature, it is *local, isolated,* and can never become a general form of *political* struggle.

Soon after the establishment of the Union (1800), began the *liberal-national* opposition of the *urban bourgeoisie* which, as in every peasant country with dwindling townlets (for example, Denmark), finds its natural leaders in *lawyers.* These also need the peasants; they therefore had to find slogans to attract the peasants. Thus *O'Connell* discovered such a slogan first in the *Catholic emancipation,* and then in the *Repeal of the Union.* Because of the infamy of the landowners, this trend has recently had to adopt a new course. While in the *social* field the *Land League* pursues more revolutionary aims (which are achievable in Ireland)—the total removal of the intruder landlords—it acts rather tamely in *political* respects and demands only Home Rule, that is, an Irish local Parliament side by side with the British Parliament and subordinated to it. This too can be achieved by constitutional means. The frightened landlords are already clamouring for the quickest possible redemption of the peasant land (suggested by the Tories themselves) in order to save what can still be saved. On the other hand, *Gladstone* declares that greater self-government for Ireland is quite admissible.

After the American Civil War, *Fenianism* took its place beside these two trends. The hundreds of thousands of Irish soldiers and officers, who fought in the war, did so with the ulterior motive of building up an army for the liberation of Ireland. The controversies between America and England after the war became the main lever of the Fenians. Had it come to a war, Ireland would in a few months have been part of the United States or at least a republic under its protection. The sum which England so willingly undertook to pay, and did indeed pay in accordance with Geneva arbitrators' decision on the Alabama affair,[337] was the *price she paid to buy off American intervention in Ireland.*

From this moment the main danger had been removed. The police was strong enough to deal with the Fenians. The treachery inevitable in any conspiracy also helped, and yet it was only *leaders* who were traitors and then became downright spies and false witnesses. The leaders who got away to America engaged there in emigrant revolution and most of them were reduced to beggary, like O'Donovan Rossa. For those who saw the European emigration of 1849-52 here, everything seems very familiar—only naturally on the exaggerated American scale.

Many Fenians have doubtless now returned and restored the old armed organisation. They form an important element in the movement and force the Liberals to more decisive action. But, apart from that, they cannot do anything but scare John Bull. Though he grows noticeably weaker on the outskirts of his Empire, he can still easily suppress any Irish rebellion so close to home. In the first place, in Ireland there are 14,000 men of the "Constabulary", gendarmes, who are armed with rifles and bayonets and have undergone military training. Besides, there are about 30,000 regulars, who can easily be reinforced with an equal number of regulars and English militia. In addition, the Navy. And John Bull is known for his matchless brutality in suppressing rebellions. *Without war or the threat of war from without, an Irish rebellion has not the slightest chance;* and *only two* powers can become dangerous in this respect: *France* and, still far more, the *United States.* France is out of the question. In America the parties flirt with the Irish electorate, make promises but do not keep them. They have no intention of getting involved in a war because of Ireland. They are even interested in having conditions in Ireland that promote a massive Irish emigration to America. And it is understandable that a land which in twenty years will be the most populated, richest and most powerful in the world has no special desire to rush headlong into adventures which could and would

hamper its enormous internal development. In twenty years
it will speak in a very different way.

However, if there should be danger of war with America,
England would grant the Irish open-handedly everything
they asked for—only not complete independence, which is
not at all desirable owing to the geographical position.

Therefore all that is left to Ireland is the constitutional
way of gradually conquering one position after the other;
and here the mysterious background of a Fenian armed conspiracy can remain a very effective element. But these
Fenians are themselves increasingly being pushed into a sort
of Bakuninism: the assassination of Burke and Cavendish[338]
could only serve the purpose of making a compromise between the Land League and Gladstone impossible. However,
that compromise was the best thing that could have happened
to Ireland under the circumstances. The landlords are evicting tens of thousands of tenants from their houses and homes
because of rent arrears, and that under military protection.
The primary need at the moment is to stop this systematic
depopulation of Ireland (the evicted starve to death or have
to emigrate to America). Gladstone is ready to table a bill
according to which arrears would be paid in the same way
as feudal taxes were settled in Austria in 1848: a third by
the peasant and a third by the state, and the other third forfeited by the landlord. That suggestion was made by the
Land League itself. Thus the "heroic deed" in Phoenix Park
appears if not as pure stupidity, then at least as pure Baku-
ninist, bragging, purposeless "*propagande par le fait*". If it
has not had the same consequences as the similar silly actions
of Hödel and Nobiling,[339] it is only because Ireland lies
not quite in Prussia. It should therefore be left to the Bakuninists and Mostians to attach equal importance to this childishness and to the assassination of Alexander II, and to
threaten with an "Irish revolution" which never comes.

One more thing should be thoroughly noted about Ireland:
never praise a single Irishman—a politician—unreservedly,
and never identify yourself with him before he is dead. Celtic
blood and the customary exploitation of the peasant (all the
"educated" social layers in Ireland, especially the lawyers,
live by this alone) make Irish politicians very responsive to
corruption. O'Connell let the peasants pay him as much as

£30,000 a year for his agitation. In connection with the Union, for which England paid out £1,000,000 in bribes, one of those bribed was reproached: "You have sold your motherland." Reply: "Yes, and I was damned glad to have a motherland to sell."

ENGELS TO EDUARD BERNSTEIN

August 9, 1882

You naturally presumed that, in view of our old friendship, Liebknecht had a perfect right to ask you to give him my letter,* and that you were obliged to give it to him. I can see nothing in that for me to complain about. You could not know that four-fifths of the many differences I have had with Liebknecht were due to such arbitrary actions on his part, to public misuse of private letters, to notes on my articles which were silly or directly contradictory to the meaning of the text, etc. This time too he has used my letter in an unjustifiable way. The letter was written with direct reference to your article. Liebknecht treated it as if it were *"my"* interpretation of the *entire* Irish question. That is terribly frivolous, particularly when speeches by Davitt are advanced against it, which *had not even been made* when the letter was written, and which have nothing to do with it, since Davitt with his state ownership of the land is so far only a *symptom*. But Liebknecht always acts frivolously when he wants to demonstrate his "superiority". I do not grudge him the fun, but he should not misuse my letters for that, and now *he* compels me to ask you in future (I want to express myself as correctly and diplomatically as possible) *de lui donner— tout au plus—lecture de mes lettres sans cependant lui abandonner l'original ni lui en laisser copie.***

---

* See preceding letter.—*Ed.*
** To give him my letters to read, at the very most, without, however, leaving him the original or a copy.—*Ed.*

Frederick Engels

THE ORIGIN OF THE FAMILY, PRIVATE PROPERTY AND THE STATE

FROM CHAPTER VII
THE GENS WITH CELTS AND GERMANS

The oldest Celtic laws that have come down to our day show the gens still in full vitality. In Ireland it is alive, at least instinctively, in the popular mind to this day, after the English forcibly blew it up. It was still in full bloom in Scotland in the middle of the last century, and here, too, it succumbed only to the arms, laws and courts of the English.

The old Welsh laws, written several centuries before the English Conquest,[340] not later than the eleventh century, still show communal field agriculture of whole villages, although only as exceptions and as the survival of a former universal custom. Every family had five acres for its own cultivation; another plot was at the same time cultivated in common and its yield divided. Judging by the Irish and Scotch analogies there cannot be any doubt that these village communities represent gentes or subdivisions of gentes, even though a reinvestigation of the Welsh laws, which I cannot undertake for lack of time (my notes are from 1869[341]), should not directly corroborate this. The thing, however, that the Welsh sources, and the Irish, do prove directly is that among the Celts the pairing family had not yet given way by far to monogamy in the eleventh century. In Wales, marriage did not become indissoluble, or rather did not cease to be subject to notice of dissolution, until after seven years. Even if only three nights were wanting to make up the seven years, a married couple could still separate. Then their property was divided between them: the woman divided, the man made his choice. The furniture was divided according to certain very funny rules. If the marriage was dissolved by the man, he had to return the woman's dowry and a few other articles; if the woman desired a separation, she received less. Of the children the man was given two, the woman one, namely, the middle child. If the woman married again after her divorce, and her first husband fetched her back, she was obliged to follow him, even if she already had *one* foot in her new husband's bed. But if two people had lived together for seven years, they were considered man and wife, even without the preliminaries of a formal marriage. Chastity among girls before marriage was by no means strictly observed, nor was it demanded; the regulations governing this subject are of an extremely frivolous nature and run counter to all bourgeois morals. When a woman committed adultery, her husband had a right to beat her—this was one of three cases when he could do so without incurring a penalty—but after that he could not demand any other redress, for

"the same offence shall either be atoned for or avenged, but not both".

The reasons that entitled a woman to a divorce without detriment to her rights at the settlement were of a very diverse nature: the man's foul breath was a sufficient reason. The redemption money to be paid to the tribal chief or king for the right of the first night (*gobr merch*, hence the medieval name *marcheta*, French *marquette*) plays a conspicuous part in the legal code. The women had the right to vote at the popular assemblies. Add to this that similar conditions are shown to have existed in Ireland; that time marriages were also quite the custom there, and that the women were assured of liberal and well-defined privileges in case of separation, even to the point of remuneration for domestic services; that a "first wife" existed by the side of others, and in dividing a decedent's property no distinction was made between legitimate and illegitimate children—and we have a picture of the pairing family compared with which the form of marriage valid in North America seems strict; but this is not surprising in the eleventh century for a people which in Caesar's time was still living in group marriage.

The Irish gens (*sept*; the tribe was called *clainne*, clan) is confirmed and described not only by the ancient law-books, but also by the English jurists of the seventeenth century who were sent across for the purpose of transforming the clan lands into domains of the King of England. Up to this time, the land had been the common property of the clan or gens, except where the chiefs had already converted it into their private domain. When a gentile died, and a household was thus dissolved, the gentile chief (called *caput cognationis* by the English jurists) redistributed the whole gentile land among the other households. This distribution must in general have taken place according to rules such as were observed in Germany. We still find a few villages—very numerous forty or fifty years ago—with fields held in so-called *rundale*. Each of the peasants, individual tenants on the soil that once was the common property of the gens but had been seized by the English conquerors, pays rent for his particular plot, but all the arable and meadow land is combined and shared out, according to situation and quality, in strips, or "*Gewanne*," as they are called on the Mosel, and each one receives a share of each *Gewann*. Moorland and pastures are used in common. As recently as fifty years ago, redivision was still practised occasionally, sometimes annually. The map of such a *rundale* village looks exactly like that of a German community of farming households [*Gehöferschaft*] on the Mosel or in the Hochwald. The gens also survives in the "factions". The Irish peasants often form parties that seem to be founded on absolutely absurd and senseless distinctions and are quite incomprehensible to Englishmen. The only purpose of these factions is apparently to rally for the popular sport of solemnly beating the life out of one another. They are artificial reincarnations, later substitutes for the blasted gentes that in their own peculiar way demonstrate the continuation of the inherited gentile instinct. Incidentally, in some localities members of the same gens still live together on what is practically their old territory. During the thirties, for instance, the great majority of the inhabitants of the county of Monaghan had only four family names, that is, were descended from four gentes, or clans.*

The downfall of the gentile order in Scotland dates from the suppression of the rebellion of 1745.[343] Precisely what link in this order the Scotch clan represents remains to be investigated; no doubt it is a link. Walter Scott's novels bring the clan in the Highlands of Scotland vividly before our eyes. It is, as Morgan says,

"an excellent type of the gens in organisation and in spirit, and an extraordinary illustration of the power of the gentile life over its members.... We find in their feuds and blood revenge, in their localisation by gentes, in their use of lands in common, in the fidelity of the clansman to his chief and of the members of the clan to each other, the usual and persistent features of gentile society.... Descent was in the male line, the children of the males remaining members of the clan, while the children of its female members belonged to the clans of their respective fathers".[344]

The fact that mother right was formerly in force in Scotland is proved by the royal family of the Picts, in which, according to Beda, inheritance in the female line prevailed.[345] We even see evidences of the punaluan family

---

* During a few days that I spent in Ireland,[342] I again realised to what extent the rural population there is still living in the conceptions of the gentile period. The landlord, whose tenant the peasant is, is still considered by the latter as a sort of clan chief who supervises the cultivation of the soil in the interest of all, is entitled to tribute from the peasant in the form of rent, but also has to assist the peasant in cases of need. Likewise, everyone in comfortable circumstances is considered under obligation to help his poorer neighbours whenever they are in distress.

Such assistance is not charity; it is what the poor clansman is entitled to by right from his rich fellow clansman or clan chief. This explains why political economists and jurists complain of the impossibility of inculcating the modern idea of bourgeois property into the minds of the Irish peasants. Property that has only rights, but no duties, is absolutely beyond the ken of the Irishman. No wonder so many Irishmen with such naïve gentile conceptions, who are suddenly cast into the modern great cities of England and America, among a population with entirely different moral and legal standards, become utterly confused in their views of morals and justice, lose all hold and often succumb to demoralisation in masses. [*Note by Engels to the 1891 edition.*]

preserved among the Scots as well as the Welsh until the Middle Ages in the right of the first night, which the chief of the clan or the king, the last representative of the former common husbands, could claim with every bride, unless redeemed.

Published in the book:
Friedrich Engels, *Der Ursprung der Familie, des Privateigenthums und des Staats,* Höttingen-Zürich, 1884

Printed according to the fourth edition published in Stuttgart in 1891
Translated from the German

From AN INTERVIEW WITH ENGELS PUBLISHED IN THE "NEW YORKER VOLKSZEITUNG"[346]

Question: What about Ireland? Is there anything—apart from the national question—which might raise the hopes of socialists?

Engels: *A purely socialist movement cannot be expected in Ireland for a considerable time.* People there want first of all to become peasants owning a plot of land, and after they have achieved that mortgages will appear on the scene and they will be ruined once more. But this should not prevent us from seeking to help them to get rid of their landlords, that is, to pass from semi-feudal conditions to capitalist conditions.

Question: What is the attitude of the English workers towards the Irish movement?

Engels: The masses are *for* the Irish. The organisations, and the labour aristocracy in general, follow Gladstone and the liberal bourgeois and do not go further than these.

Published in
New Yorker Volkszeitung
No. 226, September 20, 1888

Translated from the German

Frederick Engels

From THE PREFACE TO THE ENGLISH EDITION OF "THE CONDITION OF THE WORKING-CLASS IN ENGLAND"[347]

Again, the repeated visitations of cholera, typhus, small-pox, and other epidemics have shown the British bourgeois the urgent necessity of sanitation in his towns and cities, if he wishes to save himself and family from falling victims to such diseases. Accordingly, the most crying abuses described in this book have either disappeared or have been made less conspicuous. Drainage has been introduced or improved, wide avenues have been opened out athwart many of the worst "slums" I had to describe. "Little Ireland"* has disappeared, and the "Seven Dials"[348] are next on the list for sweeping away. But what of that? Whole districts which in 1844 I could describe as almost idyllic, have now, with the growth of the towns, fallen into the state of dilapidation, discomfort, and misery. Only the pigs and the heaps of refuse are no longer tolerated. The bourgeoisie have made further progress in the art of hiding the distress of the working-class. But that, in regard to their dwellings, no substantial improvement has taken place, is amply proved by the Report of the Royal Commission "on the Housing of the Poor", 1885. And this is the case, too, in other respects. Police regulations have been plentiful as blackberries; but they can only hedge in the distress of the workers, they cannot remove it. . . .

Free Trade meant the readjustment of the whole home and foreign, commercial and financial policy of England in accordance with the interests of the manufacturing capitalists—the class which now represented the nation. And

---

* See pp. 38-39.—*Ed.*

they set about this task with a will. Every obstacle to industrial production was mercilessly removed. The tariff and
the whole system of taxation were revolutionised. Everything was made subordinate to one end, but that end of the
utmost importance to the manufacturing capitalist: the
cheapening of all raw produce, and especially of the means
of living of the working-class; the reduction of the cost of
raw material, and the keeping down—if not as yet the
bringing down—of wages. England was to become the
"workshop of the world"; all other countries were to become
for England what Ireland already was—markets for her
manufactured goods, supplying her in return with raw
materials and food. England, the great manufacturing centre
of an agricultural world, with an ever-increasing number
of corn- and cotton-growing Irelands revolving around her,
the industrial sun. What a glorious prospect!

Published in the book:
F. Engels, *The Condition
of the Working-Class in
England in 1844*,
London, 1892

Printed according to the
text of the book

From THE PEASANT QUESTION IN FRANCE AND GERMANY[349]

The bourgeois and reactionary parties greatly wonder why everywhere among socialists the peasant question has now suddenly been placed upon the order of the day. What they should be wondering at, by rights, is that this has not been done long ago. From Ireland to Sicily, from Andalusia to Russia and Bulgaria, the peasant is a very essential factor of the population, production and political power. Only two regions of Western Europe form an exception. In Great Britain proper big landed estates and large-scale agriculture have totally displaced the self-supporting peasant; in Prussia east of the Elbe the same process has been going on for centuries; here too the peasant is being increasingly "turned out" or at least economically and politically forced into the background.

The peasant has so far largely manifested himself as a factor of political power only by his apathy, which has its roots in the isolation of rustic life. This apathy on the part of the great mass of the population is the strongest pillar not only of the parliamentary corruption in Paris and Rome but also of Russian despotism. Yet it is by no means insuperable. Since the rise of the working-class movement in Western Europe, particularly in those parts where small peasant holdings predominate, it has not been particularly difficult for the bourgeoisie to render the socialist workers suspicious and odious in the minds of the peasants as *partageux*, as people who want to "divide up", as lazy greedy city dwellers who have an eye on the property of the peasants. The hazy socialistic aspirations of the Revolution of February 1848 were rapidly disposed of by the reactionary ballots of the French peasantry; the peasant, who wanted peace of mind, dug up from his treasured memories the legend of Napoleon, the emperor of the peasants, and created the Second Empire. We all know what this one feat of the peasants cost the people of France; it is still suffering from its aftermath.

But much has changed since then. The development of the capitalist form of production has cut the life-strings of small production in agriculture; small production is irretrievably going to rack and ruin. Competitors in North and South America and in India, too, have swamped the European market with their cheap grain, so cheap that no domestic producer can compete with it. The big landowners and small peasants alike see ruin staring them in the face. And since they are both owners of land and country folk, the big landowners assume the role of champions of the interests of the small peasants, and the small peasants by and large accept them as such.

Meanwhile a powerful socialist workers' party has sprung up and developed in the West. The obscure presentiments and feelings dating back to the February Revolution have become clarified and acquired the broader and deeper scope of a programme that meets all scientific requirements and contains definite tangible demands; and a steadily growing number of socialist deputies fight for these demands in the German, French and Belgian parliaments. The conquest of political power by the Socialist Party has become a matter of the not too distant future. But in order to conquer political power this party must first go from the towns to the country, must become a power in the countryside. This party, which has an advantage over all others in that it possesses a clear insight into the interconnections between economic causes and political effects and long ago descried the wolf in the sheep's clothing of the big landowner, that importunate friend of the peasant—may this party calmly leave the doomed peasant in the hands of his false protectors until he has been transformed from a passive into an active opponent of the industrial workers? This brings us right into the thick of the peasant question.

Published in
Die Neue Zeit, Bd. 1, 10,
Stuttgart, 1894-95

Translated from the German

Frederick Engels

EXCERPTS FROM LETTERS ON THE IRISH QUESTION WRITTEN BETWEEN 1885 AND 1894

TO WILHELM LIEBKNECHT

December 1, 1885

The elections here are proceeding very nicely.[350] It is the first time that the Irish in England have voted en masse for *one* side, and in fact for the Tories. They have thus shown the Liberals the extent to which they can decide the issue even in England. The 80 to 85 Home Rulers—Liverpool, too, has elected one—who occupy the same position here as the Centre Party does in the Reichstag,[351] can wreck any government. Parnell must now show what he really is.

Incidentally, a victory has also been won by the *new* Manchester School,[352] that is, the theory of aggressive tariffs, although it is here even more absurd than in Germany, but after eight years of commercial stagnation the idea has taken possession of the young manufacturers. Then there is Gladstone's opportunist weakness and the clumsy manner of Chamberlain, who first throws his weight about and then draws in his horns; this has called forth the cry: the Church in danger! Finally, Gladstone's lamentable foreign policy. The Liberals profess to believe that the new county voters will vote for them. There is, indeed, no telling how these voters will act, but in order to obtain an absolute majority the Liberals would have to win 180 of the 300 still outstanding districts, and that will hardly happen. Parnell will almost certainly wield dictatorial powers in Great Britain and Ireland.

TO JOHANN PHILIPP BECKER

December 5, 1885

The elections in France placed the Radicals next in the running for control, thereby improving our prospects a good deal, too. The elections here have temporarily made the Irish masters of England and Scotland, for not one of the two parties can rule without them. Though the results in nearly 100 seats are not yet known they will change little. Thus the Irish problem will at last be settled, if not immediately then in the near future, and then the way will have been cleared there, too. At the same time some eight to ten workers have been elected—some are bought by the bourgeoisie, others are strict trade-unionists. They will probably make fools of themselves and hence greatly advance the formation of an independent labour party by destroying the traditional self-deception of the workers. Here history moves slowly, but it moves.

TO EDUARD BERNSTEIN

May 22, 1886

I am sending you Thursday's Parliamentary debates (*Daily News*) on the Irish Arms Bill, which restricts the right of the Irish to own and carry arms. Hitherto it was directed only against the nationalists, but now it is to be turned also against the Protestant braggarts of Ulster, who threaten to rebel.[353] There is a remarkable speech by Lord Randolph Churchill, the brother of the Duke of Marlborough, a democratising Tory; in the last Tory cabinet he was Secretary for India and is thus a member of the Privy Council for life. In face of the feeble and cowardly protestations and assurances made by our petty-bourgeois socialists regarding the peaceful attainment of the goal under any circumstances, it is indeed very timely to show that English ministers, Althorp, Peel, Morley and even Gladstone, proclaim the right to revolution as a part of constitutional theory—though only *so long as they form the opposition*, as Gladstone's subsequent twaddle proves, but even then he does not dare to deny the right as such—especially because it comes from England, the country of legality *par excellence*. A more telling repudiation could hardly be found for our Vierecks.

TO FRIEDRICH ADOLF SORGE

June 18, 1887

Yesterday evening the Irish Coercion Bill was clause by clause hurried through the House of Commons in two minutes.[354] It is a worthy counterpart of the Anti-Socialist Law and opens the door to completely arbitrary action by the police. Things regarded as fundamental rights in England are forbidden in Ireland and become crimes. This Bill is the tombstone of today's Tories, whom I did not consider so stupid, and of the Liberal Unionists,[355] whom I hardly thought so contemptible. It is moreover intended, not to last for a limited period, but indefinitely. The British Parliament has been reduced to the level of the German Reichstag. Though certainly not for long.

TO FLORENCE KELLY-WISCHNEWETZKY

February 22, 1888

The stupidity of the present Tory government is appalling—if old Disraeli was alive, he would box their ears right and left. But this stupidity helps on matters wonderfully. Home Rule for Ireland *and for London* is now the cry here; the latter a thing which the Liberals fear even more than the Tories do. The working class element is getting more and more exasperated, through the stupid Tory provocations, is getting daily more conscious of its strength at the ballot-box, and more penetrated by the socialist leaven.

TO WILHELM LIEBKNECHT

February 29, 1888

Have heard nothing of the Irish tricolour to which you refer. Irish flags in Ireland and here are simply green with a golden harp, but *without a crown* (in the British coat-of-arms the harp wears a crown). In the Fenian days, 1865-67, many were green and orange to show the Orangemen of the North[356] that they would not be destroyed, but accepted as brothers. However, no question of that any more.

TO FRIEDRICH ADOLF SORGE

December 7, 1889

I hope the next general election will be deferred for another three years—1. So that during the period of the greatest war danger Gladstone, the lackey of the Russians, should not be at the head of affairs; this might already be a sufficient reason for the Tsar* to provoke a war. 2. So that the anti-Conservative majority becomes so large that *real* Home Rule for Ireland becomes a necessity, otherwise Gladstone will cheat the Irish again, and this obstacle—the Irish question—will not be cleared away. 3. However, so that the labour movement may develop further and perhaps mature more rapidly as a result of the set-back caused by the business recession which will certainly follow the present period of prosperity. The next parliament may then comprise 20 to 40 labour deputies, and moreover of a very different kind from the Potters, Cremers and Co.

TO AUGUST BEBEL

January 23, 1890

I see no reason why we should *not* repay the Progressists for their infamous behaviour of 1887[357] and bring it home to them that they exist by our grace only. Parnell's decision of 1886 that the Irish in England should all vote against the Liberals, for the Tories, that is, for the first time since 1800 stop being a herd voting for the Liberals, transformed Gladstone and the Liberal chiefs into Home Rulers

---

* Alexander III.—*Ed.*

in a matter of six weeks.[358] *If* anything can still be made out of the Progressists, then only by showing them in the by-elections *ad oculos* that they are dependent on us.

TO NIKOLAI FRANTSEVICH DANIELSON

June 10, 1890

Here in England, *Rent* is applied as well to the payment of the English capitalist farmer to his landlord, as to that of the Irish pauper farmer, who pays a complete tribute composed chiefly of a deduction from his fund of maintenance, earned by his own labour, and only to the smallest extent consisting of true rent. So the English in India transformed the land-tax paid by the ryot (peasant) to the State into "rent", and consequently have, in Bengal at least, actually transformed the zemindar (tax-gatherer of the former Indian prince) into a landlord holding a nominal feudal tenure from the Crown exactly as in England, where the Crown is nominal proprietor of all the land, and the great nobles, the real owners, are by juridical fiction supposed to be feudal tenants of the Crown. Similarly when in the beginning of the 17th century the North of Ireland was subjected to direct English dominion, and the English lawyer Sir John Davies found there a rural community with common possession of the land, which was periodically divided amongst the members of the clan who paid a tribute to the chief, Davies <transformed> declared that tribute at once <into> to be "rent". Thus the Scotch lairds—chiefs of clans—profited, since the insurrection of 1745, of this juridical confusion, of the tribute paid to them by the clansmen, with a *"rent"* for the lands held by them, in order to transform the whole of the <common> clan-land, the common property of the clan, into their, the lairds, private property; for—said the lawyers, if they were not the landlords, how could they receive *rent* for that land? And thus this confusion of tribute and rent was the basis of the confiscation of all the lands of the Scottish Highlands for the benefit of a few chiefs of clan who very soon after drove out the old clansmen and replaced them by sheep as described in *C[apital]* chapter 24,3/ (p. 754, 3-rd edit[ion]).

TO FRIEDRICH ADOLF SORGE

February 11, 1891

The gasworkers now have the most powerful organisation in Ireland[359] and will put up their own candidates in the next election, unconcerned over either Parnell or MacCarthy. That Parnell is now so friendly with the workers, he owes to encounters with these same gasworkers, who had no compunctions about telling him the truth. Michael Davitt, too, who had at first wanted independent Irish Trades Unions, has learned from them: their constitution secures them perfectly free home rule. To them the credit for giving impetus to the labour movement in Ireland. Many of their branches consist of agricultural labourers.

TO FRIEDRICH ADOLF SORGE

August 11, 1891

Tussy's report to the Brussels Congress on behalf of the gasworkers and others, is very good. I shall send it to you. Tussy is going to Brussels with a mandate from the Dublin Congress of Gasworkers and General Labourers, thus representing 100,000.[360] Aveling, too, has 3 or 4 mandates. To all appearances, the *old* Trades Unions will be poorly represented. So much the better this time!

TO FRAU LIEBKNECHT

December 2, 1891

Nothing particularly new; Tussy has the not entirely undeserved reputation of being the leader of the Union of Gasworkers and General Labourers, and was away to agitate eight days in Northern Ireland the week before last. These gasworkers are fine fellows, their Union by far the most progressive; they are so good at "legal" agitation that eighteen months ago in Leeds they won two real battles—first against the police and then against the police and dragoons—forcing the municipality, which owns the gasworks, to capitulate.[361] As an old soldier, I can certify that I find no fault either in the strategic or tactical dispositions of Will Thorne, the General Secretary of the Union, who was in command in these battles.

TO HERMANN SCHLÜTER

March 30, 1892

Your great obstacle in America, it seems to me, lies in the exceptional position of the native workers. Up to 1848 one could only speak of the permanent native working class as an exception: the small beginnings of it in the cities in the East always had still the hope of becoming farmers or bourgeois. Now a working class has developed and has also to a great extent organised itself on trade union lines. But it still takes up an aristocratic attitude and wherever possible leaves the ordinary badly paid occupations to the immigrants, of whom only a small section enter the aristocratic trades. But these immigrants are divided into different nationalities and understand neither one another nor, for the most part, the language of the country. And your bourgeoisie knows much better even than the Austrian Government how to play off one nationality against the other: Jews, Italians, Bohemians, etc., against Germans and Irish, and each one against the other, so that differences in the standard of life of different workers exist, I believe, in New York to an extent unheard-of elsewhere. And added to this is the total indifference of a society which has grown up on a purely capitalist basis, without any comfortable feudal background, towards the human beings who succumb in the competitive struggle: "there will be plenty more, and more than we want, of these damned Dutchmen,* Irishmen, Italians, Jews and Hungarians"; and, to cap it all, John Chinaman** stands in the background who far surpasses them all in his ability to live on next to nothing.

---

* In the U.S.A. this was applied to the Germans.—*Ed.*
** A nickname for the Chinese used in the U.S.A.—*Ed.*

TO NIKOLAI FRANTSEVICH DANIELSON

June 18, 1892

Everything necessary to keep farm labourers just alive during the winter is frequently earned by women and children working in some new branch of domestic industry (see *Capital*, Vol. I, Chapter XV, Section 8, d). This is the case in southern and western England and among the small farmers of Ireland and Germany. The devastating consequences of the separation of agriculture from domestic industry carried on in the patriarchal manner are particularly marked during the transition period, and this is happening just now in your country.

TO AUGUST BEBEL

July 7, 1892

In brief, the Labour Party has declared itself clearly and unequivocally,[362] meaning that in the next election the two old parties will offer it alliance. The Tories are out of the question so long as they are led by the present dolts. But the Liberals must be considered, and likewise the Irish. Since the public outcry for that ridiculous business with adultery,[363] Parnell has suddenly become friendly to the workers, and the Irish gentlemen in Parliament will follow suit once they see that only the workers can get them Home Rule. Then there will be compromises, and the Fabians,[364] conspicuous by their absence in this election, will come forward again. But that is unavoidable in the circumstances. There is headway, as you see, and that is what matters.

TO AUGUST BEBEL

January 24, 1893

What Aveling told me confirms the suspicion I already had, namely, that Keir Hardie secretly cherishes the wish to lead the new party in a dictatorial way, just as Parnell led the Irish, and that moreover he tends to sympathise with the Conservative Party rather than the Liberal opposition. He publicly declares that Parnell's experiment, which compelled Gladstone to give in, ought to be repeated at the next election and where it is impossible to nominate a Labour candidate one should vote for the Conservatives, in order to show the Liberals the power of the party. Now this is a policy which under definite circumstances I myself recommended to the English; however, if at the very outset one does not announce it as a possible tactical move but proclaims it as tactics to be followed *under any circumstances*, then it smells strongly of Champion.

TO FRIEDRICH ADOLF SORGE

November 10, 1894

Anglo-Saxon sectarianism prevails in the labour movement, too. The Social-Democratic Federation, just like your German Socialist Labour Party,[365] has managed to transform our theory into the rigid dogma of an orthodox sect; it is narrow-mindedly exclusive and thanks to Hyndman has a thoroughly rotten tradition in international politics, which is shaken from time to time, to be sure, but which has not been broken with as yet. The Independent Labour Party is extremely indefinite in its tactics, and its leader, Keir Hardie, is a supercunning Scot, whose demagogic tricks are not to be trusted for a minute. Although he is a poor devil of a Scottish coal miner, he has founded a big weekly, *The Labour Leader*,[366] which could not have been established without considerable money, and he is getting this money from Tory or Liberal-Unionist, that is, anti-Gladstone and anti-Home Rule sources. There can be no doubt about this, and his notorious literary connections in London as well as direct reports and his political attitude confirm it. Consequently, owing to desertions by Irish and radical voters,[367] he may very easily lose his seat in Parliament at the 1895 general elections and that would be a stroke of good luck— the man is the greatest obstacle at present. He appears in Parliament only on demagogic occasions, in order to cut a figure with phrases about the unemployed—without getting anything done—or to address imbecilities to the Queen* on

---

* Victoria.—*Ed.*

the occasion of the birth of a prince, which is infinitely banal and cheap in this country, and so forth. Otherwise there are very good elements both in the Social-Democratic Federation and in the Independent Labour Party, especially in the provinces, but they are scattered; yet they have at least managed to foil all the efforts of the leaders to incite the two organisations against each other.

SUPPLEMENT

THE IRISH STATE PRISONERS.
SIR GEORGE GREY
AND THE INTERNATIONAL WORKING MEN'S
ASSOCIATION[368]

Some weeks ago Mr. J. Pope Hennessy addressed the following communication to the editor of the *Pall Mall Gazette*:

Sir,—It appears that the *Pall Mall Gazette* has thrown the Home Office into a state of vigorous activity. It is currently reported that Sir George Grey and other members of the Government have within the last few days been seen in the almost impenetrable disguise of practical and zealous citizens looking into casual wards and night refuges. Now, if this be so, I would ask you to let me point out to the transformed officials of the Home Office a rather gloomy institution where a visit or two might not be thrown away— I mean the convict prison at Pentonville. Nor should the visitors consist only of Sir George Grey and his secretaries. Pentonville has at present (or ought to have) a peculiar interest for Lord Russell and Mr. Gladstone. The political prisoners recently convicted in Ireland are undergoing within its walls the severest form of discipline next to death known to the English law—the Pentonville separate system. It is on behalf of these political prisoners especially that I venture to suggest some kind [of] inquiry. It must be admitted that Lord Russell and Mr. Gladstone in their remonstrances on the treatment of political prisoners were not always as temperate in their language as eminent statesmen in these days are expected to be. The principle they laid down, that political offenders should not be treated in all respects like common convicts, was sound enough; though to characterise the violation of that principle as a "breach of all moral law", as an "abominable persecution", as "a savage and cowardly system", was going a little too far. In borrowing Lord Russell's and Mr. Gladstone's principle, I therefore disclaim all connection with the rather violent phraseology in which they thought fit to enforce it. One reason for being somewhat more moderate than they were is self-evident. They were exposing the misconduct of foreign governments; I am endeavouring to correct the misconduct of a government in which these benevolent champions of imprisoned politicians are highly responsible members. It would be ungenerous to turn their own weapons against such champions in such a cause. Therefore, without borrowing any of the warm and indignant invectives of Lord Russell and Mr. Gladstone, I simply charge them with being parties to a breach of that well-known principle they have embodied in so many dispatches, speeches, and letters—that political convicts should not be treated like common convicts; and I also charge the present administration with treating the Irish political prisoners so severely that probably some of them will go mad. In Mr. Gladstone's famous letter to Lord Aberdeen (p. 31) he says:

I had heard that the political offenders were obliged to have their heads shaved; but this had not been done, though they had been obliged to shave away any beard they might have had. I must say I was astonished at the mildness with which they spoke of those at whose hands they were enduring these abominable persecutions.

Not many days ago Mr. Gladstone might have read how the political offenders in Ireland half an hour after they were sentenced had their heads closely cropped, their beards and whiskers shaved off; how they were then stripped of their ordinary clothes, put into the convict dress, handcuffed, and sent off to Pentonville.

"In thirty minutes," said a Government organ describing the operation, "they were so changed that their dearest friends could hardly recognise them."

In another part of his pamphlet Mr. Gladstone describes the unhappy condition of the political prisoners confined in the Bagno of Nisida after their sentence:

For one half-hour of the week, a little prolonged by the leniency of the superintendent, they were allowed to see their friends outside the prison. This was the sole view of the natural beauties with which they were surrounded. At other times they were exclusively within the walls.—P. 29.

About a fortnight ago an Irish magistrate applied to the Home Office for permission to see the political prisoners now in England. Sir George Grey refused his application on the ground that for the first six months no stranger whatever can be allowed to visit a convict undergoing the separate system at Pentonville. What is the separate system of Pentonville? It is very unlike the system so eloquently exposed by Mr. Gladstone. The prisoners are not "allowed to see their friends outside the prison", nor are they allowed to see them inside the prison; nor are they allowed to see each other. Each prisoner has a solitary world of his own, thirteen feet by seven. A portion of this cell is occupied by a water-closet, and within two yards of this he takes his solitary meals, performs his solitary task work, and rests at night. If he omits to scrub and clean out his cell every morning, or if he breaks any other law of his little world, the directors can order him to be flogged, and put on bread and water for twenty-eight days in another little world where there is no light. What is the effect of this separate system? The Blue Books of the recent Royal Commission on Transportation and Penal Servitude give us the latest and most accurate information on the subject. Sir Joshua Jebb in his evidence speaks of what he calls "the serious physical effects" of the Pentonville separate system.

When the prisoners were embarked in ships in order to go to Van Diemen's Land,[369] a number of them fell into fits, and it was only by associating them for a fortnight or so before they left Pentonville that these fits ceased on embarkation.

Earl Grey: The suddenness of the change I suppose had that effect?—Yes. The medical men could not account for them; the fits were of an anomalous character.

Sir John Pakington: What was the nature of the fits?—The medical superintendent was in dismay. He had never seen anything of the kind before. They were very peculiar.

Sir John Pakington: Did the fits affect the health of the men afterwards?—The men got better afterwards; but they were reported to be very quiet. There is reason to believe that the effect was produced by the strictness of the separation.—P. 18.

Sir John Pakington will find in Judge Therry's *Reminiscences of New South Wales* (1863) a further answer to this question. The only English convict prison to which Judge Therry refers is Pentonville. "It in a great degree unfitted them (the discharge of convicts) for domestic and general service. It imparted to them abstracted and eccentric habits." The medical profession were of opinion that the system "had seriously impaired the mental faculties of several of the Penton*villains*, as they were termed."—(P. 354.) The present practice is to send the prisoners at the termination of the Pentonville system, to Chatham or to Portland to work in gangs with other convicts. This is called letting them into the world again. It is then that the full effect of Pentonville upon the mental faculties becomes manifest. Mr. Measor, the Deputy Governor of Chatham, in his evidence before the Royal Commission, says: "I have observed when they come down to the public works' prisons that they are in a very flabby condition of mind, and a very flabby condition physically, and I believe it (the Pentonville system) produces both effects." He is asked, "You are able to state this from your own experience?" He answers:

Yes. I have seen men who have come from separate confinement to whom I should be sorry to talk upon any subject with the expectation of getting any reasonable view from them. They appear as if they had been undergoing something which had so utterly depressed their system that you would no more think of treating them as reasonable beings, capable of being strongly remonstrated with, than you would a man who was almost at the door of death. (Vol. ii, p. 446.)

The proportion of those who are driven permanently insane by the Pentonville system is by no means small. The annual report of the Directors of Convict Prisons for the same year (1863) that the Deputy Governor of Chatham gave his evidence contains a table showing the number of convicts arriving at Chatham in twelve months, and the numbers transferred from Chatham in twelve months. From this table (p. 222) the following figures are taken. They confirm Mr. Measor's evidence, though they tell a more precise and painful tale:

                                        No. of Convicts
Received into Chatham convict prison
    since the 1st of January 852
Transferred to Millbank   1
    "   to Dartmoor   2
    "   to Woking   26
    "   to Broadmoor Lunatic Asylum   85

It is only fair to state, though, that this proportion of persons who are made mad by the separate system [is smaller than when this system] was carried out with greater severity. In the same report from which these figures are taken there is a statement of the medical officer of Pentonville, in which he remarks that "since 1859 the separate system has assumed a milder character, and for the last triennial period the insanity is less than any previous one, and the suicidal cases are also less."—(P. 29.) It is evident that the Pentonville system breaks down the mind, and that the number of those who are rendered absolutely insane is in direct proportion to the severity of the treatment. After such facts, it is hardly worth while mentioning that the dietary at Pentonville is lower than in any other convict prison (Vol. i, p. 274, of the Royal Commission). In short, confinement in Pentonville is the severest punishment, except death, allowed by the law. I am not certain that I ought to say "except death", for I find the Protestant chaplain in his report saying that any one who thinks a convict is petted would change his opinion if he could visit Pentonville and "behold (a specimen of the sterner type of treatment here) a ruffian now under a sentence of life penal servitude for a savage assault committed in another prison, and ready to imbrue his hands here in the blood of any one who might come helplessly within his reach, glad to exchange his present state for the gallows."—(P. 117.) If a convict attempts to kill a warder at Portland or any other convict prison, he is punished by being sent to Pentonville. There he is left till he dies, or sent in a strait-jacket to Broadmoor. Whether those members of the Government who made themselves so very busy about political prisoners abroad will trouble themselves about political prisoners at home one can hardly venture to guess. Mr. Gladstone has before now changed his opinions, and so has Lord Russell.

But this much I think may safely be said, that the people of England will not approve of condemning political prisoners in this country to the Pentonville separate system.

I am, Sir, your faithful servant,

J. POPE HENNESSY

1, Paper-buildings, Temple,
February 2, 1866

This letter having fallen under the eye of a member of the Central Council of the International Working Men's Association,* he communicated with the wife of one of the State prisoners, and learnt from her these facts, viz., that the State prisoners now confined in Pentonville Prison were removed thither on December 23, 1865; that only one letter on either side was allowed to pass between the prisoner and his wife during the first six-months term of this mode of incarceration, and that a relaxation of this cruel rule would be a great boon to the prisoner and a consolation to his suffering family.

When these facts were laid before the Central Council of the International Working Men's Association, that body whose leading principle it is to appease national animosities and to encourage a sentiment of international fraternity— that body, which lamented the long-standing feud between English and Irish nations, and could see only a new source of hatred between the two nations in the event of the reduction to a state of mental imbecility of the Irish State prisoners—thought it its duty to take the matter into its serious consideration.

The Central Council, after full deliberation, resolved to ask Sir George Grey to receive a deputation, consisting entirely of English and Scottish members, whose prayer to the Home Secretary should be to take care of the *mental* health of the State prisoners, and in particular to allow of a more frequent correspondence between the prisoners and their nearest and dearest relatives. The aim of the Council, in resolving that the proposed deputation should consist exclusively of Britons, was to offer a pledge of amity from the dominant nation to the suffering people of Ireland. The following letter was accordingly sent to Sir George Grey:

* Fox.—*Ed.*

To the Right Hon. Sir George Grey,
Secretary of State for the Home Department

18, Bouverie Street, Fleet Street,
February 24, 1866

Sir,—A deputation, consisting exclusively of Englishmen, from the Working Men's International Association, solicit an interview with you at as early a day after next Tuesday as is convenient to you, to urge upon you the propriety of mitigating, to a very slight extent, the severity of the prison discipline now enforced at Pentonville Prison upon the Irish State prisoners.

I am, Sir, &c.,

W. R. CREMER, Hon. Sec.

To this application the Secretary has received the following reply:

Whitehall, March 1, 1866

Sir,—I am directed by Secretary Sir George Grey to acknowledge the receipt of your letter of the 24th ult., requesting him to appoint an early day for receiving a deputation from the "International Working Men's Association" on the subject of the treatment of the Irish State prisoners in Pentonville Prison, and I am to acquaint you that the Secretary of State must decline to receive a deputation on this subject.

I am, Sir, your obedient servant,

H. WADDINGTON

Mr. W. R. Cremer,
18, Bouverie Street, E. C.

The Central Council submit this correspondence to the British public, and through them to the public of both continents, without comment.

G. ODGER, President

Published in *The Commonwealth* No. 157, March 10, 1866

Printed according to the text of the book *The General Council of the First International. 1864-1866. The London Conference, 1865. Minutes,* Moscow

MEETING OF THE COUNCIL
AND MEMBERS AND FRIENDS
OF THE ASSOCIATION
NOVEMBER 19, 1867

Citizen *Weston* was unanimously elected to take the chair.

The *Secretary** read the resolution from the Minutes of the previous Council meeting, fixing the order of the day for the 19th, [it] being the discussion of the Fenian question.

The *Chairman* said: I think the Council has acted wisely in determining the discussion of this question at this time, and I have no doubt that it will receive the attention it merits.

He then called upon Citizen Jung to open the discussion.**

Mr. *Jung* said: When I proposed that this question should be discussed I thought an expression of opinion on the part of the Council of this Association was desirable. I am no abettor of physical force movement, but the Irish have no other means to make an impression. Many people seem to be frightened at the term "physical force" in this country, yet even English agitations are not free from its influence. The Reform League has accomplished much by way of moral force, but it was only under a threat that physical force might be resorted to on the occasion of the Hyde Park meetings that the Government gave way.[370] I should be sorry to find the working men of this country go wrong upon this question. They have been right upon every other. The Irish require more than simple reform. Some endeavours have

---

* Eccarius.—*Ed.*
** Here a newspaper clipping is pasted into the Minute Book.—*Ed.*

been made to divert the attention of the work-people of this country with regard to the Fenians. While they are denounced as murderers, Garibaldi is held up as a great patriot; and have no lives been sacrificed in Garibaldi's movement? The Irish have the same right to revolt as the Italians, and the Italians have not exhibited greater courage than the Irish. I may not agree with the particular way in which the Irish manifest their resistance, but they deserve to be free. (*Loud cheers.*)

Mr. *Lessner* said: Our Association is not confined to any particular nationality; we are of all nations, and the Irish question concerns us as much as any other. In the course of twenty years the Irish population has dwindled down from eight millions to five and a half millions, and this decline is in consequence of the British rule. No country can be prosperous with a declining population. Ireland declines at a rapid rate, and the Irish have a right to revolt against those who drive them out of their country; the English would do the same if any foreign power oppressed them in a similar manner. (*Cheers.*)

Mr. *Dupont*: The Council would be wanting in its duty if it remained indifferent to the Irish cause. What is Fenianism? Is it a sect or a party whose principles are opposed to ours? Certainly not. Fenianism is the vindication by an oppressed people of its right to social and political existence. The Fenian declarations leave no room for doubt in this respect. They affirm the republican form of government, liberty of conscience, no State religion, the produce of labour to the labourer, and the possession of the soil to the people. What people could abjure such principles? Only blindness and bad faith can support the contrary. We hear that those whom the English law is going to strike down for their devotedness to such a cause are exclaiming: "We are proud to die for our country and for republican principles." Let us see of what value the reproaches are that are addressed to the Fenians by the English would-be liberators. Fenianism is not altogether wrong, they say, but why not employ the legal means of meetings and demonstrations by the aid of which we have gained our Reform Bill? I avow that it is hardly possible to restrain one's indignation at hearing such arguments. What is the use of talking of legal means to a people reduced to the lowest state of misery from century to century by English oppression—to people who emigrate by thousands, to obtain bread, from all parts of the country? Is not this Irish emigration to America by millions the most eloquent legal protest? Having destroyed all—life and liberty—be not surprised that nothing should be found but hatred to the oppressor. Is it well for the English to talk of legality and justice to those who on the slightest suspicion of Fenianism are arrested and incarcerated, and subjected to physical and mental tortures which leave the cruelties of King Bomba,* of whom the would-be liberators talked so much, far behind? A citizen of Manchester, whose domicile was invaded by constables, asked one of them to show his warrant. "Here is my warrant," he replied, drawing a pistol from his pocket. This shows the conduct of the English Government towards the Irish. Without having right on their side, such conduct is enough to provoke and justify resistance. The English working men who blame the Fenians commit more than a fault, for the cause of both peoples is the same; they have the same enemy to defeat— the territorial aristocracy and the capitalists. (*Cheers.*)

Mr. *Morgan* thought it was rather unfortunate that the Irish had chosen the name of Fenians, which many Englishmen considered synonymous with all that is bad. Had they simply called themselves Republicans, they would have shut up at once all those Englishmen who profess to be in favour of Republicanism. Englishmen as a rule did not look as favourably upon things in their own country as in other countries. They applauded insurrection abroad, but denounced it in Ireland. Deeds that would be considered as heroism if committed in France, in Italy, or in Poland, would be stigmatised as crimes in Ireland. The Irish had every reason to have recourse to physical force. Moral suasion had never been used towards them by the British Government; it had always applied to the robe and the musket. The English ought at least to look as favourably upon the Irish as upon the Italians. Were they treated in the same manner by a foreign power they would revolt sooner than the Irish. (*Hear, hear.*)

---

* Ferdinand II.—*Ed.*

Mr. *Lucraft* said the question was not whether the Irish were justified in using physical force, but whether they could do any good by it. He thought they could not. He thought it rather strange that the Irish of London, for instance, had not made common cause with the English and Scotch in the reform agitation.

Mr. *Weston* thought the word Fenianism meant the heat produced by centuries of oppression, and the hatred engendered by it, which could not be cured by the concessions of reform which the English demanded for themselves. A government that had trampled upon the rights of a people could never be reached by moral suasion, but by physical force resistance. In England there was no need of bludgeons, but in Ireland moral force had not [had] fair play. The rescue of the Fenian prisoners at Manchester was an exact duplicate affair of the rescue that was now attempted by the British Government of the prisoners held in Abyssinia. If killing was murder to rescue prisoners in Manchester, it was murder in Abyssinia; if it was wrong in one place it was wrong in the other. The crime of starving the Irish was far greater than the accidental killing of one man in trying to rescue the Fenian prisoners. He did not believe in the justice of the law. The laws were made and administered by hostile partisans, and there was a possibility of finding an innocent man guilty. He thought Ireland had been governed with more heartlessness than any other country, and he was glad that the Irish question had come uppermost. The democracy of the sister kingdoms must take the matter up and redress the wrong. (*Loud cheers.*)

Mr. *William Parks* said that the Irish in Ireland, in America, and in England were all of one opinion,—they wanted Ireland for the Irish, and to govern themselves.

Citizen *Jayet* argued in a speech of some length that physical force resistance was a bounden duty for every people who was oppressed by tyrants, were they of home or foreign origin, and showed that this was laid down as a maxim in the constitution of the French Convention, of which Robespierre had been a leading member.*

---

* Jayet's speech is recorded in handwriting.—*Ed.*

Upon the proposition of Dr. *Marx*, the discussion was adjourned to Tuesday next.*

Upon the proposition of Citizen *Lucraft*, it was agreed after some discussion, and the Standing Committee with the chairman of the meeting were instructed, to draw up a memorial to the Home Secretary concerning the Fenian prisoners under sentence of death at Manchester and present it to a special meeting of the Council for adoption on Wednesday, November 20.**

Published in the book   Printed according to
The General Council of the   the text of the book
First International. 1866-1868.
Minutes, Moscow

---

* Here the following sentence is crossed out: "The Standing Committee was instructed to draw up a memorial to the Home Secretary on behalf of the Fenian prisoners now under sentence of death at Manchester." The newspaper clipping ends here.—*Ed.*

** Unsigned. For the memorial see pp. 118-19.—*Ed.*

ADDRESS OF THE LAND
AND LABOUR LEAGUE TO THE WORKING MEN
AND WOMEN OF GREAT BRITAIN AND IRELAND[371]

Fellow-Workers,

The fond hopes held out to the toiling and suffering millions of this country thirty years ago have not been realised. They were told that the removal of fiscal restrictions would make the lot of the labouring poor easy; if it could not render them happy and contented it would at least banish starvation for ever from their midst.

They rose a terrible commotion for the big loaf,[372] the landlords became rampant, the money lords confounded, the factory lords rejoiced—their will was done—Protection received the *coup de grâce*. A period of the most marvellous prosperity followed. At first the Tories threatened to reverse the policy, but on mounting the ministerial benches, in 1852, instead of carrying out their threat, they joined the chorus in praise of unlimited competition. Prepared for a pecuniary loss they discovered to their utter astonishment that the rent-roll was swelling at the rate of more than £2,000,000 a year. Never in the history of the human race was there so much wealth—means to satisfy the wants of man—produced by so few hands, and in so short a time, as since the abolition of the Corn Laws. During the lapse of twenty years the declared value of the annual exports of British and Irish produce and manufactures—the fruits of your own labour—rose from £60,000,000 to £188,900,000. In twenty years the taxable income of the lords and ladies of the British soil increased, upon their own confession, from £98,000,000 to £140,000,000 a year; that of the chiefs of trades and professions from £60,000,000 to £110,000,000 a year. Could human efforts accomplish more?

Alas! there are stepchildren in Britania's family. No Chancellor of the Exchequer has yet divulged the secret how the £140,000,000 are distributed amongst the territorial magnates, but we know all about the trades-folk. The special favourites increased from sixteen, in 1846, to one hundred and thirty-three, in 1866. Their average annual income rose from £74,300 to £100,600 each. They appropriated one-fourth of the twenty years' increase. The next of kin increased from three hundred and nineteen to nine hundred and fifty-nine individuals: their average annual income rose from £17,700 to £19,300 each: they appropriated another fourth. The remaining half was distributed amongst three hundred and forty-six thousand and forty-eight respectables, whose annual income ranged between £100 and £10,000 sterling. The toiling millions, the producers of that wealth—Britania's cinderellas—got cuffs and kicks instead of halfpence.

In the year 1864 the taxable income under schedule D[373] increased by £9,200,000. Of that increase the metropolis, with less than an eighth of the population, absorbed £4,266,000, nearly a half. £3,123,000 of that, more than a third of the increase of Great Britain, was absorbed by the City of London, by the favourites of the one hundred and seventy-ninth part of the British population: Mile End and the Tower, with a working population four times as numerous, got £175,000. The citizens of London are smothered with gold; the householders of the Tower Hamlets are overwhelmed by poor-rates. The citizens, of course, object to centralisation of poor-rates purely on the principle of local self-government.

During the ten years ending 1861 the operatives employed in the cotton trade increased 12 per cent; their produce 103 per cent. The iron miners increased 6 per cent; the produce of the mines 37 per cent. Twenty thousand iron miners worked for ten mine owners. During the same ten years the agricultural labourers of England and Wales diminished by eighty-eight thousand one hundred and forty-seven, and yet, during that period, several hundred thousand acres of common land were enclosed and transformed into private property to enlarge the estates of the nobility, and the same process is still going on.

In twelve years the rental liable to be rated to the poor in England and Wales rose from £86,700,000 to £118,300,000: the number of adult able-bodied paupers increased from one hundred and forty-four thousand five hundred to one hundred and eighty-five thousand six hundred.

These are no fancy pictures, originating in the wild speculations of hot-brained incorrigibles; they are the confessions of landlords and money lords, recorded in their own blue books. One of their experts told the House of Lords the other day that the propertied classes, after faring sumptuously, laid by £150,000,000 a year out of the produce of your labour. A few weeks later the president of the Royal College of Surgeons related to a jury, assembled to inquire into the causes of eight untimely deaths, what he saw in the foul ward of St. Pancras.

Hibernia's favourites too have multiplied, and their income has risen while a sixth of her toiling sons and daughters perished by famine, and its consequent diseases, and a third of the remained were evicted, ejected and expatriated by tormenting felonious usurpers.

This period of unparalleled industrial prosperity has landed thousands of our fellow-toilers—honest, unsophisticated, hard-working men and women—in the stone yard and the oakum room[374]; the roast beef of their dreams has turned into skilly. Hundreds of thousands, men, women and children are wandering about—homeless, degraded outcasts— in the land that gave them birth, crowding the cities and towns, and swarming the highroads in the country in search of work to obtain food and shelter, without being able to find any. Other thousands, more spirited than honest, are walking the treadmill to expiate little thefts, preferring prison discipline to workhouse fare, while the wholesale swindlers are at large, and felonious landlords preside at quarter sessions to administer the laws. Thousands of the young and strong cross the seas, flying from their native firesides, like from an exterminating plague; the old and feeble perish on the roadside of hunger and cold. The hospitals and infirmaries are overcrowded with fever and famine-stricken: death from starvation has become an ordinary every-day occurrence.

All parties are agreed that the sufferings of the labouring

poor were never more intense, and misery so widespread, nor the means of satisfying the wants of man ever so abundant as at present. This proves above all that the moral foundation of all civil government, *"that the welfare of the entire community is the highest law, and ought to be the aim and end of all civil legislation"*, has been utterly disregarded. Those who preside over the destinies of the nation have either wantonly neglected their primary duty while attending to the special interests of the rich to make them richer, or their social position, their education, their class prejudices have incapacitated them from doing their duty to the community at large or applying the proper remedies, in either case they have betrayed their trust.

Class government is only possible on the condition that those who are held in subjection are secured against positive want. The ruling classes have failed to secure the industrious wages-labourer in the prime of his life against hunger and death from starvation. Their remedies have signally failed, their promises have not been fulfilled. They promised retrenchment, they have enormously increased the public expenditure instead. They promised to lift the burden of taxation from your shoulders, the rich pay but a fractional part of the increased expenses; the rest is levied upon your necessaries—even your pawn tickets are taxed—to keep up a standing army, drawn from your own ranks, to shoot you down if you show signs of disaffection. They promised to minimise pauperism: they have made indigence and destitution your average condition—the big loaf has dwindled into no loaf. Every remedy they have applied has but aggravated the evil, and they have no other to suggest,— their rule is doomed. To continue is to involve all in a common ruin. There is but one,—and only one,—remedy. Help Yourselves! Determine that you will not endure this abominable state of things any longer; act up to your determination, and it will vanish.

A few weeks ago a score of London working men talked the matter over. They came to the conclusion that the present economical basis of society was the foundation of all the existing evils,—that nothing short of a transformation of the existing social and political arrangements could avail, and that such a transformation could only be effected by the toiling millions themselves. They embodied their conclusions in a series of resolutions, and called a conference of representative working men, to whom they were submitted for consideration. In three consecutive meetings those resolutions were discussed and unanimously adopted. To carry them out a new working men's organisation, under the title of the *"Land and Labour League"*, was established. An executive council of upwards of forty well-known representative working men was appointed to draw up a platform of principles arising out of the preliminary resolutions adopted by the conference, to serve as the programme of agitation by means of which a radical change can be effected.

After mature consideration the Council agreed to the following:

1. Nationalisation of the Land.
2. Home Colonisation.
3. National, Secular, Gratuitous and Compulsory Education.
4. Suppression of Private Banks of Issue. The State Only to Issue Paper Money.
5. A Direct and Progressive Property Tax, in Lieu of All Other Taxes.
6. Liquidation of the National Debt.
7. Abolition of the Standing Army.
8. Reduction of the Number of the Hours of Labour.
9. Equal Electoral Rights, with Payment of Members.

The success of our efforts will depend upon the pressure that can be brought to bear upon the powers that be, and this requires numbers, union, organisation and combination. We therefore call upon you to unite, organise and combine, and raise the cry throughout Ireland, Scotland, Wales and England, *"The Land for the People"*—the rightful inheritors of nature's gifts. No rational state of society can leave the land, which is the source of life, under the control of, and subject to the whims and caprices of, a few private individuals. A government elected by, and as trustee for, the whole people is the only power that can manage it for the benefit of the entire community.

Insist upon the State reclaiming the unoccupied land as a beginning of its nationalisation, and placing the unemployed upon it. Let not another acre of common land be enclosed for the private purposes of non-producers. Compel the Government to employ the army, until its final dissolution, as a pioneer force to weed, drain and level the wastes for cultivation, instead of forming encampments to prepare for the destruction of life. If green fields and kitchen gardens are incompatible with the noble sport of hunting let the hunters emigrate.

Make the Nine points of the League the Labour programme, the touchstone by which you test the quality of candidates for parliamentary honours, and if you find them spurious reject them like a counterfeit coin, for he who is not for them is against you.

You are swindled out of the fruits of your toil by land laws, money laws, and all sorts of laws. Out of the paltry pittance that is left you, you have to pay the interest of a debt that was incurred to keep your predecessors in subjection; you have to maintain a standing army that serves no other purpose in your generation, and you are systematically overworked when employed, and underfed at all times. Nothing but a series of such radical reforms as indicated on our programme will ever lift you out of the slough of despond in which you are at present sunk. The difficulty can be overcome by unity of purpose and action. We are many; our opponents are few. Then working men and women of all creeds and occupations claim your rights as with one voice, and rally round, and unite your forces under, the banner of the *"Land and Labour League"* to conquer your own emancipation!

JOHN WESTON, Treasurer
MARTIN J. BOON⎫ *Secretaries*
J. GEORGE ECCARIUS⎭

Published as a pamphlet in London in 1869

Printed according to the text of the book *The General Council of the First International. 1868-1870. Minutes,* Moscow

ARTICLES BY JENNY MARX
ON THE IRISH QUESTION[375]

I

London, February 27, 1870

The *Marseillaise* for February 18 quotes an article from
the *Daily News* in which the English paper gives information to the French press concerning the election of O'Donovan Rossa. Since this information is somewhat confused and
since partial explanations only serve to throw a false light
on the things which they are claiming to elucidate, I should
be grateful if you would kindly publish my comments on
the article in question.

Firstly, the *Daily News* states that O'Donovan was
sentenced by a jury, but it omits to add that in Ireland
the juries are composed of minions more or less directly
nominated by the government.

Then, in speaking with righteous horror of the felony
of treason, the false liberals of the *Daily News* omit to say
that this new category in the English Penal Code was
expressly invented to identify the Irish patriots with the
vilest of criminals.

Let us take the case of O'Donovan Rossa. He was one
of the editors of the *Irish People*. Like most of the Fenians
he was sentenced for having written so-called seditious
articles. Consequently the *Marseillaise* was not wrong in
drawing an analogy between Rochefort and Rossa.

Why does the *Daily News*, which aims at keeping France
informed about the Fenian prisoners, remain silent about
the appalling treatment which they have received? I trust
that you will allow me to make up for this prudent silence.

Some time ago O'Donovan was put in a dark cell with
his hands tied behind his back. His handcuffs were not
removed night or day so that he was forced to lick his

food, gruel made with water, lying on the ground. Mr.
Pigott, editor of the *Irishman*, learnt about this from Rossa
who described it to him in the presence of the prison governor and another witness, and published the information
in his newspaper, encouraging Mr. Moore, one of the Irish
members of the House of Commons, to request a parliamentary enquiry into what goes on in the prisons. The government strongly opposed this request. Thus, Mr. Moore's
motion was rejected by 171 votes to 36—a worthy supplement to the voting which crushed the right to suffrage.

And this took place during the ministry of the sanctimonious Gladstone. As you can see the great Liberal leader
knows how to mock humanity and justice. There are also
Judases who do not wear glasses.

Here is another case which also does England credit.
O'Leary, a Fenian prisoner aged between sixty and seventy,
was put on bread and water for three weeks because—the
reader of the *Marseillaise* would never guess why—because
Leary called himself a *"pagan"* and refused to say he was
Protestant, Presbyterian, Catholic or *Quaker*. He was given
the choice of one of these religions or bread and water. Of
these five evils, O'Leary, or *"pagan O'Leary"* as he is
called, chose the one that he considered the least—bread
and water.

A few days ago after examining the body of a Fenian
who died at Spike Island Prison the *coroner* expressed his
very strong disapproval of the manner in which the
deceased man had been treated.

Last Saturday a young Irishman called *Gunner Hood*
left prison after serving four years. At the age of 19 he
had joined the English army and served England in
Canada. He was taken before a military tribunal in 1866
for having written seditious articles and sentenced to two
years' hard labour. When the sentence was pronounced
Hood took his cap and threw it into the air shouting, "Long
live the Irish republic!" This impassioned cry cost him dear.
He was sentenced an extra two years in prison and fifty
strokes for good measure. This was carried out in the most
atrocious manner. Hood was attached to a plough and two
strapping blacksmiths were armed with cat-o'-nine-tails.
There is no equivalent term in French for the English

knout. Only the Russians and the English know what is meant by this. Like draws to like.

Mr. Carey, a journalist, is kept at present in the part of the prison intended for the insane, the terrible silence and the other forms of torture to which he has been subjected having turned him into a mass of living flesh deprived of all reason.

The Fenian, Colonel Burke, a man who has distinguished himself not only by his service in the American army but also as a writer and painter, has also been reduced to a pitiful state in which he can no longer recognise his closest relatives. I could add many more names to this list of Irish martyrs. Suffice it to say that since 1866, when there was a raid on the *Irish People*'s offices, *20 Fenians have died or gone mad* in the prisons of humanitarian England.

II

London, March 5

During the meeting of the House of Commons on March 3 Mr. *Stackpoole* questioned Mr. Gladstone on the treatment of Fenian prisoners. He said, among other things, that Dr. *Lyons* of Dublin had recently stated that

"the discipline, diet, personal restrictions and the other punishments were bound to cause permanent damage to the prisoners' health."

After having expressed complete satisfaction with the way in which prisoners were treated, Mr. Gladstone crowned his little *speech* with this brilliantly witty remark:

"As to the health of O'Donovan Rossa, I am glad to be able to say that during her *last* visit to her husband Mrs. O'Donovan Rossa congratulated him on *looking better*."[376]

Whereupon a burst of Homeric laughter broke out from all sides of that noble assembly. *Her last visit!* Note that Mrs. O'Donovan Rossa had not only been separated from her husband for several years, but that she had travelled all over America earning money to feed her children by giving public lectures on English literature.

And bear in mind also that this same Mr. *Gladstone,* whose quips are so pointed, is the almost sacred author of *Prayers,* the *Propagation of the Gospel, The Functions of Laymen in the Church* and the recently published homily *Ecce homo.*

Is the profound satisfaction of the head jailer shared by his prisoners? Read the following extracts from a letter written by *O'Donovan Rossa,* which by some miracle was slipped out of the prison and arrived at its destination after an incredible delay:

LETTER FROM ROSSA

I have already told you about the hypocrisy of these English masters who, after placing me in a position which forced me to get down on my knees and elbows to eat, are now depriving me of food and light and giving me chains and a Bible. I am not complaining of the penalties which my masters inflict on me—it is my job to suffer— but I insist that I have the right to inform the world of the treatment to which I am subjected, and that it is illegal to hold back my letters describing this treatment. The minute precautions taken by the prison authorities to prevent me writing letters are as disgusting as they are absurd. The most insulting method was to strip me once a day for several months and then examine my arms, legs and all other parts of my body. This took place at *Millbank* daily from February to May 1867. One day I refused, whereupon five prison officers arrived, beat me mercilessly and tore off my clothes.

Once I succeeded in getting a letter to the outside, for which I was rewarded by a visit from Messrs. *Knox* and *Pollock,* two *police magistrates.*

How ironical to send two government employees to find out the truth about the English prisons. These gentlemen refused to take note of anything important which I had to tell them. When I touched upon a subject which was not to their liking, they stopped me by saying that prison discipline was not their concern. Isn't that so, Messrs. *Pollock* and *Knox?* When I told you that I had been forced to wash in water which had already been used by half a dozen English prisoners, did you not refuse to note my complaint?

At *Chatham* I was given a certain amount of tow to pull out and told that I would go without food if I did not finish the work by a certain time.

"Perhaps you'll still punish me even if I do the job in time," I shouted. "That's what happened to me at *Millbank.*"

"How could it?" asked the jailer.

Then I told him that on July 4 I had finished my work ten minutes before the appointed time and picked up a book. The officer saw me do this, accused me of being lazy and I was put on bread and water and locked in a dark cell for forty-eight hours.

One day I caught sight of my friend *Edward Duffy*. He was extremely pale. A little later I heard that *Duffy* was seriously ill and that he had expressed the wish to see me (we had been very close in Ireland). I begged the governor to give me permission to visit him. He refused point-blank. This was round about Christmas '67—and a few weeks later a prisoner whispered to me through the bars of my cell: "*Duffy* is dead."

How movingly this would have been described by the English if it had happened in Russia!

If Mr. Gladstone had been present on such a sad occasion in Naples, what a touching picture he would have painted! Ah! Sweet Pharisees, trading in hypocrisy, with the Bible on their lips and the devil in their bellies.

I must say a word in memory of *John Lynch*. In March 1866 I found myself together with him in the exercise yard. We were being watched so closely that he only managed to say to me, "The cold is killing me." But then what did the English do to us? They took us to London on Christmas Eve. When we arrived at the prison they took away our flannels and left us shivering in our cells for several months. Yes, they cannot deny that it was they who killed *John Lynch*. But nevertheless they managed to produce officials at the enquiry who were ready to prove that *Lynch* and *Duffy* had been given very gentle treatment.

The lies of our English oppressors exceed one's wildest imagination.

If I am to die in prison I entreat my family and my friends not to believe a word of what these people say. Let me not be suspected of personal rancour against those who persecuted me with their lies. I accuse only tyranny which makes the use of such methods necessary.

Many a time the circumstances have reminded me of Machiavelli's words: "that tyrants have a special interest in circulating the Bible so that the people understand its precepts and offer no resistance to being robbed by brigands".

So long as an enslaved people follows the sermons on morality and obedience preached to them by the priests, the tyrants have nothing to fear.

If this letter reaches my fellow countrymen I have the right to demand that they raise their voices to insist that justice be done for their suffering brothers. Let these words whip up the blood that is moving sluggishly in their veins!

I was harnessed to a cart with a rope tied round my neck. This knot was fastened to a long shaft and two English prisoners received orders to prevent the cart from bouncing. But they refrained from doing this, the shaft rose up into the air and the knot came undone. If it had tightened I would be dead.

I insist that they do not possess the right to put me in a situation where my life depends on the acts of other people.

A ray of light is penetrating through the bolts and bars of my prison. This is a reminder of the day at *Newtownards* where I met *Orangemen* and *Ribbonmen* who had forgotten their bigotry!

O'Donovan Rossa
Political prisoner sentenced to hard labour

III

London, March 16, 1870

The main event of the past week has been O'Donovan Rossa's letter which I communicated to you in my last report.

The *Times* printed the letter without comment, whereas the *Daily News* published a commentary without the letter.

"As one might have expected," it says, "Mr. O'Donovan Rossa takes as his subject the prison rules to which he has been *subjected for a while.*"

How atrocious this "*for a while*" is in speaking of a man who has already been imprisoned for five years and condemned to hard labour *for life.*

Mr. O'Donovan Rossa complains among other things "of being harnessed to a cart with a rope tied round his neck" in such a way that his life depended on the movements of English convicts, his fellow prisoners.

But, exclaims the *Daily News*, "is it really unjust to put a man in a situation where his life depends on the acts of others? When a person is in a car or on a steamer does not his life also depend on the acts of others?"

After this brilliant piece of arguing, the pious casuist reproaches O'Donovan Rossa for not loving the *Bible* and preferring the *Irish People*, a comparison which is sure to delight its readers.

"Mr. O'Donovan," it continues, "seems to imagine that prisoners serving sentences for *seditious writing* should be supplied with cigars and daily newspapers, and that they should above all have the right to correspond freely with their friends."

Ho, ho, virtuous Pharisee! At last you have admitted that O'Donovan Rossa has been sentenced to hard labour for life for *seditious writing* and not for an attempted *assassination* of Queen Victoria, as you vilely insinuated in your first address to the French press.

"After all," this shameless newspaper concludes, "O'Donovan Rossa is simply being treated for what he is, that is, an ordinary convict."

After Mr. Gladstone's special newspaper, here is a different angle from the "liberal" press, the *Daily Telegraph*, which generally adopts a rougher manner.

"If we condescend," it says, "to take note of O'Donovan Rossa's letter, it is not because of the Fenians who are incorrigible, but exclusively for the well-being of France.
"Let it be known that only a few days ago in the House of Commons Mr. Gladstone made a formal denunciation of all these outrageous lies, and there cannot be any intelligent Frenchmen of whatever party and class who would dare doubt the word of an English gentleman."

But if, contrary to expectation, there were parties or people in France perverse enough not to believe the word of an English gentleman such as Mr. Gladstone, France could not at least resist the well-meant advice of Mr. Levy who is not a gentleman and who addresses you in the following terms:

"We advise our neighbours, the Parisians, to treat all the stories of cruelties committed on political prisoners in England as so many insolent lies."

With Mr. Levy's permission, I will give you a new example of the value of the *words* of the gentlemen who make up Gladstone's Cabinet.

You will remember that in my first letter I mentioned Colonel *Richard Burke,* a Fenian prisoner who has gone insane thanks to the humanitarian methods of the English government. The *Irishman* was the first to publish this news, after which Mr. Underwood sent a letter to Mr. Bruce, the Home Secretary, asking him for an enquiry into the treatment of political prisoners.

Mr. Bruce replied in a letter which was published in the English press and which contained the following sentence:

"With regard to Richard Burke at Woking Prison, Mr. Bruce is bound to refuse to make an enquiry on the grounds of *such ill-founded and extravagant insinuations* as those contained in the extracts from the *Irishman* which you have sent me."

This statement by Mr. Bruce is dated January 11, 1870. Now in one of its recent issues the *Irishman* has published the same Minister's reply to a letter from Mrs. Barry, Richard Burke's sister, who asked for news about her brother's "alarming" condition. The ministerial reply of *February* 24 contains an official report dated January 11 in which the prison doctor and Burke's special guard state that he has become insane. Thus, the very day when Mr. Bruce publicly declared the information published by the *Irishman* to be false and ill-founded, he was concealing the irrefutable official proof in his pocket! It should be mentioned incidentally that Mr. Moore, an Irish member in the House of Commons, is to question the Minister on the treatment of Colonel Burke.

The *Echo*, a recently founded newspaper, takes an even stronger liberal line than its companions. It has its own principle which consists of selling for one penny, whereas all the other newspapers cost twopence, fourpence or sixpence. This price of one penny forces it on the one hand to make pseudo-democratic professions of faith so as not to lose its working-class subscribers, and on the other hand to make constant reservations in order to win over respectable subscribers from its competitors.

In its long tirade on O'Donovan Rossa's letter it finished up by saying that "perhaps even those Fenians who have received an amnesty will refuse to believe the exaggerations of their compatriots", as if Mr. Kickham, Mr. Costello and others had not already published information on their suffering in prison totally in accordance with Rossa's letter! But after all its subterfuge and senseless evasions the *Echo* touches on the sore point.

The "publications by the *Marseillaise*," it says, "will cause a scandal and this scandal will spread all round the world. The continental mind is perhaps too limited to be able to discern the difference between the crimes of a Bomba and the severity of a Gladstone! So it would be better to hold an enquiry" and so on.

The *Spectator*, a "liberal" weekly which supports Gladstone, is governed by the principle that all genres are bad except the boring one.[377] This is why it is called in London the journal of the seven wise men. After giving a brief account of O'Donovan Rossa and scolding him for his aversion to the Bible, the journal of the seven wise men pronounces the following judgment:

"The Fenian O'Donovan Rossa does not appear to have suffered anything more than the ordinary sufferings of convicts, but we confess that we should like to see changes in this regime. It is very right and often most advisable to shoot rebels. It is also right to deprive them of their liberty as the most dangerous type of criminals. But it is neither right nor wise to degrade them."

Well said, Solomon the Wise!

Finally we have the *Standard*, the main organ of the Tory party, the Conservatives. You will be aware that the English oligarchy is composed of two factions: the landed aristocracy and the plutocracy. If in their family quarrels one takes the side of the plutocrats against the aristocrats one is called a liberal or even radical. If, on the contrary, one sides with the aristocrats against the plutocrats one is called a Tory.

The *Standard* calls O'Donovan Rossa's letter an apocryphal story probably written by A. Dumas.

"Why," it says, "did the *Marseillaise* refrain from adding that Mr. Gladstone, the Archbishop of Canterbury and the Lord Mayor were present each morning while O'Donovan Rossa was being tortured?"

In the House of Commons a certain member once referred to the Tory party as the "stupid party". Is it not a fact that the *Standard* well deserves its title as the main organ of the stupid party!

Before closing I must warn the French not to confuse the newspaper rumours with the voice of the English proletariat which, unfortunately for the two countries, Ireland and England, has no echo in the English press.

Let it suffice to say that more than 200,000 men, women and children of the English working class raised their voices in Hyde Park to demand freedom for their Irish brothers, and that the General Council of the *International Working Men's Association*, which has its headquarters in London and includes well-known English working-class leaders among its members, has severely condemned the treatment of Fenian prisoners and come out in defence of the rights of the Irish people against the English government.[378]

P.S. As a result of the publicity given by the *Marseillaise* to O'Donovan Rossa's letter, Gladstone is afraid that he may be forced by public opinion to hold a parliamentary public enquiry into the treatment of political prisoners. In order to avoid this again (we know how many times his corrupt conscience has opposed it already) this diplomat has just produced an official, but anonymous denial of the facts quoted by Rossa.[379]

Let it be known in France that this denial is nothing more than a copy of the statements made by the prison jailer, police magistrates Knox and Pollock, etc., etc. These gentlemen know full well that Rossa cannot reply to them. He will be kept under stricter supervision than ever, but ... I shall reply to them in my next letter with *facts* the verification of which does not depend on the goodwill of jailers.

IV

London, March 18, 1870

As I announced in my last letter Mr. Moore, an Irish member of the House of Commons, yesterday questioned the government on the treatment of Fenian prisoners. He referred to the request made by Richard Burke and four other prisoners held in Mountjoy Prison (in Dublin) and asked the government whether it considered it honourable to hold the bodies of these men after having deprived them of their senses. Finally, he insisted on a "full, free and public enquiry".

So here was Mr. Gladstone with his back to the wall. In 1868 he gave an insolent, categorical refusal to a request to hold an enquiry made by the same Mr. Moore. Since then he has always replied in the same fashion to repeated demands for an enquiry.

Why give way now? Perhaps it would not be a bad idea

to admit to being alarmed by the uproar on the other side of the Channel. As to the charges levelled against our governors of prisons, we have asked them to give a full explanation in this connection.

The latter have unanimously replied that all this is sheer nonsense. Thus, our ministerial conscience is naturally satisfied. But after the explanations given by Mr. Moore (these are his exact words) it appears "that the point in question is not exactly satisfaction. That the satisfaction of the minds of the government derives from its confidence in its subordinates and, therefore, it would be both political and just to conduct an enquiry into the truth of the jailers' statements".[380]

> One day he says this, and the next day says that,
> His yesterday's views today he will shelve,
> He now wears a helmet, and now a top hat,
> A nuisance to others, a bore to himself.*

But he does not give way at last without making reservations.

Mr. Moore demanded a "full, free and public enquiry". Mr. Gladstone replied that he was responsible for the "form" of the enquiry, and we already know that this will not be a "parliamentary enquiry", but one conducted by means of a Royal Commission. In other words the judges in this great trial, in which Mr. Gladstone appears as the main defendant, are to be selected and appointed by Mr. Gladstone himself.

As for Richard Burke, Mr. Gladstone states that the government had learnt of his insanity as early as January 9. Consequently, his honourable colleague Mr. Bruce, the Home Secretary, lied outrageously by declaring in his open letter of January 11 that this information was untrue. But, Mr. Gladstone continues, Mr. Burke's mental disturbance had not reached a sufficiently advanced stage to justify his release from prison. It must not be forgotten that this man was an accessory to the blowing up of Clerkenwell Prison.[381] Really? But Richard Burke was already detained in Clerkenwell Prison when a number of other people took it into their heads to blow up the prison in order to free him. Thus he

---

* Boileau, *Satires*, Satire 8.—*Ed.*

was an accessory to this ridiculous attempt which, it is thought, was instigated by the police and which, if it had succeeded, would have buried him under the ruins! Moreover, concludes Mr. Gladstone, we have already released two Fenians who went mad in our English prisons. But, interrupts Mr. Moore, I was talking about the four insane men detained in Mountjoy Prison in Dublin. Be that as it may, replies Mr. Gladstone. There are still two madmen less in our prisons.

Why is Mr. Gladstone so anxious to avoid all mention of Mountjoy Prison? We shall see in a moment. This time the facts are verified not by letters from the prisoners, but in a Blue Book published in 1868 by order of Parliament.

After the Fenian skirmish[382] the English government declared a state of general emergency in Ireland. All guarantees of the freedom of the individual were suspended. Any person "being suspected of Fenianism" could be thrown into prison and kept there without being brought to court as long as it pleased the authorities. One of the prisons full of suspects was Mountjoy Convict Prison in Dublin, of which John Murray was the inspector and Mr. M'Donnell the doctor. Now what do we read in the *Blue Book* published in 1868 by order of Parliament?

For several months Mr. M'Donnell wrote to Inspector Murray protesting against the cruel treatment of suspects. Since the inspector did not reply, Mr. M'Donnell then sent three or four reports to the prison governor. In one of these letters he referred to

"certain persons who show unmistakable signs of insanity." He went on to add: "I have not the slightest doubt that this insanity is the consequence of the prison regime. Quite apart from all humane considerations, it would be a serious matter if one of these prisoners, who have not been sentenced by a court of law but are merely suspects, should commit suicide."

All these letters addressed by Mr. M'Donnell to the governor were intercepted by John Murray. Finally, Mr. M'Donnell wrote direct to Lord Mayo, the First Secretary for Ireland. He told him for example:

"There is no one, my Lord, as well informed as you yourself are on the harsh discipline to which the 'suspect' prisoners have been subjected for a considerable time, a more severe form of solitary confinement than that imposed on the convicts."

What was the result of these revelations published by order of Parliament? The doctor, Mr. M'Donnell, was dismissed!!! Murray kept his post.

All this took place during the Tory ministry. When Mr. Gladstone finally succeeded in unseating Lord Derby and Mr. Disraeli by fiery speeches in which he denounced the English government as the true cause of Fenianism, he not only confirmed the savage Murray in his functions but also, as a sign of his special satisfaction, conferred a large sinecure, that of "Registrar of habitual criminals", on his post of inspector.

In my last letter I stated that the anonymous reply to Rossa's letter, circulated by the London newspapers, emanated directly from the Home Office.

It is now known to be the work of the Home Secretary, Mr. Bruce. Here is a sample of his "ministerial conscience!"

> As to Rossa's complaint that he is obliged "to wash in water which has already been used for the convicts' ablutions, the police magistrates Knox and Pollock have declared that after their careful enquiry it would be superfluous to consider such nonsense", says Mr. Bruce.

Luckily the report by police magistrates Knox and Pollock has been published by order of Parliament. What do they say on page 23 of their report? That in accordance with the prison regime a certain number of convicts use the same bath one after the other and that "the guard cannot give priority to O'Donovan Rossa without offending the others. It would, therefore, be superfluous to consider such nonsense".

Thus, according to the report by Knox and Pollock, it is not O'Donovan Rossa's allegation that he was forced to bathe in water which had been used by convicts which is nonsense, as Mr. Bruce would have them say. On the contrary, these gentlemen find it absurd that O'Donovan Rossa should have complained about such a disgrace.

During the same meeting in the House of Commons in which Mr. Gladstone declared himself ready to hold an enquiry into the treatment of Fenian prisoners, he introduced a new Coercion Bill for Ireland, that is to say, the suppression of constitutional freedoms and the proclamation of a state of emergency.

Theoretical fiction has it that constitutional liberty is the rule and its suspension an exception, but the whole history of English rule in Ireland shows that a state of emergency is the rule and that the application of the constitution is the exception. Gladstone is making agrarian crimes the pretext for putting Ireland once more in a state of siege. His true motive is the desire to suppress the independent newspapers in Dublin. From henceforth the life or death of any Irish newspaper will depend on the goodwill of Mr. Gladstone. Moreover, this Coercion Bill is a necessary complement to the Land Bill recently introduced by Mr. Gladstone which consolidates landlordism in Ireland whilst appearing to come to the aid of the tenant farmers.[383] It should suffice to say of this law that it bears the mark of Lord Dufferin, a member of the Cabinet and a large Irish landowner. It was only last year that this Dr. Sangrado published a large tome[384] to prove that the Irish population has not yet been sufficiently bled, and that it should be reduced by a third if Ireland is to accomplish its glorious mission to produce the highest possible rents for its landlords and the largest possible quantities of meat and wool for the English market.

V

London, March 22

There is a London weekly with a wide circulation among the mass of the people which is called *Reynolds's Newspaper*. This is what it has to say about the Irish question:

"Now we are regarded by the other nations as the most hypocritical people on earth. We blew our own trumpets so loudly and so joyfully and exaggerated the excellence of our institutions so much, that now when our lies are being exposed one by one it is not at all surprising that other peoples should ridicule us and ask themselves whether it can be possible. It is not the people of England who have brought about such a state of affairs, because the people also have been tricked and deceived—the blame lies with the ruling classes and a venal, parasitic press."[385]

The Coercion Bill for Ireland which was introduced on Thursday evening* is a detestable, abominable, execrable

---

* March 17, 1870.—*Ed.*

measure. This Bill extinguishes the last spark of national liberty in Ireland and silences the press of this unhappy country in order to prevent its newspapers from protesting against a policy which is the crying disgrace of our time. The government wants its revenge on all those newspapers which did not greet its wretched Land Bill with transports of delight, and will get it. In effect the *Habeas Corpus Act* will be suspended, because from now onwards it will be possible to imprison for six months or even for life any person who cannot explain his behaviour to the satisfaction of the authorities.

Ireland has been put at the mercy of a band of well-trained spies who are euphemistically referred to as "detectives".

Not even Nicholas of Russia ever published a crueller ukase against the unfortunate Poles than this Bill of Mr. Gladstone's against the Irish. It is a measure which would have won Mr. Gladstone the good favour of the famous King of Dahomey. Nevertheless, Mr. Gladstone had the colossal effrontery to boast in front of Parliament and the nation of the generous policy which his government is proposing to adopt with regard to Ireland. At the end of his speech on Thursday Gladstone even went as far as producing expressions of regret pronounced with a sanctimonious, lachrymose solemnity worthy of the reverend Mr. Stiggins. But snivel as he may, the Irish people will not be deceived.

We repeat that the Bill is a shameful measure, a measure worthy of *Castlereagh,* a measure which will invoke the condemnation of all free nations on the heads of those who invented it and those who sanction and approve it. Finally, it is a measure which will bring well-deserved opprobrium to Mr. Gladstone and, we sincerely hope, lead to his swift defeat. And how has the demagogic minister Mr. Bright been able to keep silent for forty-eight hours?

We state without hesitation that Mr. Gladstone has proved to be the most savage enemy and the most implacable master to have crushed Ireland since the days of the notorious *Castlereagh.*

As if the cup of ministerial shame were not already full to overflowing, it was announced in the House of Commons on Thursday evening, the same evening as the Coercion Bill was introduced, that Burke and other Fenian prisoners had been tortured to the point of insanity in the English prisons, and in the very face of this appalling evidence Gladstone and his jackal Bruce were protesting that the political prisoners were treated with all possible care. When Mr. Moore made this sad announcement to the House he was constantly interrupted by hoots of bestial laughter. Had such a disgusting and revolting scene taken place in the American Congress, what a howl of indignation would have gone up from us!

Up till now the *Reynolds's News,* the *Times,* the *Daily News, Pall Mall,* the *Telegraph,* etc., etc., have greeted the Coercion Bill with shouts of wild joy, particularly the measure for the destruction of the Irish press. And all this is taking place in England, the acknowledged sanctuary of the press. But isn't there a case after all for wanting revenge on these new writers. You will agree that it was too hard to watch the *Irishman* each Saturday demolish the tissue of lies and calumny which these Penelopes worked on for six days of the week with sweat on their brows, and that it is quite natural that the latter should give a frantic welcome to the police who come to tie the hands of their formidable enemy. At least these fine fellows realise their own collective worth.

A characteristic exchange of letters has taken place between Bruce and Mr. M'Carthy Downing concerning Colonel Richard Burke. Before reproducing it I should like to remark in passing that Mr. Downing is an Irish nember of the House of Commons. This ambitious advocate joined the ministerial phalanx with the noble aim of making a career. Thus, we are not dealing here with a suspect witness.

February 22, 1870

Sir,

If my information is correct, Richard Burke, one of the Fenian prisoners formerly held in Chatham Prison, has been transferred to Woking in a state of insanity. In March 1869 I took the liberty of bringing his state of apparent ill-health to your notice, and in the following July Mr. Blake, former member for Waterford, and I informed you of our opinion that if the system of his treatment were not changed, the worst consequences were to be feared. I received no reply to this letter. My object in writing to you is the cause of humanity and the hope of obtaining his release so that his family may have the consolation of seeing to his needs and mitigating his suffering. I have in my hand a letter from the prisoner to his brother dated December 3 in which he says that he has been systematically *poisoned,* this being, I imagine, one of the phases of his disease. I sincerely trust that the kind sentiments for which you are known will urge you to grant this request.

Yours, etc.,

M'Carthy Downing

Home Office,
February 25, 1870

Sir,

Richard Burke was transferred from Chatham as a result of his illusion that he was poisoned or cruelly treated by the prison medical officers. At the same time, without him being positively ill, his health deteriorated. Consequently I gave orders for him to be moved to Woking and had him examined by Dr. Meyer from Broadmoor Asylum, who was of the opinion that his illusion would disappear when his health improved. His health did, in fact, improve rapidly and an ordinary observer would not have noted any signs of his mental weakness. I should very much like to be in a position to give you an assurance of his early release, but am not able to do so. His crime and the consequences of the attempt to free him are too serious for me to be able to give you such an assurance. Meanwhile all that medical science and good treatment can do to restore his mental and physical health will be done.

H. A. Bruce

February 28, 1870

Sir,

After receiving your letter of the 25th in reply to my request that Burke should be handed over to the care of his brother, I hoped to find an occasion to talk to you on this matter in the House of Commons, but you were so busy on Thursday and Friday that an interview was out of the question. I have received letters from a number of Burke's friends. They are waiting anxiously to hear whether my request has been successful. I have not yet informed them that it has not. Before disappointing them I felt "justified" in writing to you again on the matter. I thought that as a person who has invariably and at some risk denounced Fenianism, I could permit myself to give a word of impartial, friendly advice to the government.

I have no hesitation in saying that the release of a political prisoner who has become mentally unbalanced would not be criticised and certainly not condemned by the general public. In Ireland people would say: "Well, the government is not as cruel as we thought." Whereas

if, on the other hand, Burke is kept in prison this will provide new
material for the national press to attack it as being even crueller than
the Neapolitan governors in their worst days. And I confess that I
cannot see how men of moderate views could defend the act of refusal
in such a case....

M'Carthy Downing

Sir,
I regret that I am unable to recommend Burke's release.
It is true that he has shown signs of insanity and that in ordinary
cases I would be "justified" in recommending him to the mercy of the
Crown. But his case is not an ordinary one, because he was not only
a hardened conspirator, but his participation in the attempt to blow
up Clerkenwell which, if it had succeeded, would have been even more
disastrous than it was, makes him an improper recipient of pardon.

H. A. Bruce

Could anything be more infamous! Bruce knows perfectly
well that if there had been the slightest suspicion against
Colonel Burke during the trial concerning the attempt to
blow up Clerkenwell, Burke would have been hung next to
Barrett who was sentenced to death on the testimony of a
man who had previously given false testimony against three
other men, and in spite of the evidence of eight citizens
who made the journey from Glasgow to prove that Barrett
had been there when the explosion had taken place. The
English have no scruples (Mr. Bruce can confirm this) when
it is a question of hanging a man—especially a Fenian.

But all this spate of cruelty cannot break the iron spirit
of the Irish. They have just celebrated their national holiday, St. Patrick's Day, more demonstratively than ever in
Dublin. The houses were decorated with flags saying:
"Ireland for the Irish!", "Liberty!" and "Long live the political prisoners!" and the air rang with the sound of their
national songs and—the *Marseillaise*.

VI

AGRARIAN OUTRAGES IN IRELAND

London, April 2, 1870

In Ireland the plundering and even extermination of the
tenant farmer and his family by the landlord is called the
property right, whereas the desperate farmer's revolt

against his ruthless executioner is called an agrarian outrage. These agrarian outrages, which are actually very few in number but are multiplied and exaggerated out of all proportion by the kaleidoscope of the English press in accordance with orders received, have, as you will know, provided the excuse for reviving the regime of white terror in Ireland. On the other hand, this regime of terror makes it possible for the landowners to redouble their oppression with impunity.

I have already mentioned that the Land Bill consolidates landlordism under the pretext of giving aid to the tenant farmers. Nevertheless, in order to pull the wool over people's eyes and clear his conscience, Gladstone was compelled to grant this new lease of life to landlord despotism subject to certain legal formalities. It should suffice to say that in the future as in the past the landlord's word will become law if he succeeds in imposing on his tenants at will the most fantastic rents which are impossible to pay or, in the case of land tenure agreements, make his farmers sign contracts which will bind them to voluntary slavery.

And how the landlords are rejoicing! A Dublin newspaper, the *Freeman,* publishes a letter from Father P. Lavelle, the author of *The Irish Landlord since the Revolution,* in which he says:

> "I have seen piles of letters addressed to tenants by their landlord, the brave captain, an "absentee" living in England, warning them that from now on their rents are to be raised by 25%. This is equivalent to an eviction notice! And this from a man who does nothing for the land except live off its produce!"

The *Irishman* on the other hand publishes the new tenure agreements dictated by Lord Dufferin, the member of Gladstone's Cabinet who inspired the Land Bill and introduced the Coercion Bill in the House of Lords. Add the rapacious shrewdness of an expert moneylender and the despicable chicanery of the advocate to feudal insolence and you will have a rough idea of the new land tenure agreements invented by the noble Dufferin.

It is now easy to see that the rule of terror has arrived just in time to introduce the rule of the Land Bill! Let us suppose, for example, that in a certain Irish county the farmers refuse either to allow a 25% rent increase or to sign *Dufferin's* land tenure agreements! The county's landlords will then get their valets or the police to send them anonymous threatening letters, as they have in the past. This also counts as an "agrarian outrage". The landlords inform the Viceroy, Lord Spencer, accordingly. Lord Spencer then declares that the district is subject to the provisions of the Coercion Act which is then applied by the same landlords, in their capacity as magistrates, against their own tenants!

Journalists who are imprudent enough to protest will not only be prosecuted for sedition, but their printing presses will be confiscated without the semblance of legal proceedings!

It should, perhaps, now be obvious why the head of your executive* congratulated Gladstone on the improvements which he had introduced in Ireland, and why Gladstone returned the compliment by congratulating your executive on its constitutional concessions. "A Roland for an Olivier",[386] those of your readers who know Shakespeare will say. But others who are more versed in the *Moniteur* than in Shakespeare will remember the letter sent by the head of your executive to the late Lord Palmerston containing the words "Let us not act like knaves!"

Now I shall return to the question of political prisoners, not without good cause.

The publication of Rossa's first letter in the *Marseillaise* produced a great effect in England—the result is to be an enquiry.

The following dispatch was printed by all the newspapers in the United States:

"The *Marseillaise* says that O'Donovan Rossa was stripped naked once a day and examined, that he was starved, that he was locked in a dark cell, that he was harnessed to a cart, and that the death of his fellow prisoners was caused by the cold to which they were exposed."

The *Irishman's New York* correspondent says:

"The Rochefort *Marseillaise* has placed the suffering of the Fenian prisoners before the eyes of the American people. We owe a debt of gratitude to the *Marseillaise* which, I trust, will be promptly paid."

---

* Napoleon III.—*Ed.*

Rossa's letter has also been published by the German press.

From now onwards the English government will no longer be able to commit its outrages in silence. Mr. Gladstone will gain nothing from his attempt to silence the Irish press. Each journalist imprisoned in Ireland will be replaced by a hundred journalists in France, Germany and America.

What can Mr. Gladstone's narrow-minded, out-of-date policies do against the international spirit of the nineteenth century?

VII

THE DEATH OF JOHN LYNCH

Citizen Editor,

I am sending you extracts from a letter written to the *Irishman* by an Irish political prisoner during his detention (he is now at liberty) in a penal colony in Australia.

I shall limit myself to translating the episode concerning John Lynch.

Letter from John Casey

The following is a brief, impartial report of the treatment to which my brother exiles (twenty-four in number) and I were subjected during our incarceration in that pit of horrors, that living tomb which is called Portland Prison.

Above all it is my duty to pay a tribute of respect and justice to the memory of my friend John Lynch who was sentenced by an extraordinary tribunal in December 1865 and died at Woking Prison in April 1866.

Whatever may be the cause to which the jury has attributed his death, I confirm, and am able to furnish proof, that his death was accelerated by the cruelty of the prison warders.

To be imprisoned in the heart of winter in a cold cell for twenty-three hours out of twenty-four, insufficiently clad, sleeping on a hard board with a log of wood as a pillow and two worn blankets weighing barely ten lbs. as one's only protection against the excessive cold, deprived through an inexpressibly fine stroke of cruelty of even covering our frozen limbs with our clothes which we were forced to put outside our cell door, given unhealthy, meagre nourishment, having no exercise apart from a daily walk lasting three-quarters of an hour in a cage about 20 ft. long by 6 ft. wide designed for the worst type of criminals: such privation and suffering would break even an iron constitution. So it is not surprising that a person as delicate as Lynch should succumb to it almost immediately.

On arrival at the prison Lynch asked for permission to keep his flannels on. His request was rudely refused. "If you refuse I shall be dead in three months," he replied on that occasion. Ah, little did I suspect that his words would come true. I could not imagine that Ireland was to lose one of her most devoted, ardent and noble sons so soon, and that I myself was to lose a tried and tested friend.

At the beginning of March I noticed that my friend was looking very ill and one day I took advantage of the jailer's brief absence to ask him about his health. He replied that he was dying, that he had consulted the doctor several times, but that the latter had not paid the slightest attention to his complaints. His cough was so violent that although my cell was a long distance from his, I could hear it day and night resounding along the empty corridors. One jailer even told me, "Number 7's time will soon be up—he should have been in hospital a month ago. I've often seen ordinary prisoners there looking a hundred times healthier than him."

One day in April I looked out of my cell and saw a skeleton-like figure dragging itself along with difficulty and leaning on the bars for support, with a deathly pale face, glazed eyes and hollow cheeks. It was Lynch. I could not believe it was him until he looked at me, smiled and pointed to the ground as if to say: "I'm finished."

This was the last time I saw Lynch.

This statement of Casey's corroborates Rossa's testimony about Lynch. And it should not be forgotten that Rossa wrote his letter in an English prison whilst Casey was writing in an Australian penal colony, making any communication between the two of them quite impossible. However, the government has just stated that Rossa's assertions are lies. Bruce, Pollock and Knox even declare "that Lynch was given flannels before he asked for them".

On the other hand Mr. Casey insists as firmly as Mr. Bruce denies it that Lynch complained that "even when he was incapable of walking and was forced to remain in the terrible solitude of his cell his request was refused".

But as Mr. Laurier said in his beautiful speech:

"Let us leave aside human testimony and turn to the testimony that does not lie, the testimony that does not deceive, the *silent testimony*."[387]

The fact remains that Lynch entered Pentonville blooming with life, full of hope and vitality, and, three months later, this young man was a corpse.

Until Messrs. Gladstone, Bruce and his cohort of police can prove that Lynch is not dead, they are wasting their time in vain oaths.

VIII
LETTER FROM ENGLAND

London, April 19, 1870

"No priests in politics" is the watchword which can be heard all over Ireland at the moment.

The large party which has been opposing with all its might the despotism of the Catholic Church, ever since the "*disestablishment*" of the Protestant Church, is growing daily with remarkable rapidity and has just dealt the clergy a crushing blow.

At the Longford election the clerical candidate, Mr. Greville-Nugent, beat the people's candidate, John Martin, but the nationalists challenged the validity of his election because of the illegal means by which it had been won, and got the better of their opponents. The election of Nugent was annulled by Judge Fitzgerald who declared Nugent's agents, that is to say the priests, guilty of having bribed the voters by flooding the country not with the Holy Spirit, but with spirits of a different kind. It appeared that in the single month from December 1 to January 1 alone the reverend fathers had spent £3,500 on whisky!

The *Standard* allows itself to make some most peculiar comments on the Longford election:

"With regard to their scorning of the intimidation by the clergy," writes the mouthpiece of the "stupid party",* "the nationalists deserve our praise.... The great victory which they have won will encourage them to put up new candidates against Mr. Gladstone and his ultra-montane allies."

The *Times* writes:

"From the Papal Bull issued in the eternal city to the intrigues of the country priests, all ecclesiastical power was lined up against Fenianism and the nationalists. Unfortunately this ardour was not accompanied by prudence, and will result in a second battle at Longford."

---

* This refers to the Tories.—*Ed.*

The *Times* is right. The battle of Longford will break out
again and be followed by those of Waterford, Mallow and
Tipperary, the nationalists in these three counties also
having presented petitions requesting the annulment of the
election of the official members. In Tipperary it was
O'Donovan Rossa who first won the election, but since Parliament stated that he was incapable of representing
Tipperary the nationalists proposed Kickham in his place,
one of the Fenian patriots who has just finished a spell
in English prisons. Kickham's supporters are now declaring
that their candidate has been duly elected in spite of the
fact that Heron, the government and clerical candidate,
gained a majority of four votes.

Bear in mind, however, that one of these four voters for
Heron is a wretched *maniac* who was taken to the poll by
a reverend father—you know the weakness which priests
have for the poor in spirit, for theirs is the kingdom of
heaven. And that the second voter is a *corpse*! Yes, the
honest and moderate party actually dared to profane the
name of a man who died a fortnight before the election by
making him vote for a Gladstonian. Apart from this, patriotic voters say that eleven of their votes were discounted
on the grounds that the first letter of Kickham's name was
illegible, that their telegrams were not delivered, that the
authorities were bribing electors right and left and that a
base system of intimidation was practised.

The pressure which was brought to bear in Tipperary was
unprecedented even in the history of Ireland. The bailiff
and the policeman, who stand for eviction warrants, besieged the tenants' hovels in order to terrify wives and children first. The booths in which the voting took place were
surrounded by police, soldiers, magistrates, landlords and
priests.

The latter hurled stones at people who were putting up
posters for Kickham. On top of all this, the moneylender
was present in the booths, his eyes resting hungrily on his
wretched debtor during the voting. But the government got
nothing for all its pains. One thousand six hundred and
sixty-eight small tenants braved it out and, unprotected by
secret ballot, gave their votes openly for Kickham.

This brave act reminds us of the heroic struggle of the Poles.

Faced with the battles waged in Longford, Mallow, Waterford and Tipperary, will anyone still dare to say that the Irish are the abject slaves of the clergy.

Published in the newspaper
La Marseillaise Nos. 71, 79, 89,
91, 99, 113, 118, 125 for March 1,
9, 19, 21 and 29, and April 12,
17 and 24, 1870

Translated from the
French

DECLARATION BY THE GENERAL COUNCIL OF THE INTERNATIONAL WORKING MEN'S ASSOCIATION[388]

POLICE TERRORISM IN IRELAND

The national antagonism between English and Irish working men, in England, has hitherto been one of the main impediments in the way of every attempted movement for the emancipation of the working class, and therefore one of the mainstays of class dominion in England as well as in Ireland. The spread of the International in Ireland, and the formation of Irish branches in England, threatened to put an end to this state of things. It was quite natural then that the British Government should attempt to nip in the bud the establishment of the International in Ireland by putting into practice all that police chicanery which the exceptional legislation and the practically permanent state of siege there enable it to exercise. How Ireland is governed in a truly Prussian way, under what is called the Free British Constitution, will appear from the following facts.

In Dublin, at the meeting of the International, a sergeant and private of the police, in full uniform, were stationed at the door of the place of meeting, the owner of which asked them whether they were sent officially, and the sergeant said he was, the International having a dreaded name.

In Cork the same trick is practised. Two constables of the "Royal Irish Constabulary" are placed opposite the house door of the secretary of the local section, during the day, and four after dark, and the name of every one is noted down who calls upon him. A sub-inspector has recently called upon several persons by whom members of the Cork section were employed, and demanded the addresses of the latter, and many persons have been warned by the "Constabulary" that if they are seen speaking to the secretary their names will be sent to "The Castle"—a name of horror to the working class of Ireland.[389]

In the same city, according to a letter received,

"The magistrates have held several special meetings, extra police have been drafted in, and on Easter Sunday the constables were all under arms, with ten rounds of ball cartridge each. They expected we were going to have a meeting in the park; the magistrates are trying all they can to provoke a riot."

If the British Government continues in this way they may be sure that the last shreds of the mask of liberalism will be torn from their faces. In the International papers all over the world, the name of Mr. Gladstone will be coupled week after week with those of Sagasta, Lanza, Bismarck, and Thiers.

By order of
the General Council

R. APPLEGARTH, M. BARRY, M. J. BOON, F. BRADNICK, G. H. BUTTERY, É. DELAHAYE, EUGÈNE DUPONT, W. HALES, G. HARRIS, HURLIMAN, JULES JOHANNARD, C. KEEN, HARRIETT LAW, F. LESSNER, LOCHNER, C. LONGUET, C. MARTIN, ZÉVY MAURICE, H. MAYO, G. MILNER, CH. MURRAY, PFÄNDER, J. ROACH, RÜHL, SADLER, COWELL STEPNEY, A. TAYLOR, W. TOWNSHEND, E. VAILLANT, J. WESTON, YARROW.

Corresponding Secretaries:

LEO FRANKEL, for Austria and Hungary; *A. HERMAN*, Belgium; *T. MOTTERSHEAD*, Denmark; *A. SERRAILLIER*, France; *KARL MARX*, Germany and Russia; *C. ROCHAT*, Holland; *J. P. McDONNELL*, Ireland;

F. ENGELS, Italy and Spain; *WALERY WRÓBLEWSKI*, Poland; *HERMANN JUNG*, Switzerland; *J. G. ECCARIUS*, United States; *LE MOUSSU*, for French branches of United States; *J. HALES*, General Secretary

Published as a leaflet in April 1872

Printed according to the text of the book *The General Council of the First International.   1871-1872. Minutes,* Moscow

COUNCIL MEETING*
MAY 14th, 1872

Citizen *Serraillier* in the chair.

Members present: Citizens *Boon, Barry, Cournet, Dela-*
haye, Eccarius, Engels, Arnaud, Frankel, Hales, Jung,
Lessner, Mayo, Martin, McDonnell, Milner, Mottershead,
Murray, Le Moussu, Rühl, Serraillier, Townshend, Vaillant
and *Yarrow*.

The Minutes of the preceding meeting having been read
and confirmed, the *Secretary*** read a declaration from
Citizen Weston, to the effect that his name had been appended to the document, purporting to be the rules of the
Universal Federalist Council of the International,[390] without
his knowledge. This document was signed by Weston in the
presence of Eccarius, Roach and himself, and he told them
that he visited one of the meetings of the dissentients, but
he went upon invitation, unofficially, and knew nothing of
any intention to publish. He disagreed with the publication,
though he did consider a competent tribunal had a right
to arraign the General Council; what he meant by a competent tribunal was a body of men still within the International. He had a complaint against the Council himself, and
that was that his name had been used without letting him
know of any intention to publish. He knew the Council
had a right to use the name of the members, but he thought
that out of courtesy information ought to be sent to all, so

---

* The Minutes are in Hales's hand on pp. 448-53 of the Minute
Book.—*Ed.*
** Hales.—*Ed.*

as to give them an opportunity to be present if they wished. He felt very strongly upon the point because when he did attend the Council he was treated very cavalierly by certain members if he happened to disagree with them. He did write a letter in answer to the second one, but found he was too late for post when he had finished. No slight whatever was intended.

A motion was carried accepting the reply as satisfactory and it was ordered to be sent for publication in the report.

Citizen *Engels* reported that the seat of the new Federal Council of Spain had been fixed at Valencia; he had received the first letter; Lorenzo was the new secretary. He asked for the addresses of all the other Federal Councils.

Citizen *McDonnell* reported that the movement was progressing in Cork and Dublin. He read a letter from a correspondent in Dublin, which expressed a hope that the journals of the Association would avoid any articles expressing atheistical opinions, or condemnation of Catholicism, as anything of the kind would do great damage in Ireland, which opinion Citizen McDonnell endorsed.

Citizen *Yarrow* announced that the Alliance Cabinet-Makers (who were members) had formed an amalgamation with the East [End] London Cabinet-Makers, and it had been resolved that all fresh jobs should be taken as daywork and the prices based upon time.

Citizen *Hales* reported that Messrs. Shaen and Roscoe had sent a letter informing the Council that Mr. Wilkinson of St. George's Hall had consented to pay the damages asked for, upon production of receipts.

A suggestion was made by Citizen *Barry* that the Council should celebrate the fall of the *Commune*; but as no proposition was made, the matter fell through.

Citizen *Hales* proposed "That in the opinion of the Council the formation of *Irish* nationalist branches in England is opposed to the General Rules and principles of the Association". He said he brought forward the motion in no antagonism to the Irish members, but he thought the policy being pursued [is] fraught with the greatest danger to the Association, besides being in antagonism to the Rules and principles. The fundamental principle of the Association was to destroy all semblance of the nationalist doctrine, and remove all barriers that separated man from man, and the formation of either Irish or English branches could only retard the movement instead of helping it on. The formation of *Irish* branches in England could only keep alive that national antagonism which had unfortunately so long existed between the people of the two countries. Misunderstandings would arise—nay, had arisen, and there was almost certain to be conflicts between the different sections upon important matters of policy. The Secretary for Liverpool* wrote and said he understood an Irish section had been formed in Liverpool, but he didn't know where it was, nor what it was doing; did that savour of international harmony? A section had been formed in Middlesbrough based upon the section which previously existed in that town—and it had decided that it should not be called an Irish section but simply the Middlesbrough section. Yet when Citizen Roach wrote and asked the section to correspond with the Federal Council, he received an answer telling him virtually to mind his own business and informing him that if he wanted to know anything about the section he could apply to Citizen McDonnell. So that jealousy had already arisen. No one knew what the Irish branches were doing, and in their rules they stated that they were republican, and their first object was to liberate Ireland from a foreign domination. Now he contended that the International had nothing to do with liberating Ireland, nor with the setting up of any particular form of government, either in England or Ireland, and it was the duty of the Council to prevent any mistake upon the subject by passing the resolution he proposed. If such was not done they would have splits which perhaps could not be healed.

Citizen *Mayo* seconded.

Citizen *Mottershead* could not escape from the logic of the motion, but he deprecated the spirit in which it was made. The speech of Citizen Hales showed the animus with which he was actuated, and, seeing that, he could not vote for the motion. He would rather vote for a motion recommending our English members to cultivate a spirit of fra-

* George Gilroy.—*Ed.*

ternity with the Irish members. He unfortunately knew too well the domineering spirit with which Englishmen of the ignorant class treated their Irish brethren. They had been treated as aliens in a foreign land, and were looked down upon by the English workers much the same as the mean Whites of the South looked down upon Negroes. He objected to the style and manner of the Secretary's speech and he hoped the Council would show its feeling upon the matter by rejecting the motion.

Citizen *McDonnell* quite agreed with Mottershead that it was desirable that Englishmen should cultivate a fraternal feeling with the Irish, and he thought such speeches as that delivered by Citizen Hales were the most injurious it was possible to conceive. Why, the speech he made when he gave notice of motion, had it been reported, would have prevented the establishment of the Association in Ireland and would have destroyed all hopes of doing so. It seemed very strange that the General Secretary should, at the moment when there were dangers and difficulties attending the work of propaganda in Ireland, come forward with a motion which would virtually destroy the work that had been done. It looked suspicious. Why, to ask Irishmen to give up their nationality was to insult them. He was proud to say that he had worked for the redemption of Ireland and would continue to do so; it was impossible to crush out the aspirations of the Irish people. The only effect of the passing of the resolution proposed would be to prevent Irishmen joining. He would ask what had been done before he joined the Council to extend the Association among Irishmen. Nothing! And now [that] he had done something it was proposed to undo it.

Citizen *Boon* was sorry that the motion should have been brought on, though he was not surprised that the Secretary should have done so. The Normans conquered Ireland and held her in subjection by the aid of their Saxon serfs, and the motion made meant that the rule of the Saxon should still continue. The same spirit of domination was still rampant in the minds of some of the English working men. He approved of the nationalist character of the Irish people's organisations and he hoped they would still continue and not be coerced into giving up their rights either by the English Government or the English working class. He was strongly of an opinion that Hales did not understand the Irish character; he would protest against the passage of the motion.

Citizen *Engels* said the real purpose of the motion, stripped of all hypocrisy, was to bring the Irish sections into subjection to the British Federal Council, a thing to which the Irish sections would never consent, and which the Council had neither the right nor the power to impose upon them. According to the Rules and Regulations, the Council had no power to compel any section or branch to acknowledge the supremacy of any Federal Council. It was certainly bound, before admitting or rejecting any new branch within the jurisdiction of a Federal Council, to consult that Council, but he maintained that the Irish sections in England were no more under the jurisdiction of the British Federal Council than the French, German, Italian or Polish sections in this country. The Irish formed a distinct nationality of their own, and the fact that [they] used the English language could not deprive them of their rights. Citizen Hales had spoken of the relations of England and Ireland being of the most idyllic nature—breathing nothing but harmony. But the case was quite different. There was the fact of seven centuries of English conquest and oppression of Ireland, and so long as that oppression existed, it would be an insult to Irish working men to ask them to submit to a British Federal Council. The position of Ireland with regard to England was not that of an equal, but that of Poland with regard to Russia. What would be said if the Council called upon Polish sections to acknowledge the supremacy of a Council sitting in Petersburg, or the North Schleswig and Alsatian sections to submit to a Federal Council in Berlin? Yet that was asked by the motion. It was asking the conquered people to forget their nationality and submit to their conquerors. It was not Internationalism, but simply prating submission. If the promoters of the motion were so brimful of the truly international spirit, let them prove it by removing the seat of the British Federal Council to Dublin and submit to a Council of Irishmen. In a case like that of the Irish, true Internationalism must necessarily be based upon a distinct national organisation, and they were under the necessity to state in the preamble to their rules that their first and most pressing duty as Irishmen was to establish their own national independence. The antagonism....*

Citizen *Murray* didn't regret the discussion though it had all been on one side. Citizen Hales seemed to imagine that unity could be obtained by putting down the Irish branches. He thought that a mistake. The Irish could not forget all at once 700 years of English misrule and it must be remembered that the English workmen had not treated the Irish as they ought to have done. It was only yesterday that the columns of the newspapers used to contain the stereotyped advertisement "That no Irish need apply" and the passage of the resolution would be virtually saying no Irish need apply.

Citizen *Hales* said all the speeches made in opposition really proved his case. It was admitted that the Irish did not understand the principles of the International, for all the speakers urged that if the word "Irish" was struck out of the names of the branches, the Irish would not join, which was only saying that they were national and not international. He had been told he didn't understand the Irish character—well, he thought he did, and that was the reason he brought on his motion. He believed the majority of the members of the Irish branches did not understand the principles of the Association; as the correspondent of the *Standard* said: They were only Fenians under another name, and they became members of the International because they saw that it would be a convenient cloak under which to prosecute their special designs—and he objected to that not because he had any objection to Fenianism, but because he wanted the Association [to be] free from special sects or cliques. He had advocated Fenianism for he held that the Irish like other people had a right to govern themselves; the right of self-government was inalienable, and no people could be deprived of that right; he should like to see Ireland ruling herself tomorrow for he was convinced that the Irish themselves would then wake from their enchantment and find that nationalism was no remedy for the ills of society.

---

* The record breaks off here; 15 lines are left blank in the Minute Book. For the full text of Engels's speech see pp. 302-04.—*Ed.*

He asked them to pass the motion and thus prevent future mischief.

The motion was put and lost, only one voting in favour.

A short discussion then took place on the advisability of reporting the discussion and it was decided that Citizen Hales should draw up a report to be submitted on Saturday.*

The Council adjourned at 11.30.**

Published in the book
*The General Council of
the First International.
1871-1872. Minutes,*
Moscow

Printed according to the text of the book

---

* May 18.—*Ed.*
** Unsigned.—*Ed.*

WILLIAM THORNE AND ELEANOR MARX-AVELING TO SAMUEL GOMPERS
JANUARY 25, 1891[394]

Mr. Samuel Gompers
for the American Federation of Labour

Dear Comrade,

During the recent visit of Comrades Bebel, Liebknecht and Singer on the occasion of Frederick Engels' 70-th birthday, they met representatives of the Gasworkers and General Labourers Union (comprising about 100,000 men and women belonging to over seventy trades) and of several other Unions and Organisations, besides John Burns, Cunninghame Graham, M. P. and others. At this meeting the feeling was very strong that the time had come to bring about a close and organised relation between the labour parties of the different countries. The most immediate question is that of preventing the introduction from one country to another of unfair labour, i.e., of workers who not knowing the conditions of the labour-struggle in a particular country, are imported into that country by the Capitalists, in order to reduce wages, or lengthen the hours of labour, or both. The most practical way of carrying this out appears to be the appointing in each country of an International Secretary, who shall be in communication with all the other International Secretaries. Thus, the moment any difficulty between capitalists and labourers occurs in any country, the International Labour Secretaries of all the other countries should be at once communicated with, and will make it their business to try to prevent the exportation from their particular country of any labourers to take the place on unfair terms of those locked-out or on strike in the country where the difficulty has occurred. Whilst this is the most immediate and most obvious matter to be dealt with, it is hoped that an arrangement of the kind proposed, will in every way facilitate the interchange of ideas on all questions between the workers of every nation that is becoming every day and every hour the most pressing necessity of the working-class movement.

If your organisation agrees with the views of the Gas-workers and General Labourers Union, will you at once communicate with us, and give us the name of the Secretary appointed by it to take part in this important movement?

Yours fraternally

W. Thorne
(General Secretary)

Eleanor Marx-Aveling
(On the behalf of the Executive Committee)

Published in:
Marx and Engels,
Collected Works,
second Russian ed.,
Vol. 38, Moscow,
1965

Printed according to
the original

NOTES

[1] *Letters from London* is a series of four reports written by Engels in May and June 1843 during his stay in England, where he had been sent by his father, a German textile manufacturer, in November 1842 to learn the business in the Manchester branch of Ermen & Engels. In England, where he remained till August 1844, Engels rejected his idealism and revolutionary democratism and arrived at a materialist and communist world outlook. The reports were written for the *Schweizerischer Republikaner,* a radical Swiss journal published in Zurich.   p. 33

[2] *The Anglo-Irish Union* was imposed on Ireland by the English Government after the suppression of the Irish rebellion of 1798. The Union, which became valid as of January 1, 1801, abrogated the autonomous Irish Parliament and made Ireland even more dependent on England. In the 1820s Repeal of the Union became the most popular slogan in Ireland. However, the Irish bourgeois liberals (O'Connell and others) who headed the national movement wanted to use the agitation for Repeal of the Union solely as means for exerting pressure on the English Government to make it grant small concessions to the Irish bourgeoisie and landowners. In 1835, O'Connell made an agreement with the Whigs and stopped this agitation altogether. In 1840, after the Tories assumed office, the Irish liberals were compelled, under pressure from the mass movement, to set up the Repeal Association (Repealers) but endeavoured to make it take the road of compromises with the English ruling classes.   p. 33

[3] *Principal tenant*—person renting land directly from the landowner and then leasing it in small lots to subtenants. The subtenants also often broke up their lots and leased them out to even smaller tenants. Thus, a hierarchy of intermediaries often formed between the landowner and the direct producer.   p. 34

[4] Marx uses the term *juste-milieu,* literally "golden mean"—a half-hearted, in-between position, attempting to avoid extremes. Speaking below of O'Connell's friends from the *juste-milieu,* Engels was referring to the English Whigs, whom, from 1835, the Irish liberals supported against the Tories.   p. 35

⁵ The second Chartist petition for a People's Charter (a programme of six points providing for the introduction of universal suffrage and other reforms of the English political system) included the demand for Ireland to be allowed to annul the enforced Union with England of 1801. The petition was drafted by the Executive of the National Charter Association (founded in 1840), the first mass party of the working class in the history of the English labour movement. On May 2, 1842, the petition was submitted to Parliament, but even though it had been signed by about three and a half million people, it was rejected by the House of Commons. The Irish liberals headed by O'Connell did not approve of the Chartist agitation. p. 35

⁶ The book *The Condition of the Working-Class in England* was written by Engels between September 1844 and March 1845 in Barmen on the basis of material he had collected during his two-year stay in England. The work is one of the first in socialist literature to substantiate the historic role of the working class as a social force called upon to carry out the socialist transformation of society in the interests of all working people. Engels's description of the living and working conditions of the English proletariat also contains an account of the conditions of Irish immigrant labourers, and of the working people in Ireland. Two relevant excerpts are given in this collection. The book was first published in Leipzig in 1845. In Engels's lifetime two editions of the authorised English translation appeared—one in New York in 1887, the other in London in 1892 (passages from the Preface to the latter are also included in this collection, see pp. 344-45). In the appendix to the American edition of 1887, the author warns the reader that the book was written in the period when the theory of scientific communism was only taking shape and was therefore not mature in all respects, containing here and there some old views he had not yet completely repudiated when the book was written. p. 37

⁷ Engels is referring to the book *The Moral and Physical Condition of the Working Classes, employed in the Cotton Manufacture in Manchester* by James Ph. Kay, Dr. Med., 2nd ed., 1832. In the Chapter "The Great Towns", Engels's descriptions of the workers' districts in Manchester are often based on Kay's book. p. 39

⁸ The *Tithes Commutation Bill* was passed by the English Parliament in 1838. Its passing was preceded by a stubborn struggle of the Irish peasants against the tithe, which was a heavy burden and emphasised their religious inequality (the tithe was collected in favour of the Anglican Church, which was alien to the Irish Catholics) and social disparity (the landlord stock-breeders were exempt from the tithe). Beginning with 1831 the resistance against this levy turned into "the war against the tithe" and led to armed clashes between the peasants and the police and troops. Afraid that it might arouse even greater discontent, the English ruling circles agreed to some concessions. According to the Commutation Bill tithes were lowered by 25 per cent and were levied not

directly on the harvest but as a tax on rent, which was paid by the landlords. The latter increased rents correspondingly, and the Irish peasants, though in different form, continued to pay for the maintenance of the Anglican Church in Ireland. p. 41

9 Engels is referring to "Irish Immigration", an earlier chapter in his book, where he cited a view on the Irish working people from Thomas Carlyle's well-known pamphlet *Chartism* (published in 1839). Engels points out that Carlyle's views were both one-sided and exaggerated. In the quoted passage he says that these views reflect the Anglo-Saxon nationalist prejudices of the author and his bias against the Irish people. On the whole, in his *Condition of the Working-Class in England* Engels was not critical enough of Carlyle, particularly as regards his opinion of the Irish people. In their later works—the *Manifesto of the Communist Party* (1848) and the review of Thomas Carlyle's book *Latter-Day Pamphlets* (1850)—Marx and Engels fully exposed and strongly condemned the reactionary views of Carlyle, an ideologist of feudal socialism and an enemy of the working-class and national liberation movements. p. 41

10 See Note 2. p. 42

11 In 1843 the movement for the Repeal of the Union assumed a large scale in Ireland. Mass meetings in support of this demand were held all over the country. O'Connell proclaimed 1843 "repeal year". The Tory Government decided to resort to repressive measures: it outlawed the meeting set for October 5 in Clontarf, near Dublin, and concentrated troops there. Afraid of losing control over the movement, if it should come to clashes, the liberal leaders of the Repeal Association cancelled the meeting. Encouraged by this capitulation, the English authorities brought O'Connell and eight other Irish leaders to trial, and the hearing took place in January and February 1844. O'Connell was sentenced to 12 months' imprisonment, the other accused to somewhat shorter terms. Three months later, however, following widespread protests and general ferment, the House of Lords quashed the sentence. p. 43

12 This was the first of a series of articles by Engels in *La Réforme*. In addition to strictly English subjects—the trade crisis, the upsurge of the Chartist movement, etc., Engels also devoted much attention to Irish issues, notably to the advance of the national liberation movement in Ireland.

La Réforme—a democratic daily newspaper published between 1843 and 1850 in Paris by a group of petty-bourgeois republicans and socialists—Flocon, Ledru-Rollin, Louis Blanc and others. p. 44

13 In the spring of 1846, the English Prime Minister Robert Peel submitted to the House of Commons a Bill legalising police terror in Ireland under the pretext of a ban on the carrying of arms. The Bill was defeated by the opposition party, the Whigs, who used it to topple Peel's Cabinet. On coming to power, however, the

Whigs passed, in 1847, a Coercion Bill for Ireland, which launched
a wave of new cruel reprisals against the Irish people.   p. 45

[14] *King's County* was the name given by the English conquerors in
the 16th century to the Irish county of Offaley, in honour of the
husband of the English Queen Mary Tudor—Philip II of Spain.
p. 45

[15] See Note 2.   p. 46

[16] The three kingdoms referred to are England, Scotland and Ireland.
p. 47

[17] Engels's article "Feargus O'Connor and the Irish People" was published in the *Deutsche-Brüsseler Zeitung*, the mouthpiece of the
Communist League, the first international revolutionary organisation of the proletariat. Marx's and Engels's contributions to that
newspaper, founded by German political emigrants in Brussels,
determined its trend, reflecting the consistent revolutionary-democratic and communist views of the League's members. The paper
appeared between January 1847 and February 1848.   p. 48

[18] The newspaper *Northern Star*, founded in 1837, was published up
to 1852, first in Leeds and from November 1844 in London. Its
founder and editor was Feargus O'Connor, but it was Julian
Harney, a leader of the revolutionary Chartists, who determined
its revolutionary trend. Under his guidance the paper became a
militant proletarian organ, which was greatly esteemed by the
masses and exerted a major influence on them. Harney enlisted
Frederick Engels as permanent contributor and Engels's articles
appeared regularly in the paper between 1843 and 1850. Marx
and Engels highly valued the *Northern Star* as a militant organ
of the proletarian democrats. When O'Connor deserted the proletarian movement and took up petty-bourgeois democratic positions,
Harney was compelled to leave the paper in 1850 and it lost its
revolutionary trend.   p. 48

[19] *Repealers*—participants in the movement for the Repeal of the Union
between England and Ireland and for the setting up of an independent Irish Parliament. The leadership of the movement was
exercised by liberals. See also Note 23.   p. 48

[20] *Conciliation Hall*, one of the biggest halls in Dublin, in which the
Repeal Association held public meetings. They were frequently
addressed by Daniel O'Connell and later by his son John, who
became the head of the Association after his father's death. Both
father and son opposed a genuinely revolutionary struggle against
English colonial rule, although they constantly swore that they
would achieve Ireland's independence, if necessary, by means of
armed uprising.   p. 49

[21] See Note 5.   p. 49

[22] The excerpt given below is from a speech Marx made at a meeting
in Brussels dedicated to the second anniversary of the Polish up-

rising against Austrian rule in Cracow. The meeting was called by the Association démocratique in Brussels, an international organisation founded in the autumn of 1847, which united in its ranks proletarian revolutionaries and progressive bourgeois and petty-bourgeois democrats. Marx was its Vice-Chairman. p. 51

23 The upsurge of the national liberation struggle in Ireland widened the already existing differences between the moderate and revolutionary wings of the Repeal Association. The liberal landowners, making up its Right wing, wanted the movement to confine itself to "legal means". The revolutionary wing, whose most consistent champions were John Mitchel and Thomas Lalor, were for armed struggle against English colonial rule and the setting up of an Irish Republic, for giving the land to the Irish peasants, for an alliance with the Chartists and the implementation of democratic reforms. In January 1847, the Repeal Association split up and its revolutionary-democratic wing formed an organisation of its own— the Irish Confederation—which began to prepare an uprising (see also Note 54).

After the uprising was suppressed in 1848, the Irish Confederation fell to pieces and the majority of its active members were either banished or gaoled; the survivors emigrated, mainly to the U.S.A., where they later joined the Fenian movement. p. 51

24 The article "Köln in Gefahr" (Cologne Is in Danger) appeared in *Neue Rheinische Zeitung*, the militant organ of the proletarian revolutionaries published from 1848 to 1849. Marx was the editor-in-chief and the other members of the editorial board were Frederick Engels, Wilhelm Wolff, Georg Weerth, Ferdinand Wolff, Ernst Dronke, Ferdinand Freiligrath and Heinrich Bürgers. The newspaper became a headquarters of the proletariat's revolutionary forces, its organisational centre. Its editorials, generally written by Marx and Engels, defined the working class's attitude towards all important revolutionary issues and events. The newspaper educated the masses in a revolutionary spirit, in the spirit of proletarian internationalism. It supported the Irish national liberation movement and called on the Chartists to ally themselves with revolutionary Ireland. Reports on events in England and Ireland appeared regularly on its pages. p. 52

25 The beginning of Goethe's poem "Reineke Fuchs" (Reineke the Fox). The reference to the confusion of cases applies to the Berlin dialect. p. 52

26 Engels is referring to the repressive measures the English Government launched against the Chartists and the participants in the Irish national liberation movement. At the end of May the leaders of the armed uprising under preparation in Ireland were arrested; John Mitchel, the leader of the revolutionary wing of the Irish nationalists, was deported to the Bermudas for 14 years. Many members of the clubs were subjected to repressive measures. Ernest Jones, a leader of the revolutionary Chartists, and his collaborators

were arrested early in June and accused of high treason. Mass
arrests of Chartists began on June 4, 1848. p. 52

27 *The Committee of Fifty* was elected by the Preparliament, the meeting
of the public functionaries of the German states held in Frankfurt
am Main between March 31 and April 4, 1848. Most of its members
were constitutional monarchists. The Committee rejected the proposal
of the Bundestag to set up a triumvirate as an interim central
power of the German Union. On June 28, 1848, the Frankfurt National Assembly adopted a decision to set up a government body
comprising the Imperial Regent and the Imperial Ministry. p. 52

28 A reference to the return of the Prince of Prussia to his Berlin
palace, from which he had fled when the revolution broke out.
Berlin's insurgent workers traced the inscription: "Property of the
entire nation", on the palace wall. p. 52

29 This article by Marx was published in *The New-York Daily Tribune*
of which Marx became a regular correspondent in August 1851.
The newspaper was founded in 1841 by Horace Greeley, a
prominent American journalist and political figure, and up to the
mid-fifties reflected the views of the Left wing of the American
Whigs. Later it became the organ of the Republican Party. During
the forties and fifties the paper stood on progressive positions and
campaigned against slavery in America. In the early sixties, during
the Civil War, the champions of a compromise with the Southern
States gained the upper hand in the paper, producing a corresponding change in its political trend.

  Marx contributed to the paper for more than ten years, up to
March 1861. Many articles were written by Engels at his request.
Marx's and Engels's articles dealt with the most diverse questions
of international and domestic policy, of the working-class and
democratic movements, economics, etc.

  The facts on the expropriation of the land from the rural
population by the Sutherland family given in this article were
later used by Marx in Chapter XXVII of the first volume of his
Capital. p. 53

30 J. Dalrymple, *An Essay towards a General History of Feudal
Property in Great Britain*, London, 1759. p. 53

31 D. Ricardo, *On the Principles of Political Economy, and Taxation.*
First edition appeared in London in 1817. p. 55

32 J.-C.-L. Simonde de Sismondi, *Nouveaux principes d'économie politique, ou de la richesse dans ses rapports avec la population.* I-II,
Paris, 1819. p. 55

33 Marx's term for the excitement following the discovery of gold in
Australia and California. Among the people who rushed to these
gold-fields were many young, politically active English workers,
who emigrated there to look for work and to escape persecution
for their participation in the Chartist movement. The mass emigra-

tion had a baneful effect on the composition of the English working class. p. 56

34 In his *Republic* Plato expressed the conviction that the ideal state must rely on a strict division of labour and that poets must be banished from it because, he said, they were of no use. p. 57

35 Marx realised this intention in the article "The Future Results of the British Rule in India", printed in *The New-York Daily Tribune* on August 8, 1853. p. 59

36 The *Coalition Ministry* (1852-55), headed by Aberdeen, consisted of representatives of both ruling parties: the Whigs and the Tories and a group of Peelites (moderate Tories), to whom the Premier himself belonged. Whigs predominated in the Ministry. Aberdeen's Government was ironically called the "ministry of all talents". p. 59

37 A draft Bill submitted by Aberdeen's Government to the House of Commons in June 1853. The government expected to normalise the relations between landlords and tenants by giving the latter some rights and thereby mitigating the class struggle in the country. After more than two years of debates Parliament rejected the Bill. p. 59

38 The article referred to was printed in *The Times* on June 25, 1853. p. 60

39 With the introduction of the Union in 1801 the English Parliament abolished the tariffs which had protected the emergent Irish industry against European competition since the end of the eighteenth century. The abrogation of the tariffs dealt a mortal blow to Irish manufacture, which was unable to compete with the far more powerful English industry. Cotton and wool manufacture died out altogether and Ireland became an agrarian appendage of England. p. 61

40 *Free traders*—champions of unencumbered trade and non-intervention by the state in the economy. The centre of the free traders was in Manchester, where the so-called Manchester School emerged —a trend in economic thought reflecting the interests of the industrial bourgeoisie. The movement was headed by the textile manufacturers Cobden and Bright, who in 1838 organised the Anti-Corn Law League. In the forties and fifties the free traders were a separate political grouping of bourgeois radicals, who at the end of the fifties amalgamated with the emerging English Liberal Party. p. 63

41 Marx means the Irish Sea. p. 66

42 Jonathan Swift bequeathed his entire fortune to the building of a lunatic asylum in Dublin. It was opened in 1757. p. 66

43 In 1853, Parliament adopted a Bill on the encumbered estates in Ireland belonging to the Irish nobility. At that time there were many manorial estates in Ireland which had been mortgaged and

mortgaged anew because their owners were unable to make ends meet. Moreover, according to English legislation, they were obliged to help the poor residing on their lands. According to the 1853 Act, these manorial estates (the remnants of the indigenous Irish landed estates) were to be quickly sold to the highest bidder and the proceeds used to pay off creditors. This was one of the measures that helped English landlords to take possession of Irish lands and to use them as pastures. p. 67

[44] See Note 36. p. 68

[45] Karl Marx's *Lord Palmerston* appeared as a series of articles in the Chartist *People's Paper* and in abridged form in *The New-York Daily Tribune*. In addition, some of the articles were printed in England as a separate pamphlet. The pamphlet gives an accurate and witty portrait of Lord Palmerston, England's major statesman and a typical representative of the bourgeois-aristocratic oligarchy. In his person Marx gave a portrait and appraisal of English diplomacy in general, of the country's entire official foreign policy. By giving concrete examples of Palmerston's attitude to the national liberation struggle of the Polish, Irish, Hungarian and Italian people, Marx exposed the counter-revolutionary essence of English foreign policy, the constant support it gave to reactionary regimes in all countries, the deeply anti-popular policy of hypocrisy and cynicism. Marx convincingly demonstrated that Palmerston's policy was typical of the English ruling circles, whose interests he defended. p. 70

[46] *Emancipation of Catholics*—the abolition by the English Government in 1829 of restrictions placed on the political rights of Catholics. Catholics, most of whom were Irish, were granted the right to stand for election to Parliament and to hold some government offices. Simultaneously, the property census was raised fivefold. The 1829 Act was introduced after several decades of struggle by the Irish Catholic bourgeoisie, landowners and Catholic clergy, into which they had drawn the peasantry. The Act was to some extent a concession by the English Government, which at the same time expected that this manoeuvre would split and weaken the national movement and bring the élite of the Irish bourgeoisie and the landowners over to its side. p. 70

[47] The *Irish Brigade*—the name given by Marx to the faction of Irish deputies in the British Parliament. In the 1830s-1850s it was made up mainly of representatives of the Right wing in the national movement, who were reflecting the interests of the élite of the Irish bourgeoisie, the landlords and the Catholic clergy. Among them there were also Irish liberal functionaries who were relying on support from well-to-do tenants. Owing to the balance between the Tories and the Whigs in the House of Commons, the Irish Brigade, alongside with representatives of the free trader bourgeoisie, was able to tip the scale in the House of Commons and to influence the struggle in it, sometimes even to decide the fate of the government. p. 72

⁴⁸ In February 1835, Daniel O'Connell, the leader of the Irish bourgeois nationalists, signed an agreement with representatives of the
Whigs according to which he was to support them in the House
of Commons in return for some concessions; in particular, Irish
political leaders were promised posts in the administrative appa-
ratus after the Whigs came to office. For his part, O'Connell undertook to stop the Repeal of the Union campaign. The agreement was
negotiated in Lord Lichfield's London house and became known
as the Lichfield-House Contract. It meant that the liberal circles
of the Irish bourgeoisie and the medium landowners had reached
a compromise with the English politicians and had renounced
consistent struggle for Ireland's independence.   p. 73

⁴⁹ The article "Ireland's Revenge" was published in the *Neue Oder-
Zeitung*—a bourgeois-democratic daily that appeared between 1849
and 1855 in Breslau. The most radical newspaper at the time, it
was often persecuted by the ruling circles. Max Friedländer, a
German journalist, and a cousin of Lassalle, invited Marx to co-
operate on it, and as from December 1854 Marx became the London
correspondent of the paper, contributing two or three articles a
week.   p. 74

⁵⁰ See Note 2.   p. 74

⁵¹ See Note 40.   p. 74

⁵² See Note 46.   p. 75

⁵³ In 1845-47 a grievous famine blighted Ireland due to the ruin
of farms and the pauperisation of the peasants, who were cruelly
exploited by the English landlords. Although there was a great
dearth of potatoes, the principal diet of the Irish peasants, the
English landlords continued to export food from the country,
condemning the poorest sections of the population to starvation.
About a million people starved to death and the new wave of
emigration caused by the famine carried away another million.
As a result large districts of Ireland were depopulated and the
abandoned land was turned into pastures by the Irish and English
landlords.   p. 76

⁵⁴ In 1848, a popular uprising was being prepared in Ireland. Its aim
was the national liberation of the country and the establishment of
a republic. The preparations for the uprising were directed by
the Left wing of the Confederation (Mitchel, Lalor, Reilly and
others), who set up armed clubs throughout the country, which
trained units of the national guard and manufactured arms. Mitchel
and his friends established contacts with the Left wing of the
Chartists (Jones and others), who planned to rise simultaneously
with the Irish. At the end of May 1848, the English authorities
arrested Mitchel and other active leaders of the clubs. Mitchel
was deported to the Bermudas. More troops were sent to Ireland
and the inviolability of the person guaranteed by the Constitution
was revoked. After long hesitation, late in June 1848, the surviving
leaders of the Irish Confederation (Smith O'Brien and others) called

upon the Irish to revolt. But they had missed the moment. The uprising took the form of uncoordinated actions in several counties which were easily put down by the troops. The English Government was supported by the Catholic clergy and the landowner élite. p. 76

55 See Note 43. p. 76

56 Marx is referring to the new offensive begun by the English and French troops in the spring of 1855 during the Crimean War (1853-56). Marx and Engels believed that it could have led to the rout of tsarism, if the Allied troops had taken energetic action. Marx sharply censured the foreign policy pursued by the English and French governments, who were striving to consolidate their positions in the Balkans and oust Russia while simultaneously trying to preserve the tsarist autocracy as an instrument for the suppression of revolutionary and national liberation movements. In the articles describing the war, Marx and Engels paid tribute to the skill of the Russian soldiers who defended Sevastopol. p. 78

57 Marx's *Lord John Russell*, consisting of six articles, is a vivid document revealing the essence of the two-party system in England. It was directed against John Russell, a typical Whig, and exposed the policies of that party. Marx showed that the struggle between Whigs and Tories did not affect any crucial issues of domestic and foreign policy, that the attacks by the opposition on the government were a component of the two-party mechanism, and that the efforts of both parties were aimed at preserving the power in the hands of the aristocracy and bourgeois élite. p. 79

58 *Anti-Jacobin war*—the war England waged in alliance with European absolutist states against republican France in 1793. p. 79

59 In April 1833, Parliament adopted an Act designed to suppress the peasant movement in Ireland and introducing a state of siege in the country. p. 79

60 See Note 11. p. 79

61 See Note 48. p. 80

62 *Corn Laws*—the high import tariffs on corn, aimed at limiting or prohibiting the import of corn to England—were introduced in 1815 in the interests of the big landlords. The struggle over the Corn Laws between the industrial bourgeoisie and the landed aristocracy ended in 1846 with the passing by the Peel Government of a Repeal Bill. This was a heavy blow to the landed aristocracy and promoted the development of capitalism in England. p. 81

63 *Habeas Corpus Act* was adopted by the English Parliament in 1679; it was a guarantee against police arbitrariness, for it required that the authorities should state reasons for taking persons into custody and release them if they were not brought before a court within a limited period. However, Parliament was entitled to suspend the Act, and the English ruling classes constantly abused it in Ireland. p. 82

[64] Clarendon, Lord Lieutenant in Ireland, cruelly suppressed the uprising of the Irish peasants in the summer of 1848 (see Note 54).
p. 82

[65] A reference to the 1848 uprising. See Note 54.
p. 82

[66] Engels was accompanied on his tour of Ireland in May 1856 by his wife Mary Burns.
p. 83

[67] A reference to the 1845-47 famine. See Note 53.
p. 84

[68] An error has crept into this passage: the English wars of conquest began in 1169.
p. 84

[69] See Note 43.
p. 85

[70] The article *"The Question of the Ionian Islands"* was written by Marx in connection with the English Government's policy aimed at obstructing the liberation of the islands from the English protectorate, established in 1815, and their cession to Greece. The decision on the cession of the islands to Greece was adopted by the Legislative Assembly of Corfu, the main island. Gladstone went on a special mission to the Ionian Islands in November 1858. The English Government succeeded in delaying a solution of the problem up to 1864.
p. 86

[71] The editorial offices of *The Times* are on Printing-House Square in London.
p. 86

[72] *Orange Lodges or Orangemen* (the Orangeist Order), named after William III, Prince of Orange—a terrorist organisation, set up by the landlords and Protestant clergy in Ireland in 1795 to fight against the national liberation movement of the Irish people. The Order united ultra-reactionary English and Irish elements from all layers of society and systematically incited Protestants against the Irish Catholics. The Orangemen had a particularly great influence in Northern Ireland, where the majority of the population were Protestants.
p. 87

[73] *Dublin Castle* was built by the English conquerors in the thirteenth century and became the seat of the English authorities, a stronghold against the Irish population. Dublin Castle was a symbol of English colonial rule.
p. 87

[74] *Phoenix Club*—an Irish secret society formed of the revolutionary clubs smashed after 1848, and uniting mainly small employees, sales-assistants and workers. The society was connected with Irish revolutionary émigrés in the U.S.A. In 1858, most of the club members joined the secret Fenian society, and shortly after the Phoenix Club was broken up by the English police.
p. 89

[75] *Ribbonism*—an Irish peasant movement that emerged in Northern Ireland at the end of the eighteenth century. Its members were united in secret societies and wore a green ribbon as an emblem. The Ribbonmen movement was a form of popular resistance to the arbitrary rule of the English landlords and the forcible eviction

of tenants from the land. The Ribbonmen attacked estates, organised attempts on the lives of hated landlords and managers. The
activities of the Ribbonmen had a purely local, decentralised character and they had no common programme of action.   p. 89

76 The English ruling circles and reactionary Irish landlords did everything they could to foment religious strife between Catholic and
Protestant Irishmen, which substantially weakened the national
liberation movement in Ireland. In the 1780s they helped to set
up secret terrorist Protestant organisations in Northern Ireland,
the "Peep-o'-Day Boys" society among them. The members of
these societies generally broke into the houses of Catholics at day-
break and, pretending to search for arms, which Catholics were
not allowed to possess, destroyed their property.
    Defenders—the members of an organisation of Irish Catholics,
which emerged in the 1780s in defence against the "Peep-o'-Day
Boys".   p. 89

77 The reference is to the setting up in England in 1854 of corrective
schools to which juvenile delinquents, aged from 12 to 16, were
sent for crimes which according to former laws were punishable
by short-term imprisonment.   p. 92

78 On the 1845-47 famine in Ireland see Note 53.   p. 93

79 *"The Crisis in England"* was one of 52 articles published by Marx
in the Viennese newspaper *Die Presse* between October 1861 and
the end of 1862.
    Die Presse was an Austrian bourgeois daily of a liberal trend.
It appeared between 1848 and 1896 with small interruptions. In
the early sixties it took an anti-Bonapartist stand in foreign policy
questions and opposed the reactionary course of the quasi-constitu-
tionalist government of the Austrian Empire.   p. 95

80 See Note 62.   p. 95

81 Marx is referring to the sluggishness of the English officials and
members of both Houses as regards the Bills on landlords and
tenants in Ireland when these were debated in Parliament in 1853-
55. On the nature and fate of these Bills, see Marx's articles "The
Indian Question—Irish Tenant Right" and "From Parliament: Bulwer's Proposal—the Irish Question" (pp. 59-65, 77-78) and also
Note 37.   p. 97

82 In 219 B.C., the ancient Spanish town of Saguntum, which was in
alliance with Rome, was besieged by Hannibal of Carthage. After
eight months the resistance of the citizens of Saguntum was broken,
but many of them burned themselves rather than surrender and
be enslaved.   p. 97

83 *Unionist troops*—troops of the Northern States fighting the armies of
the slave-owning Southern Confederation during the American
Civil War (1861-65). The seizure of New Orleans in April 1862
by General Butler with the support of the Unionist navy was one
of the most important events in the spring and summer offensive

of the Northern armies. However, the successes achieved in this offensive by Grant in the west and Butler in the south were brought to naught by the defeat of the main Unionist forces on the central front near Richmond. The tide did not turn in favour of the Northern armies until 1863. p. 97

84 *Statute Law*—legal norms based on statutes, the legislative acts of the English Parliament. p. 98

85 After the suppression of the 1798 national liberation uprising in Ireland, in 1801, on the instance of Castlereagh, State Secretary for Ireland, the English Parliament adopted reactionary laws introducing a stage of siege in Ireland and suspending the Habeas Corpus Act, according to which reasons had to be stated for every arrest. p. 98

86 The *December coup d'état* was carried out by Louis Bonaparte, the French President, on December 2, 1851, and led to the establishment of the Bonapartist dictatorship in France. Immediately after the coup Palmerston, who was at that time English Foreign Secretary, approved of Louis Bonaparte's actions in a talk with the French ambassador. Palmerston made this statement without consulting the other cabinet members and was forced to resign because of it. In principle, however, the English Government supported this line and was the first in Europe to recognise the Bonapartist regime in France. p. 98

87 *Zouaves*—light-armed French infantry, originally recruited from the Algerian Kabyle tribe of Zouaoua. p. 98

88 Chapter XXV of Volume I of *Capital* is called "The General Law of Capitalist Accumulation". In the first five subsections of Section 5 Marx illustrates the operation of that law by examples from the position of different categories of the English working class, and in subsection *f*, given in this collection, by the example of the working people's social conditions in agrarian Ireland dependent on England. p. 99

89 The reference is to the plague which ravaged Western Europe in 1348-49, decimating the population. It caused an acute labour shortage, which in England led to a temporary rise in the wages of workers in both town and country. In this the followers of the reactionary economist Malthus saw proof of his thesis that the pauperisation of the masses was due not to social causes but to a natural disproportion between the excessive growth of the population and the production of means of consumption. p. 106

90 *Thirty Years' War* (1618-48)—a general European war resulting from the struggle between Protestants and Catholics. The Spanish and Austrian Hapsburgs and Catholic princes of Germany joined under the banner of Catholicism and with the support of the Pope waged war against the Protestant countries: Bohemia, Denmark, Sweden, the Dutch Republic and several Protestant German states. The Protestant bloc was supported by the Catholic French kings,

opponents of the Hapsburgs. Germany was the main battleground and bore the brunt of the devastation and havoc.   p. 106

91 Marx refers to Section 8 of Chapter XV of the first volume of *Capital*—"Revolution Effected in Manufacture, Handicrafts and Domestic Industry by Modern Industry". Subsection *d* of that section deals with "Modern Domestic Industry".   p. 108

92 *Fenians*—Irish petty-bourgeois revolutionaries, who took their name from the warriors of ancient Erin. The first Fenian organisations were founded in 1857 in the U.S.A. where they united Irish immigrants; later they emerged also in Ireland. In the early sixties the Fenians set up a secret Irish Revolutionary Brotherhood, which unfolded the struggle for an independent Irish republic, oriented on armed revolt. The Fenians, who belonged to the revolutionary wing of the Irish national movement, voiced the protest of the Irish people against colonial oppression and against the eviction of the Irish peasants from the land. At the same time they made mistakes of a sectarian and nationalistic character, resorted to conspiratorial methods, and failed to understand the importance of an alliance with England's democratic and proletarian circles. As a result, they failed to win the support of the mass of the Irish people. Most of the Fenians were urban petty bourgeois or intellectuals. The Fenians were very active in the latter half of the sixties, but the movement declined in the seventies, giving way to more massive and effective forms of national liberation struggle.
                    p. 115

93 Chapter XXXVII opens Part VI of Volume III of *Capital*, which deals with the transformation of surplus-profit into ground-rent. The passage is taken from the Introduction to this part.   p. 117

94 The reference is to the struggle over the Bills on landlords and tenants in Ireland that unfolded when they were debated in Parliament in 1853-55. See Marx's articles in this collection, "The Indian Question—Irish Tenant Right" and "From Parliament: Bulwer's Proposal—the Irish Question" (pp. 59-65, 77-78) and also Note 37.   p. 117

95 This memorial of the General Council of the International Working Men's Association written by Marx was adopted in connection with the conclusion in Manchester of the trial of the Irish Fenians, who had made an armed attack on a prison van in an attempt to liberate Kelly and Deasy, two Fenian leaders. The attack took place on September 18, 1867. Kelly and Deasy managed to escape but during the clash a policeman was killed. Five Irishmen charged with murder were brought to trial and, although there was no direct evidence and they were splendidly defended by Counsel Ernest Jones, a former Chartist leader, they were sentenced to death. Mac-Guire was subsequently pardoned, and Condon's sentence was commuted to life imprisonment, but Larkin, Allen and O'Brien were executed. The Fenian trial in Manchester aroused a storm of protest in Ireland and England. On the instance of Marx, the

General Council of the International started, on November 19, a discussion on the Irish question (the minutes of the meeting will be found on pp. 368-72), during which the leaders of the international proletarian organisation expressed their solidarity with the struggle of the Irish people for independence and condemned the position of the reformist trade union leaders, who in the wake of the English bourgeois radicals denied the right of the Fenians to resort to revolutionary methods in the struggle. The discussion was scheduled to continue on November 26, but when the news of the conviction was received, the General Council convened an emergency meeting on November 20 and addressed a memorial to the Home Secretary asking for the commutation of the death sentence. The English Government ignored the memorial of the International Working Men's Association. Because of the opposition set up by the trade union leaders, the memorial was not published in the English labour press in its original wording. The French translation was published by *Le courrier français,* a weekly which appeared in Paris and was linked with the International.   p. 118

[96] A hint at the extensive amnesty granted by President Lincoln in 1863 and President Johnson in 1865 to persons who participated in the American Civil War on the side of the South.   p. 118

[97] These notes were written by Marx as a conspectus for his speech to be made at the meeting of the General Council of the International Working Men's Association on November 26, 1867, when the discussion on the Irish question begun on November 19 was to continue. In view of the immense excitement caused by the execution of the three condemned Fenians (Larkin, Allen and O'Brien) on November 23, Marx considered this speech as no longer suitable. Feeling that at such a moment it would be more appropriate for one of the English members of the General Council to express sympathy with the Irish revolutionaries, he gave the floor to Peter Fox, who was known for his support of the Irish national liberation movement. Marx described the meeting in great detail in his letter to Engels of November 30, 1867 (see pp. 146-48). Later, preparing for a report on the Irish question in the German Workers' Educational Association in London (see pp. 126-39), Marx used this draft and the materials he had compiled for it.   p. 120

[98] A reference to the *Act of Settlement* adopted by the Long Parliament on August 12, 1652, during the English bourgeois revolution, following the suppression of the 1641-52 national liberation uprising in Ireland. The Act legalised the reign of terror and violence established by the English colonialists in Ireland and sanctioned the wholesale plunder of Irish lands in favour of the English bourgeoisie and the "new" bourgeoisified nobility. This Act declared the majority of Ireland's indigenous population "guilty of revolt". Even those Irishmen who had not been directly involved in the uprising but had failed to show the proper "loyalty" to the English Crown were considered "guilty". Those declared "guilty" were classified into categories, depending on the extent of their involve-

ment in the uprising, and subjected to brutal reprisals: execution, deportation, confiscation of property. On September 26, 1653, the Act of Settlement was supplemented by the Act of Satisfaction which prescribed the forcible resettlement of Irish people whose property had been confiscated to the barren province of Connaught and to Clare County and defined the procedure for allotting the confiscated land to the creditors of Parliament, the officers and men of the English army. Both Acts consolidated and extended the economic foundations of English landlordism in Ireland.   p. 120

99 See Note 63.   p. 120

100 Marx uses an appraisal of the Fenian movement from Queen Victoria's address to Parliament of November 19, 1867, to describe the brutal policy of the English Government towards the Irish Fenians.
                                                                             p. 120

101 During an abortive coup in Boulogne in 1840, Prince Louis Bonaparte wounded an officer of the government troops. This crime did not stop the English Government from obsequiously recognising the Bonapartist regime after the usurpation of power by Louis Bonaparte in 1851. In 1867, however, three Irish Fenians were sent to the gallows only on the suspicion of having made an attempt on the life of a policeman while attacking a prison van in Manchester.   p. 120

102 The *corn-acre system*—the subletting to the poorest peasants of small plots (of an area of up to half an acre) by middlemen on fettering terms, which was extensively practised in Ireland. The term came into use in the 18th century, after the adoption of a law decreeing that corn be sown on these small holdings.   p. 122

103 Following the repeal of the Corn Laws in 1846, which led to a drop in grain prices due to the fall in the demand for Irish grain in England, and the rise in the demand for wool and other stock-breeding products from Ireland, landlords and rich farmers switched to extensive pasture farming which led to the mass eviction of small Irish tenants from the land ("clearing of estates") in the mid-19th century.   p. 123

104 A reference to the forcible eviction from the land of the population of the Scottish Highlands (the Gaels) by the Anglo-Scottish nobility in the 18th and the beginning of the 19th centuries, a process similar to the "clearing of estates" in Ireland. Marx describes this process in Chapter XXVII of the first volume of *Capital*.   p. 123

105 The *roundheads*—the name given to the supporters of Parliament during the English bourgeois revolution in the 17th century because of their puritan custom of cutting their hair close, while the *cavaliers*—supporters of the King—wore their hair long.   p. 123

106 See Note 2.   p. 123

107 In the first decades of the 19th century the Irish national movement developed under the slogan of the abolition of political restrictions

for the Catholic population and the granting to Catholics (who formed the majority of the population) of the right to stand for election to Parliament (see Note 46 on the Catholic Emancipation Act of 1829). After the thirties the struggle was waged under the banner of Repeal of the Anglo-Irish Union of 1801 (see Note 2). O'Connell and his supporters championed moderate, peaceful means of struggle ("moral force"). In the mid-forties, however, the supporters of the liberation of Ireland by revolutionary methods, up to and including armed uprising against English rule ("Young Ireland" group, John Mitchel and his friends), gained ground in the Repeal Association headed by O'Connell. The differences between O'Connell and those advocating the use of "physical force" led to a split in the Repeal Association and the formation of the more radical Irish Confederation (see Note 23). p. 124

[108] A reference to the reactionary foreign policy pursued by Castlereagh, the British Foreign Secretary (1812-22). He supported the efforts of the Holy Alliance aimed at strengthening the reactionary feudal monarchies in Europe, notably the measures against the revolutionary movements in Italy and Spain. The counter-revolutionary Tory policy of Castlereagh was continued by Palmerston, the Whig leader, who relied on the support of the Right wing of that party. He, however, masked the real nature of this policy in liberal phrases and hypocritical expressions of sympathy with the oppressed peoples. In his *Lord Palmerston* (an excerpt from which is published in this collection, see pp. 70-71), Marx showed that in his capacity of Foreign Secretary Palmerston played an ignoble role with regard to the Polish struggle for independence during the general uprising of 1830-31 and the uprising in the free city of Cracow in 1846. While inciting the Poles to action by his false promises of assistance, Palmerston sanctioned the suppression of the Polish movement by tsarist Russia, Austria and Prussia. p. 125

[109] The *Reform League*—an organisation set up in London in the spring of 1865 on the initiative and with the participation of the General Council of the International. It was to be a political centre for the guidance of the mass movement of workers for a second electoral reform (the first, carried out in 1832, fully preserved the political privileges of the ruling classes and denied rights to the workers). By advancing the slogan of universal suffrage, the League won considerable influence among the proletarian masses and set up branches in many English towns. However, due to the vacillations of the bourgeois radicals in the League's leadership, who were frightened by the mass movement, and because of the policy of compromise pursued by the trade union leaders on the Council and Executive Committee, the Reform League acted inconsistently and half-heartedly. This enabled the English ruling classes to make the 1867 electoral reform a moderate one and to extend franchise only to the petty bourgeoisie and the upper crust of the working class.

The leadership of the Reform League committed a grave error in the Irish question by refusing to give any real support to the

Irish national liberation movement, although many of its rank-and-file members expressed sympathy with it. The meeting of the League's Council on November 1, 1867, adopted a resolution condemning Fenianism, tabled by bourgeois radicals. When the Irish question came up for discussion in the General Council of the International in November 1867, the speeches were spearheaded against this chauvinistic and anti-revolutionary position of the Reform League and its supporters among the liberal trade unionists.

p. 125

110 This outline is a draft conspectus for a report on the Irish question Marx was to make at the meeting of the German Workers' Educational Association in London on December 16, 1867. "Yesterday I read in our German Workers' Association (but three other German workers' associations were represented there, about 100 people in all) a one-and-a-half hour long report on Ireland," Marx wrote in this connection to Engels on December 17, 1867. Some members of the General Council of the International also attended the meeting. Eccarius, a Council member, who attached great importance to this report, which explained the attitude of the General Council towards the Irish national liberation movement, took notes in order to prepare them for publication (see pp. 140-42). A copy of these notes was sent to Johann Philipp Becker, the editor of *Vorbote*, a monthly magazine in Geneva, which was the mouthpiece of the German sections of the International Working Men's Association in Switzerland; but it was not published.

The *London German Workers' Educational Association* was founded in February 1840 by German revolutionary emigrants. After the founding of the Communist League—the first international communist organisation of the working class—the leading role in the Association was assumed by the local sections of the League. Marx and Engels took an active part in the Association's activities (except when sectarian elements temporarily gained the upper hand). At the end of the fifties, Friedrich Lessner, a pupil and comrade-in-arms of Marx and Engels, became one of the leaders of the Association. The Association, which was linked with English workers' organisations, participated in the inauguration of the International Working Men's Association in 1864 and began to act as its German section in London. The Association continued to exist up to 1918. p. 126

111 A reference to the three biggest national liberation uprisings in Ireland.

The *1641-52 uprising* was provoked by the colonialist policy which the English absolute monarchy pursued in Ireland, and which was continued during the English bourgeois revolution by the English bourgeoisie and the "new" nobility. The majority of the insurgents were Irish peasants led by the expropriated clan chiefs and the Catholic clergy. The Anglo-Irish nobility, descendants of the first English conquerors who had become related to the Irish clan élite and adopted many Irish customs and habits, also participated in the uprising. In October 1642, the insurgents formed the

Irish Confederation in Kilkenny. A struggle went on within it between the indigenous Irish, who stood for Ireland's independence and action both against the Long Parliament and the English Royalists, and the Anglo-Irish aristocrats, who endeavoured to come to terms with Charles I on the condition that they would be allowed to keep their estates and receive a guarantee of freedom of worship for Catholics. The latter gained the upper hand and a treaty was signed with a representative of Charles I. After the rout of the Royalists in England, Oliver Cromwell, the Lord Protector of the new bourgeois republic, organised an expedition to Ireland on the pretext of suppressing a Royalist revolt there but in fact with the aim of reducing her to colonial submission and plundering the land. He hoped that by confiscating Irish lands he would solve the problem of paying the creditors of the republic, the officers and men in the army. In 1649-52, the Irish uprising was brutally suppressed; the garrisons and population of entire towns were destroyed, the Irish were sold en masse into slavery in the West Indies, and Irish lands were confiscated and handed over to new English landlords. These actions of Cromwell and his successors did much to prepare the ground for the restoration of the monarchy in England in 1660.

The *1689-91 uprising* followed in the wake of the 1688-89 coup d'état in England (known as the Glorious Revolution), involving the overthrow of James II Stuart and the establishment of a bourgeois-aristocratic constitutional monarchy in England under William III of Orange. The Catholic nobility in Ireland, supported by the masses who were dissatisfied with the colonial regime, rose against William. Under the banner of defence of the Stuarts the insurgents fought for the abolition of Ireland's political and religious inequality and the return of the confiscated estates. James II, who had taken refuge in Ireland and was endeavouring to use the Irish movement to regain the crown, became its official head and recognised the demands of the Irish people. But the differences between the reactionary Jacobites and the Irish patriots weakened the insurgents. Despite their stubborn resistance, they were finally defeated.

The *1798 uprising* was the result of the upsurge of national sentiments in Ireland, caused by the growth of the liberation movement and the impact of the American and French bourgeois revolutions at the end of the 18th century. It was prepared by Irish bourgeois revolutionaries (Theobald Wolfe Tone, Edward Fitzgerald), who in 1791 founded the patriotic society "The United Irishmen" in Belfast (the chief town of the Northern Irish province of Ulster) and proclaimed a fight for an independent Irish republic. On the eve of the uprising, however, most of the society's leaders were tracked by government spies and arrested. The uprising broke out on May 23 and lasted until June 17, 1798. It flared up in a number of counties in South-East and Northern Ireland and was particularly strong in County Wexford. The majority of the insurgents were peasants and urban poor. In August and September 1798, after the landing of a French force in support of the Irish patriots, the uprising spread to a number of places in Connaught. The English authorities launched savage reprisals against the rebels (almost all the leaders were executed) and passed the Act of Anglo-Irish Union in 1801.   p. 126

112 About 1155 Pope Adrian IV issued a bull which conferred on the English King Henry II the title of Supreme Ruler of Ireland in exchange for the promise to subject the Irish Church to Rome. Henry II used this "gift" to launch an aggressive expedition against Ireland in 1171.

In 1576, in connection with the exacerbation of relations between Protestant England and the Catholic powers, Pope Gregory XIII declared that Queen Elizabeth I had forfeited the right to the Irish crown.   p. 127

113 *English Pale*—the medieval English colony in South-East Ireland founded by the Anglo-Norman barons in the 1170s. The term came into use in the second half of the 14th century. The boundaries of the English Pale changed during the continual wars of the conquerors against the hitherto unsubdued population. Castles and fortifications were built in the border areas. At the end of the 15th century the Pale included only part of the present counties of Louth, Meath, Dublin and Kildare, but it served as a bridgehead for the complete subjection of Ireland by the English in the 16th century. Dublin was the centre of the Pale and the seat of the English Lord Deputy.   p. 127

114 The *Anglo-Irish Parliament*, convoked at the end of the 13th century, was initially made up of representatives of the big barons and dignitaries of the Church of the English colony in Ireland (the Pale). With the extension of the power of the English crown to the entire island (16th-early 17th centuries) the Parliament became a representative body of the English and Anglo-Irish aristocracy under the English Lord Deputy. The competency of that Parliament was limited; according to the Act passed by Lord Deputy Poynings in 1495, it could be convoked only with the sanction of the Royal Privy Council. Under the impact of the growing national liberation movement, in the 1780s the English Government was compelled to extend the rights of the Irish Parliament. In 1801, however, the Irish Parliament was abolished under the Act of Union.   p. 128

115 A reference to the Act of Settlement (1652) and the Act of Satisfaction (1653). For details, see Note 98.   p. 128

116 A reference to the capitulation at Limerick, an agreement signed in October 1691, between the Irish insurgents and representatives of the English command, and approved by King William III. The surrender terms were honourable: the insurgents were given permission to serve either in foreign armies or in the army of William III; the people were promised an amnesty, the preservation of their property, suffrage and religious freedom. The Limerick terms, however, were soon flagrantly violated by the English authorities.
                                                                                        p. 128

[117] *Absentees*—landlords who owned estates in Ireland but lived permanently in England. Their estates were managed by realty agents who robbed the Irish peasants, or were leased to speculator-middlemen, who subleased small plots to the peasants.   p. 129

[118] A reference to the book: W. Molyneux, *The Case of Ireland's Being Bound by Acts of Parliament in England Stated,* Dublin, 1698.   p. 129

[119] *Penal Code* or *penal laws*—a set of laws passed by the English for Ireland at the end of the 17th and in the first half of the 18th centuries on the pretext of struggle against Catholic conspiracies. These laws deprived the indigenous Irish, the majority of whom were Catholics, of all civil and political rights. They limited the right of Catholics to inheritance, to the acquisition and alienation of property, and introduced the practice of confiscating property for petty offences. The Penal Code was used as an instrument for the expropriation of the Irish who still owned land. It established unfavourable lease terms for Catholic peasants, promoting their dependence on the English landlords. The ban on Catholic schools, the stern punishment meted out to Catholic priests, and other measures were intended to stamp out Irish national traditions. The penal laws were abrogated, and then only in part, at the end of the 18th century under the influence of the growing national liberation struggle in Ireland.   p. 129

[120] Catholics were officially deprived of voting rights by the Act on the Regulation of Elections passed in 1727. Irish Catholics had not enjoyed the right to stand for election to Parliament from the end of the 17th century, following the introduction of an oath to be taken by M.P.s involving an abjuration of Catholic dogmas. The latter restriction was only lifted in 1829. Voting rights were restored to the Catholic population somewhat earlier, in 1793, since the English landlords themselves often needed the votes of their Catholic tenants.   p. 130

[121] *Freehold*—a category of small landownership which had come down from medieval England. The freeholder paid the lord a comparatively small rent in cash and was allowed to dispose of his land as he saw fit.   p. 130

[122] The war England waged against Napoleonic France ended in 1815.   p. 131

[123] *Cottiers*—a category of the rural population consisting of land-hungry or landless peasants. In Ireland cottiers rented small plots of land and cottages from landlords or real estate agents on extremely onerous terms. Their position resembled that of farm-hands.   p. 132

[124] See Note 48.   p. 133

[125] See Note 102.   p. 136

[126] *The Irishman*—an Irish bourgeois weekly published between 1858

and 1885, first in Belfast, later in Dublin. It supported the national liberation movement and came out in defence of the Fenians. At the same time it was subject to class and national limitations (refusing to publish the documents of the International in support of the Irish revolutionaries).   p. 139

127 The Reformation begun in England under King Henry VIII (Act of Supremacy, which declared the King the head of the Church in place of the Pope, and other Acts) was completed under Elizabeth I (the adoption, in 1571, of the "39 articles" of the Anglican Church—a variety of Protestantism). The introduction of the Reformation to Catholic Ireland was a means of subjecting her to the English absolute monarchy and expropriating her population in favour of the English colonists on the pretext of struggle against Catholicism.   p. 140

128 A reference to the Restoration of the Stuart dynasty in England in 1660. The restored Stuarts (Charles II and James II) continued to rule up to the Glorious Revolution of 1688-89. The Restoration was the result of a compromise between the bourgeois élite and the "new nobility", which had grown rich during the revolution, with the aristocrats supporting the Stuarts. The adherents of the Stuarts, many of whom had lost their estates in England, now received title to confiscated Irish lands by way of compensation. The representatives of the new regime satisfied complaints and petitions for the return of property to Irish owners only in rare cases; and following the 1665 Act such complaints were no longer considered. Thus, the sweeping expropriation of the Irish population implemented during the English bourgeois revolution was sanctioned by the restored monarchy.   p. 140

129 See Note 114.   p. 140

130 On the Fenian trial in Manchester, see Note 95.   p. 143

131 A meeting of the Reform League Council was called on October 23, 1867, to discuss the letter in which the League's President, the bourgeois radical Beales, sharply censured the Fenian movement. Odger and Lucraft, English trade union leaders who were members of the League Council (and also members of the General Council of the First International), objected to the publication of the letter and expressed solidarity with the Irish liberation movement. This was evidence of the influence on trade union leaders of the internationalist ideas of Marx and his closest collaborators in the General Council of the International Working Men's Association. However, at the meetings of the League Council on October 30 and November 1, Odger and Lucraft, on whom pressure had in the meantime been brought to bear by bourgeois radicals, denied their former position and said that they had been misunderstood. p. 143

132 The reference is to *Agricultural Statistics, Ireland. Tables showing the Estimated Average Produce of the Crops for the Year 1866; and the Emigration from Irish Ports, from 1st January to 31st*

December, 1866; also the Number of Mills for scutching Flax in each County and Province", Dublin, 1867. p. 143

133 Despite the negative attitude to the Fenian movement of the Reform League leaders, the Manchester trial of the Fenians aroused among many League members a feeling of sympathy for the fighters for Ireland's liberation who had become victims of police reprisals. The General Council of the International did everything in their power to strengthen this sentiment. Marx mentions here two meetings of the Reform League's branches in London—one held on October 31, precisely at the time the League Council debated the anti-Fenian resolution, the other on November 5, 1867. The watchword at both meetings was solidarity with the Irish national liberation movement and protest against the persecution of the Fenians by the judiciary and the police. p. 144

134 In his letter to Engels of November 28, 1867, Marx pointed out that he had to "behave diplomatically" with respect to Fenianism. He implied that refusal at that particular juncture to publicly condemn the mistakes of the Fenians, their conspiratorial tactics, the manifestations of petty-bourgeois nationalist ideology, etc., in the interests of strengthening the solidarity of the International with the Irish national liberation movement, should not mean condoning these mistakes in general. In the letter published in this collection Engels expresses his complete agreement with this opinion. p. 145

135 Engels hints at the warm sympathy for the Irish national liberation movement felt by Lizzy Burns, his wife, who was a working-woman of Irish descent, and shared the revolutionary proletarian convictions of her husband. Lizzy Burns regarded the execution of the three Fenians in Manchester as a personal tragedy (that is why Engels writes of black and green in his house—green being the Irish national colour). p. 146

136 The reference is to the article "London Meeting", which appeared in *The Times* on November 21, 1867. p. 146

137 Marx is referring to a disagreement between Peter Fox and Hermann Jung, the Corresponding Secretary of the General Council of the International for Switzerland. At the Council meeting of November 5, 1867, Jung criticised Fox for his intention to refuse the post of Corresponding Secretary for America. p. 147

138 The resolution moved by Peter Fox essentially slurred over the question of Ireland's national self-determination, proposing instead, in very hazy terms, self-administration for Ireland; it recommended to the English people "to accord an unprejudiced hearing to the arguments advanced on behalf of Ireland's right to autonomy". At Marx's initiative the Standing Committee shelved the draft resolution.

The *Standing Committee* or *Sub-Committee* was the executive body of the General Council of the International, which generally assembled once a week and drafted many of the decisions which were later adopted by the Council. The Standing Committee evolved from a commission, elected when the International Working Men's Association was set up, to draft its programme documents—the Rules and the Inaugural Address. On the Committee were the President of the General Council (until this office was abolished in September 1867), its General Secretary and the corresponding secretaries for the different countries. Marx took an active part in the work of the Standing Committee as Corresponding Secretary for Germany.   p. 147

139 Marx refers to the demands advanced by the military patriotic organisation, the Irish Volunteers, in the early 1780s. The Volunteers demanded above all Ireland's independence in the field of legislation, the responsibility of the administration to the autonomous Irish Parliament, the reorganisation of the anti-democratic Parliament, representing a narrow clique of Anglo-Irish landlords and obeying the English Government, into a genuinely national representative body. In view of the upsurge of national sentiment in Ireland, the English Government was compelled, temporarily, to meet some of these demands. The Act of 1782 abolished the right of the English Parliament to pass laws for Ireland, and the Irish Parliament could now be convened without authorisation from the English Government. In 1783 the autonomy of the Irish Parliament was reaffirmed by the Act of Renunciation. However, the reform of the Irish Parliament was obstructed by the English authorities. Moreover, with the enactment of the Anglo-Irish Union, Irish parliamentary autonomy was abolished.   p. 148

140 Marx apparently did not make a speech on the Irish question in the General Council as planned. In December 1867, the Council met twice, on the 17th and 31st, and as of January 1868, illness prevented Marx from attending the Council meetings for several months. His view on the Irish question, which reflected the position of the revolutionary proletarian wing of the General Council, was set forth in the detailed report he made on December 16 in the London German Workers' Educational Association (see pp. 126-39).
                                                                       p. 148

141 On December 13, 1867, a group of Fenians set off an explosion in London's Clerkenwell Prison in an unsuccessful attempt to free the gaoled Fenian leaders. The explosion destroyed several neighbouring houses causing the death of several people and wounding 120. The Fenian attempt was used by the bourgeois press to incite chauvinistic anti-Irish feelings among the English population. p. 149

142 Richard Pigott, the publisher of *The Irishman* (see Note 126) and Alexander Sullivan, the owner of the Irish bourgeois radical *Weekly News* appearing in Dublin from 1858, received prison sentences in 1867 and 1868 respectively for publishing articles in defence of the Fenians.   p. 149

143 Marx is referring to the elections to the English Parliament which were to be held in November 1868, on the basis of the 1867 Act

on Household Suffrage, which extended the franchise to the tenants of flats and cottages, that is, to the petty bourgeoisie and the working-class élite. Before the elections, Gladstone, the leader of the Liberal Party, made many promises to settle the Irish question in the hope of winning votes among the new categories of voters. Even before the election campaign got under way, he proposed the separation of the Anglican Church from the state in Ireland, thereby depriving it of state support and subsidies. He expected that this would win him popularity with the Irish Catholic voters. After winning the elections and assuming office at the end of 1868, Gladstone passed a Bill through Parliament in March 1869 which placed the Anglican Church in Ireland on an equal footing with the Catholic Church. Gladstone and the Liberals hoped that their policy of moderate reform would weaken the revolutionary movement in Ireland.   p. 150

[144] On October 24, 1869, a mass demonstration was held in London in support of the demand for an amnesty for Irish political prisoners. The General Council of the International helped organise the demonstration. From various parts of the capital columns of demonstrators marched to Trafalgar Square, whence an impressive procession moved to Hyde Park, where the mass meeting took place. The demonstration was held under the slogan "Justice for Ireland!" It was part of the amnesty campaign conducted in Ireland and England, which grew in intensity when Gladstone, despite his pre-election promises, insisted on humiliating terms for the Irish prisoners as a condition for granting them an amnesty.   p. 151

[145] After Marx's report at the General Council meeting on October 26, 1869, in which he said that the bourgeois press had given a distorted picture of the demonstration of solidarity with the Irish people held in London on October 24, the General Council of the International passed a decision on adopting an address to the English people. However, on the instance of Marx, the Sub-Committee or Standing Committee (see Note 138) decided to refrain from such a general address and to pass resolutions on concrete items of the agenda for a discussion of the Irish question proposed by Marx. Eccarius, the Secretary of the Council, informed the Council of this decision on November 9. On November 12 Marx wrote to Engels:

"Instead of the address on the Irish question, for which there was no real occasion, I put on the agenda for next Tuesday's meeting (to adopt resolutions) the following items:

"1) The behaviour of the British Government over the Irish amnesty question.

"2) The attitude of the English working class towards the Irish question."   p. 151

[146] See Note 96.   p. 152

[147] The reference is to the amnesty granted to the participants in the Hungarian national liberation movement following the reorganisation of the Austrian Empire into Austria-Hungary in 1867, on the

basis of an agreement between the Austrian Government and the Hungarian aristocratic opposition. This action was the result of Austria's defeat in the Austro-Prussian war of 1866 and the growth of national contradictions within the multinational Austrian state.

p. 152

148 Before they assumed office in December 1868, when the election campaign was in full swing, Gladstone and the Liberals sharply criticised in the House of Commons the Conservative Government's policy in Ireland, especially the reprisals against the participants in the Fenian movement. The Liberals compared the actions of the Conservatives with the conquest of England by William the Conqueror in the 11th century.

The *Fenian uprising* was prepared by the Fenian Irish Revolutionary (republican) Brotherhood early in 1867 with the aim of winning independence for Ireland. It was to start on March 5. The organisers planned to form several mobile columns of insurgents who were to conduct guerilla warfare from bases in woods and mountainous areas. However, weak military leadership and the fact that the authorities got to know the insurgents' intentions prevented the plan being brought to fruition. Armed revolt broke out only in some eastern and southern counties. The insurgents seized several police barracks and stations and for a short time gained control of the town of Killmalock (County Limerick). There were also clashes with the police in the suburbs of Dublin and Cork. The uprising failed because of the conspiratorial tactics of the Fenians and their weak ties with the masses. Half of the 169 participants in the uprising brought to trial were sentenced to hard labour. p. 152

149 A reference to Gladstone's negative reply to the petitions for an amnesty for Irish prisoners adopted at mass meetings in Ireland, including the one in Limerick on August 1, 1869. Gladstone endeavoured to justify his refusal in his letters to O'Shea and Butt, two Irish functionaries, which were published in *The Times* on October 23 and 27, 1869. Marx criticises the motives given by Gladstone in those letters. p. 153

150 Marx refers to *The New-York Irish People,* an Irish newspaper published in the U.S.A., which in one of its articles said that by his refusal to grant an amnesty to the participants in the Fenian movement, Gladstone was only promoting the movement (this remark was quoted by *The Irishman* in its issue of November 13, 1869). The comparison of Gladstone with the Head Centre of the plot is a touch of irony, since this was the title of the leader of the secret Fenian organisation—the Irish Revolutionary Brotherhood. p. 154

151 In a speech made on October 7, 1862, in Newcastle, Gladstone (then Finance Minister) greeted the Confederation of the Southern States in the person of its President Jefferson Davis, justifying the rebellion of the southern slaveowners against Lincoln's lawful government. The speech was published in *The Times* on October 9, 1862. p. 154

¹⁵² *Dissenters,* or dissidents—people who disagree with official religious doctrine. The reference here is to the various sectarians who dissented from the official Anglican Church. See Note 143 on Gladstone's promises to establish equality between the Anglican and Catholic churches in Ireland in order to win the support of the Irish Catholic élite, and on his Church Bill.   p. 154

¹⁵³ In 1840, a single Parliament was set up in England's Canadian possessions. The 1867 Act transformed them into the self-governing Canadian Confederation and granted it Dominion status.   p. 155

¹⁵⁴ On October 30, 1869, *The Irishman* carried a report which said that in his letter to the Dublin branch of the Ancient Order of Foresters (a mutual-assistance society founded in England as early as 1745, which took part in the movement for the amnesty for Irish prisoners), Gladstone had refused to acknowledge his pre-election promises to improve Ireland's position.   p. 155

¹⁵⁵ At the by-elections to Parliament in County Tipperary (South-Eastern Ireland), the candidacy was advanced of O'Donovan Rossa, a prominent Fenian who in 1865 had been sentenced to penal servitude for life. On November 25, 1869, Rossa was elected M.P. Even though the elections were quashed, the fact of his election testified to growing protest against English policy among the Irish masses.   p. 155

¹⁵⁶ *Political Register*—an abbreviation for *Cobbett's Weekly Political Register* which appeared between 1802 and 1835 in London. In it W. Cobbett and other English radicals sharply criticised the policy of the English Government, notably its police measures in Ireland.   p. 157

¹⁵⁷ See Note 148.   p. 157

¹⁵⁸ See Note 111.   p. 157

¹⁵⁹ A reference to the book: Ledru-Rollin, *The Decline of England,* London, 1850.   p. 157

¹⁶⁰ On November 23, 1869, during the debate of the draft resolution of the General Council of the International on the English Government's policy towards the Irish prisoners, Odger, a trade union leader, proposed to delete the word "deliberately" from the sentence "Mr. Gladstone deliberately insults the Irish nation". At the next meeting on November 30, 1869, he made new attempts to subdue the revolutionary and anti-government tone of the resolution. p. 158

¹⁶¹ At the meeting of the General Council held on November 30, Odger proposed several fresh amendments to the draft resolution on the English Government's policy towards Irish prisoners. This was an attempt of the reformist trade union leaders to reduce the resolution, which was a document exposing English policy and expressing solidarity with the fighters for Ireland's independence, to a humble appeal to the mercy of the ruling circles. In his speech, Odger justified and defended Gladstone's policy. Odger's proposal was rejected by the Council. Of all the amendments proposed by

him only the one to omit the word "deliberately", advanced at the former meeting, was accepted. p. 158

162 The reference is to item 2 in Marx's draft for the debate on the Irish question—the attitude to it of the English working class. Marx explained the stand of the General Council of the International on this question in the "Confidential Communication" published in this collection. In this report and a number of letters, including that written to Meyer and Vogt on April 9, 1870 (see pp. 292-95), Marx deals with the international importance of the problem. p. 159

163 This document is an answer to the attacks against the General Council made by Bakunin, the anarchist leader, and his supporters. After his unsuccessful attempt at the Basel Congress in 1869 to win the leadership in the International by transferring the General Council to Geneva, Bakunin changed his tactics and resorted to open attacks against the Council. Bakunin's supporters gained control over the editorial board of the Swiss organ of the International, the newspaper *Égalité*, and in the autumn of 1869 brought a number of accusations against the Council in its columns, one of the main ones being that by its statements on the Irish question the Council was diverting the attention of the international workers' organisation from its direct task—the solution of social problems.

The fact that the General Council was simultaneously fulfilling the function of the British Federal Council was also basely slandered. These and similar attacks by the anarchist sectarians revealed their inert, nihilistic attitude to the question of the proletariat's support of the national liberation movement, their failure to understand its role as an ally of the working class, and their denial of the need for workers to take part in political struggle.

Somewhere around January 1, 1870, Marx wrote a circular letter—"The General Council to the Federal Council of Romance Switzerland"—in which he gave a strong rebuff to the Bakuninists and explained, in particular, the International's position on the Irish question. The circular letter was sent to the various sections of the Association. On March 28, 1870, Marx appended some new information on the intrigues of the Bakuninists and sent the document in the form of a "Confidential Communication" to the Executive of the German Social-Democratic Workers' Party.

Early in 1870, even before the receipt of the circular letter, the Federal Council of Romance Switzerland succeeded in removing the Bakuninists from the editorial board of *Égalité* and the newspaper resumed its former revolutionary proletarian trend. p. 160

164 In November 1866, J. Hales proposed the reorganisation of the branch of the International in England so that it would rely not on the trade unions affiliated with it, but on the newly organised sections, formed according to the territorial principle and headed by a special Federal Council. Similar proposals were advanced at the end of 1869. Marx and other leaders of the Council considered the moment inopportune since this would have isolated the Inter-

national from the workers' mass organisations. Only after the events of 1871 (the Paris Commune), when the situation in England and in the world had changed radically and reformist trends had gained supremacy in the trade unions, did Marx and his supporters consider it advisable to form the British Federation of the International with a special Council at its head.   p. 160

165 *The Land and Labour League* was founded in London in October 1869 with the participation of the General Council of the International. Ten members of the Council were on the League's Executive Committee. The programme of the League, drafted by Eccarius with Marx's help, included the demands for the nationalisation of the land and the banks, for a shorter working day, universal suffrage and the abolition of the standing army. On the basis of these demands, which transcended purely bourgeois-democratic reforms and expressed proletarian interests, the organisers of the League endeavoured to rally the working people not only of England, but also of Ireland, Scotland and Wales (see Address of the League published in the Supplement to this collection, pp. 373-78). Marx regarded the League as a means of setting up an independent workers' party in England. However, reformist elements soon gained ground in the League's leadership and it eventually lost its connections with the International.   p. 161

166 The position of the International on the Irish question as expounded in this document essentially anticipated the resolution on item 2 of the programme for the debate on this question in the General Council, proposed by Marx early in November 1869, namely, on the item defining the attitude of the English working class to the liberation struggle of the Irish people (see p. 151). Even though other official documents of the International were soon to remove the need for a special resolution on this issue, Marx stuck to his idea of continuing the debate on the Irish question in the General Council for a long time. Circumstances prevented this, notably Marx's protracted illness which stopped him from regularly attending Council meetings in the winter and spring of 1870. Later more urgent matters arose, and in July 1870, the Franco-Prussian war broke out, which absorbed the attention of the Council. Therefore, the Council confined itself to the decisions already adopted on the Irish question.   p. 163

167 This article was sent by Marx to the organ of the Belgian sections of the International Working Men's Association, the weekly *L'Internationale*, which appeared between 1869 and 1873 in Brussels. It was sent as a private letter to the editor César De Paepe. Marx expected that the letter would be edited by De Paepe before it was printed. The editors, however, printed it almost without changes, only adding some explanations in brackets and dividing it into two parts. A small editorial comment was appended, which is not published in this edition.   p. 164

168 The reference is to the book: Garibaldi, *The Rule of the Monk, or Rome in the Nineteenth Century,* London, 1870.   p. 164

¹⁶⁹ *The Irish People*—an Irish weekly, the main organ of the Fenians, appearing in Dublin between 1863 and 1865. It was banned by the English Government, the members of its editorial board were arrested and sentenced to long terms of hard labour. O'Donovan Rossa, its publisher, was sentenced to penal servitude for life.   p. 164

¹⁷⁰ On O'Donovan Rossa's election to Parliament, see Note 155.   p. 166

¹⁷¹ See Note 63.   p. 167

¹⁷² A reference to Gladstone's pamphlet *Two Letters to the Earl of Aberdeen on the State Persecution of the Neapolitan Government,* published in London in 1851, in which Gladstone exposed the cruel treatment by the Government of the Neapolitan King Ferdinand II (nicknamed "Bomba" for the bombardment of Messina in 1848) of political prisoners arrested for their part in the 1848-49 revolutionary movement.   p. 167

¹⁷³ The *Land Bill* for Ireland was discussed in the English Parliament in the first half of 1870. Submitted by Gladstone on behalf of the English Government on the pretext of assisting Irish tenants, it contained so many provisos and restrictions that it actually left the basis of big landownership by the English landlords in Ireland intact. It also preserved their right to raise rents and to drive tenants off the land, stipulating only that the landlords pay a compensation to the tenants for land improvement, and instituting a definite judicial procedure for this. The Land Act was passed in August 1870. The landlords sabotaged the implementation of the Act in every way and found various ways round it. The Act greatly promoted the concentration of farms in Ireland into big estates and the ruination of small Irish tenants.   p. 168

¹⁷⁴ Marx is referring to Gladstone's speech in the House of Commons on February 15, 1870, which was published in *The Times* on February 16, 1870.   p. 168

¹⁷⁵ The report on the coroner's inquest on the body of Michael Terbert was published in *The Irishman* on February 19, 1870.   p. 169

¹⁷⁶ *The Bee-Hive-Newspaper*—an English weekly published by the trade unions in London between 1861 and 1876. In November 1864, the paper became the official organ of the International Working Men's Association, and took to printing the documents of the International and reports on the meetings of the General Council. However, the general reformist trend of the newspaper and its chauvinistic position on the Irish question were in sharp contrast with the revolutionary principles of the International. The editors of the paper often deferred the publication of the International's documents, sometimes falsified them, and handled reports on the meetings of the General Council in a most arbitrary fashion. At the beginning of 1870, bourgeois radicals and liberals began to exert an even greater influence on the newspaper and Samuel Morley, a liberal businessman, became its owner. Marx believed it essential to break with the *Bee-Hive* and to make this break

public, since on the Continent the paper was still considered an organ of the International. In his speech printed in this collection Marx gives the reasons for the break. The resolution on the break with the *Bee-Hive* proposed by Marx was adopted by the General Council early in May 1870. p. 170

177 The reference is to the *Coercion Bill* submitted by Gladstone to the House of Commons on March 17, 1870. Aimed against the national liberation movement, the Bill provided for the suspension of constitutional guarantees in Ireland, the introduction of a state of siege and the granting of extraordinary powers to the English authorities for the struggle against Irish revolutionaries. The Bill was passed by the English Parliament. p. 170

178 *History of Ireland* is a fragment of a voluminous work Engels intended to write and on which he worked at the end of 1869 and during the first half of 1870. Engels studied a vast selection of literary and historical sources: the works of antique and medieval writers, annals, collections of ancient law codes, legislative acts and legal treatises, folklore, travellers' notes, numerous works on archaeology, history, economics, geography, geology, etc. Engels's bibliography, embracing over 150 titles, is selective and includes but a fraction of the sources he studied.

The draft (see p. 210) shows that Engels's work was to consist of four long chapters, the last two being subdivided into sections. Engels actually succeeded in finishing only the first chapter— "Natural Conditions". The second chapter—"Ancient Ireland"— is unfinished. The manuscript breaks off where Engels intended to throw light on the social structure of Irish society before the invasion of the English conquerors in the second half of the 12th century. Engels did not begin writing the last two chapters, which were to describe the development of the country up to the events of his own day, although he had compiled most of the material for them. In his letter to Sigismund Borkheim in 1872 (see pp. 299-300), Engels mentioned that the Franco-Prussian war, the Paris Commune, the clash with the Bakuninists in the International, etc., interrupted his work. Engels used the results of his research in his theoretical works, including *The Origin of the Family, Private Property and the State*, and in his letters to various correspondents.

The fragment *History of Ireland* and some preparatory material Engels collected for this work were first published in 1948 in Russian in the *Marx-Engels Archives*, Vol. X. p. 171

179 Engels is referring to the formation of a centralised feudal state in England after her conquest in 1066 by William, Duke of Normandy. The reforms carried out in the 12th century by Henry II Plantagenet were particularly instrumental in strengthening the King's power. One of the objects of the English monarchy's aggressive designs was Ireland, a country at an earlier stage of social and political development than England, and still in a state of feudal decentralisation. Between 1169 and 1171 part of the island was conquered by the Anglo-Norman barons, who founded a colony there known as the Pale (see Note 113). p. 171

180 A reference to County Laoighis (Leix) in Central Ireland, which, in 1557, following the confiscation by the Tudors of the lands of local tribal communities (the clans), was renamed Queen's County in honour of Mary Tudor, the English Queen. The neighbouring Offaley County, the population of which had also fallen victim to the expropriation policy of the English colonial authorities, was renamed King's County in honour of Mary's husband, Philip II of Spain.   p. 173

181 In modern terms—deposits of the Mesozoic and Cainozoic periods.
                                                                    p. 174

182 A. Stieler, *Handatlas*, Gotha, 1868.   p. 176

183 A reference to the period of cruel reprisals against the Irish population and their wholesale expropriation, which began soon after the suppression of the Irish national liberation uprising of 1641-52 by the troops of the English bourgeois republic. According to the Acts of the English Parliament of 1652 and 1653, some of the Irish landowners, who were declared guilty of revolt (see Note 98), were to be forcibly moved to the barren province of Connaught and the swampy southern County of Clare. Resettlement was carried out under pain of execution.

On the eve of the 1798 Irish uprising, Connaught, and to an even greater extent the bordering counties of the province of Ulster in the north, became the scene of widespread terrorism by the English mercenaries and Protestant gangs hired by the landlords from among their menials (Ancient Britts, Orangemen, etc.), against the local Catholic population and its self-defence units. Under the pretext of confiscating arms from the population and billeting, soldiers and the Orangemen committed all kinds of outrages, torturing and murdering Irish people who fell into their hands and burning down their homes. Many Catholic peasants were evicted from Ulster after receiving threatening notes reading: "Go to the devil or Connaught."   p. 182

184 A reference to the Repeal of the Corn Laws in 1846 (see Note 62), leading to the inflow of cheap corn to England and creating conditions which from the point of view of the landlords and bourgeoisie favoured the development of stock-breeding in Ireland.   p. 185

185 G. Boate, *Ireland's Natural History*, London, 1652. Engels, like Wakefield, gives an earlier date of publication.   p. 186

186 The reference is to England's participation in the war against Napoleonic France and the European countries depending on her (in 1812 England fought Napoleon in alliance with Russia, Spain and Portugal), and to the Anglo-American war which broke out in the same year because the English ruling classes had refused to recognise the sovereignty of the U.S.A. and attempted to re-establish colonial rule there. The war was won by the United States in 1814.
                                                                    p. 188

187 The third volume of this publication, comprising the conclusion of the collection *Senchus Mor* (The Great Book of Old), appeared

in 1873, after Engels had written the passage in this book. *Senchus Mor* is one of the most detailed written records of the laws of the Brehons, the guardians of and commentators on laws and customs in Celtic Ireland. p. 189

188 Engels is referring to the collection *Rerum Hibernicarum Scriptores Veteres* (Ancient Annalists of Ireland), published in four volumes in 1814, 1825 and 1826 by Charles O'Conor in Buckingham.

The collection contains the first publication of part of the *Annales IV Magistrorum*, the *Annales Tigernachi*, which were written between the 11th and 15th centuries and described events from the close of the third century, the *Annales Ultonienses* (compiled by various chroniclers between the 15th and 17th centuries and describing events beginning with the mid-5th century), and the *Annales Inisfalensis* (generally assumed to have been compiled from 1215 onwards, and treating events up to 1318), mentioned by Engels. p. 191

189 Arthur O'Connor was one of the few leaders of the United Irishmen society, which prepared the 1798 uprising (see Note 111), who managed to escape execution. After his release from gaol in 1803 O'Connor was banished to France, where he stayed to the end of his days. p. 192

190 *Saerrath* and *Daerrath*—two forms of tenancy in ancient Ireland, whereby the tenant, generally an ordinary member of the community, was given the use of stock and later also of land by the chief of the clan or tribe and by other representatives of the tribal élite. They involved partial loss of personal freedom (especially in the case of Daerrath) and various onerous duties. These forms of dependence were typical of the period of the disintegration of tribal relations in ancient Irish society and of the early stages of feudalisation. At this time land tenure was on the whole still communal, while stock and farming implements were already private property, and private landownership already existed in embryonic form.

Engels's "see below" refers to the section of this chapter which remained unwritten. p. 196

191 S. Bernard, *Vita S. Malachiae.* p. 196

192 The works of Giraldus Cambrensis on Ireland, *Topographia Hibernica* and *Expugnatio Hibernica* (in Engels's manuscript *Hibernia Expugnata*), were included in the 5th volume of the *Giraldi Cambrensis Opera*, mentioned by Engels, the publication of which was begun by J. S. Brewer. The 5th volume published by J. F. Dimock appeared in 1867. p. 196

193 A reference to the following works: M. Hanmer, *The Chronicle of Ireland;* E. Campion, *History of Ireland;* E. Spencer, *A View of the State of Ireland,* published in *Ancient Irish Histories. The Works of Spencer, Campion, Hanmer and Marleburrough,* vols. I-II, Dublin, 1809, and also to: John Davies, *Historical Tracts,* London, 1786; W. Camden, *Britannia,* London, 1637; F. Moryson, *An Itinerary Containing Ten Years Travels Through the Twelve Dominions*

of Germany, Bohmerland, Switzerland, Netherland, Denmark, Poland, Italy, Turkey, France, England, Scotland and Ireland, London, 1617.

p. 197

[194] Engels is referring to Huxley's public lecture on the subject "The Forefathers and Forerunners of the English People", read in Manchester on January 9, 1870. A detailed account of the lecture was published in the *Manchester Examiner* and *The Times* on January 12, 1870.

p. 198

[195] Diodorus Siculus, *Bibliothecae historicae*, Vol. 5.

p. 198

[196] Strabo, *Geographie*, translated by K. Kärcher, Buch 7, Tübingen, 1835.

p. 198

[197] Ch. Fourier, *Le nouveau monde industriel et sociétaire ou invention du procédé d'industrie attrayante et naturelle distribuée en series passionnées*. The first edition appeared in Paris in 1829. For the passage mentioned by Engels see p. 399 of that edition.

p. 199

[198] Claudius Ptolemaeus, *Geographia*, Book II, Chapter 2.

p. 199

[199] A reference to *The Poems of Ossian* written by the Scottish poet James Macpherson, who published them in 1760-65. He ascribed them to Ossian, the legendary Celtic bard. Macpherson's poems are based on an ancient Irish epos in a later Scottish interpretation.

p. 200

[200] S. Eusebius Hieronymus, *Commentariorum in Jeremiam Prophetam libri sex*. Prologus.

p. 200

[201] Gennadius, *Illustrium virorum catologus*.

p. 201

[202] The references are to the following medieval works: Claudianus, *De IV consulatu Honorii Augusti panegiricus*; Isidorus Hispalensis, *Etymologiarum libri XX*; Beda Venerabilis, *Historiae Ecclesiasticae libri quinque*; Anonymus Ravenatis, *De Geographiae libri V*; Eginhard, *Vita et gesta Karoli Magni*; Alfred the Great, *Anglo-Saxon Version of the Historian Orosius*. In all probability Engels used extracts from the above-mentioned works contained in K. Zeuss, *Die Deutschen und die Nachbarstaemme*. See pp. 568-69 of the edition published in Munich in 1837.

p. 201

[203] Ammianus Marcellinus, *Rerum gestarum libri XXXI*, liber XX.

p. 201

[204] Nennius, *Historia Brittonum*, with an English Version by Gunn, London, 1819, § 15.

p. 201

[205] *Triads*—medieval Welsh works written in the form characteristic of the poetry of the ancient Celts of Wales, with persons, things, events, etc., arranged in sets of three. As regards their content the *Triads* are historical, theological, judicial, poetical and ethical. The early *Triads* were composed not later than the 10th century, but the extant manuscripts of these works relate to the period from the 12th to the 15th century.

p. 201

206 G.W.F. Hegel, *Vorlesungen über die Geschichte der Philosophie* (Lectures on the History of Philosophy), Bd. 3. In: *Werke*, Bd. XV, Berlin, 1836, S. 160. p. 202

207 *Alexandrian Neoplatonic school*—a trend in ancient philosophy originating in the 3rd century A. D. in Alexandria during the decline of the Roman Empire. The source of neoplatonism was Plato's idealism, and the idealistic aspect of Aristotle's teaching, interpreted in a mystical spirit by the neoplatonic philosophers. In the 5th century A. D. an unknown adherent of this school, who attempted to combine the Christian teaching with neoplatonism, signed his works with the name of Dionysius the Areopagite, the first Christian Bishop of Athens. p. 202

208 *Haraldsaga* was written early in the 13th century by the Icelandic poet and chronicler Snorri Sturluson. He tells of the life and exploits of the Norwegian King Harald (9th-10th centuries), founder of the Hårfagr dynasty. p. 204

209 *Krâkumâl* (Song of Krâka)—a medieval Scandinavian poem, composed as the death-song of Ragnar Lodbrôk (9th century), a Danish Viking taken prisoner and put to death by Ella, the King of Northumberland. According to the legend Krâka, Ragnar's wife, sang the song to her children to inspire in them the desire to avenge their father's death. Engels used the text of the song as given in the reader: F. E. Ch. Dietrich, *Altnordisches Lesebuch*, Leipzig, 1864, S. 73-80. p. 204

210 J. Johnstone, *Lodbrokar-Quida; or, the Death Song of Lodbroke*, London, 1782. p. 205

211 *Niâlssaga*—an Icelandic saga which according to recent research was recorded at the end of the 13th century from oral tradition and ancient written monuments. The central theme is the life story of Gunnar, an Icelandic Hawding (a member of the clan nobility) and his friend Bond Niâl (a free community member), an expert on and commentator of ancient customs and laws. The saga tells of the battle of the Norsemen against the Irish King Brian Boru, and is an authentic source for the study of a major event in Irish history—the Irish victory over the Norse invaders in 1014 at the battle of Clontarf. Engels quoted the excerpt from the *Niâlssaga* according to the text of the reader: F. E. Ch. Dietrich, *Altnordisches Lesebuch*, Leipzig, 1864, S. 103-08. p. 205

212 Modern scholars transcribe the name of King Brian's residence in Munster as Kankaraborg, or Kincora. p. 208

213 The Cimbri and Teutons, Germanic tribes, invaded Southern Gaul and Northern Italy in 113-101 B.C. In 101 B.C. these tribes were routed by the Roman General Marius in the battle of Vercelli (Northern Italy). The battle of the Romans against the Cimbri and Teutons was described by Plutarch in his biography of Marius, by Tacitus in *Germania*, and by other ancient historians. p. 209

214 *Beowulf*—a poem about the legendary hero Beowulf is supposed

to have been recorded in the 8th century and ranks as the finest known work of Anglo-Saxon poetry. The poem is based on folk sagas about the life of the Germanic tribes of the early 6th century.

Hildebrandslied—an 8th century German epic poem, of which only some passages have survived.

Edda—a collection of epic poems and songs about the lives and deeds of the Scandinavian gods and heroes. It has come down to us in a manuscript dating from the 13th century, discovered in 1643 by the Icelandic Bishop Sveinsson—the so-called *Elder Edda*—and in a treatise on the poetry of the scalds compiled in the early 13th century by Snorri Sturluson (*Younger Edda*).   p. 209

215 *Leges barbarorum*—records of the common law of various Germanic tribes, compiled between the 5th and 9th centuries.   p. 209

216 The preparatory material for Engels's uncompleted *History of Ireland* is vast. Passages copied from various sources fill the better part of 15 large exercise-books. In addition, there are numerous notes and fragments on separate pages and a large number of newspaper cuttings. The material is extremely varied, including analyses of sources (ancient laws, medieval chronicles, legal and historical treatises of the 16th and 17th centuries, travel notes, etc.), passages from books, notably, those relating to Irish history from ancient times to the 1860s, and jottings of Engels's own thoughts. Some of the notes represent Engels's own synthesis of data drawn from several sources. Engels generally made remarks, sometimes sharply critical ones, on the excerpts taken from the works of various authors.

Only a small part of Engels's manuscript has been published to date (in Russian, in the edition: *Marx-Engels Archives*, Vol. X, Moscow, 1948). The materials chosen for this volume show Engels's own creative contribution to the study of Irish history. They include the plan for his book, containing also in general outline his own division into periods of Irish history, the most complete and significant fragments, a chronological review of events from ancient times to the mid-17th century and the work "Varia on the History of the Irish Confiscations".   p. 210

217 See Note 119.   p. 210

218 A reference to the uprising of the Scottish highlanders in 1745. The rebellion was the result of oppression and eviction from the land carried out in the interests of the Anglo-Scottish landed aristocracy and bourgeoisie. Part of the nobility in the Scottish Highlands, who supported the claims to the English crown of the overthrown Stuart dynasty (the official aim of the insurgents was to enthrone Charles Edward, the grandson of James II), took advantage of the dissatisfaction of the highlanders. The suppression of the rebellion put an end to the clan system in the Scottish Highlands and brought about increased evictions.   p. 211

219 *The Island of Heligoland* (North Sea) was in early times settled by a Germanic tribe, the Frisians. Having become a Danish possession in the 18th century, it was captured by the English in 1807 and

ceded to England in 1814 by the Treaty of Kiel. In 1890, England gave Heligoland to Germany in exchange for Zanzibar.   p. 211

220 The Prussians defeated the Austrians in the Austro-Prussian war on July 3, 1866, near the village of *Sadowa*, in the vicinity of the town of Königgraetz in Bohemia (now Hradec Králové).

North-German Confederation—a federal German state established in 1867 under the leadership of Prussia after her victory over Austria in 1866. It existed until the formation, in January 1871, of the German Empire, incorporating in addition to the North-German Confederation the South-German states.   p. 211

221 The name given in Ireland to those who took part in the movement against the colonial authorities and landlords in the latter half of the 17th and early 18th centuries. The name was derived from the original meaning of the word—a bully, a ruffian. The Tories were mostly peasants, their leaders—expropriated Irish noblemen. At the end of the 17th century there emerged detachments made up of peasants alone—the *rapparees*. The authorities used extremely brutal methods in the fight against the Tories and rapparees. Those caught were hung, drawn and quartered. People giving information leading to their capture received high rewards. In England the nickname Tory was given by the Whigs to their opponents—the representatives of the conservative aristocratic circles, supporting the absolutist claims of the Stuarts, who were restored in 1660.   p. 212

222 A reference to the trial held in Dublin in the autumn of 1865 of the prominent participants in the Fenian movement, accused of organising an anti-government plot. The main accused were O'Leary, Luby, Kickham and O'Donovan Rossa, the publishers and editors of *The Irish People*, the Fenian newspaper suppressed by the police on September 15 (see Note 169). Many other Fenians were also arrested on denunciation by *agents provocateurs* and traitors. The picked jury was composed of supporters of English rule hostile to the Irish revolutionaries. The sentences were extremely severe: O'Leary, Luby and Kickham were sentenced to twenty years of penal servitude and O'Donovan Rossa to penal servitude for life.   p. 212

223 "*Chronology of Ireland*" was compiled by Engels mainly according to the book by Thomas Moore, outstanding Irish poet and historian, *The History of Ireland*, vols. I-IV, Paris, 1835-46. Engels admired this book for wealth of facts, literary merits and the author's deep sympathy with the oppressed people. Apart from the "Chronology", Engels used also other passages from the book. Scientifically Moore's *The History of Ireland* did not excel other works on Irish history written in the first half of the 19th century, and reflected many of the shortcomings of Irish romantic historiography of that period. This largely explains Engels's wish to make the information he drew from it fuller and more precise by turning to other sources, references to which crop up frequently in the "Chronology of Ireland". Engels, however, did not have the opportunity at that time to make all the necessary corrections to Moore's dating of events.

Yet, the general line of Ireland's historical development in his work, and his appraisals of events and people are extremely valuable and have been corroborated by later historical research. "Chronology of Ireland" ends, as does Moore's book, with 1646, the climax of the 1641-52 Irish uprising. Engels traced the subsequent course of this uprising in his excerpts from other books.   p. 213

224 The Anglo-Saxon King Athelstan defeated the Danes of Northumberland, and the Normans and Irish who came to their assistance, in the battle of Brunanburh (Central England) in 937.   p. 213

225 In the "Chronology of Ireland" Engels refers to this important landmark in Ireland's history only in general outline; a   detailed description of the beginning of the conquest of Ireland by the English is given in his other excerpts and notes. The Anglo-Norman barons from South Wales were the organisers of the first aggressive campaigns. The most influential among them, Richard de Clare, Earl of Pembroke (nicknamed Strongbow), consented to return the crown to Dermot, the King of Leinster, who had been banished from Ireland, on condition that the latter would give him his daughter in marriage and appoint him his successor. In May 1169, troops under the Anglo-Norman barons Fitzstephen and Prendergast landed on the south-eastern coast of Ireland. The next spring, troops under Maurice Fitzgerald and Raymond Le Gros invaded Ireland, and in August of the same year Pembroke himself captured Dublin. More and more feudal adventurers landed in Ireland in later years in search of booty. In October 1171, King Henry II invaded Ireland at the head of an army. Henry not only wanted to subjugate Ireland, but also to make the Anglo-Norman barons amenable to his wishes and foil their intention of creating a kingdom of their own. Henry forced the barons and the Irish chiefs to recognise him as the "supreme ruler" of Ireland, and placed his garrisons in the strongpoints of Wexford, Waterford and Dublin. He left Ireland in April 1172, leaving a Governor behind (Hugh de Lacy was the first).

In the fierce battles against the Anglo-Norman invaders, the Irish clans suffered defeat because of the lack of unity among their leaders and the enemy's superiority in arms and tactics. The establishment of the Anglo-Norman colony in Ireland marked the beginning of the age-long struggle between conquerors and local population.
                                                                                    p. 217

226 *Magna Carta Libertatum* (the Great Charter of Liberties)—a deed the insurgent barons of England, supported by the knights and townspeople, forced King John Lackland to sign on June 15, 1215. Magna Carta introduced certain limitations to the royal prerogative primarily in the interests of the big feudal lords, and made the latter's privileges secure. Some concessions were also granted to the knights and townspeople.   p. 218

227 *Geraldines*—Anglo-Irish aristocratic family descending from the first conquerors of Ireland, the Anglo-Norman nobles from South Wales. In Ireland the Geraldines became related with the clan chiefs, thereby acquiring considerable connections and influence. At the

same time they participated in the wars of conquest against the indigenous Irish. From the beginning of the 14th century, two branches of the Geraldine family—the Earls of Desmond and the Earls of Kildare—played a particularly prominent role. Both were descendants of Maurice Fitzgerald, the leader of one of the first armies of the Anglo-Norman barons to invade Ireland in 1169-71.

p. 219

228 Engels is referring to his excerpts from Thomas Moore's *The History of Ireland*. Regarding the 1295 Acts of Parliament, they say the following: In 1295 Irish *Parliament*. Acts:

"1) ...a new division of the kingdom into counties....

"2) ...all such marchers as neglected to maintain their necessary wards should forfeit their lands....

"3) ...all absentees should assign [*thus, already so early!*], out of their Irish revenues, a competent portion for that purpose [for the maintenance of a military force.—*Ed.*]

"4) ...no lord should wage war but by licence of the chief governor, or by special mandate of the king....

"5) ...an effort was made at this time to limit the number of their retainers, by forbidding every person, of whatever degree, to harbour more of such followers than he could himself maintain; and for all exactions and violences committed by these idle men ... their lords were to be made answerable."

Engels's remark (in italics) notes a feature typical of later times: the English owners of Irish estates did not reside in Ireland (see Note 117 on lords absentees).

p. 221

229 In 1286, following the death of the Scottish King Alexander III, King Edward I of England laid claim to the Scottish crown and succeeded in annexing Scotland. In 1297, an uprising flared up against English rule, and in 1306 it developed into a full-scale war of independence. The revolt was headed by Robert Bruce, a remarkable soldier. In 1314, the army of Edward II was defeated and Scotland once again became an independent kingdom.

p. 221

230 On July 24, 1314, the Scots led by Robert Bruce defeated the far bigger English army at Bannockburn, thereby liberating Scotland from English rule.

p. 221

231 In 1367, the Parliament of the English colony in Kilkenny adopted the famous Statute of Kilkenny—a code of prohibitions designed to protect the colonists from the spread among them of Irish customs and habits. The adoption of the Statute was prompted by the desire of the English authorities to intensify their policy of conquest in Ireland and to legalise the inequality of the Irish population in the vanquished part of the island, as well as to counteract the separatist tendencies of the Anglo-Irish nobility, whose strength lay in their ties with the Irish clan chiefs. The racialist, colonialist Statute demanded that the Irish be treated as enemies and their laws (the laws of the Brehons, the keepers and commentators of ancient Irish law) as the customs of an inferior race. In the excerpts from Thomas Moore's *The History of Ireland* Engels interprets the content of this

Statute as follows (Engels's own remarks are italicised): "The Statute of Kilkenny, 1367, directed against Irelandisation. Intermarriages with the natives, or any connection with them *in the way of fostering or gossipred* (see E. Spencer, [*A View of the State of Ireland*]) should be considered and punished as high treason:—that any man of English race, *assuming an Irish name, or using the Irish language, apparel, or customs, should forfeit all his lands and tenements*:—that to adopt or *submit* to the *Brehon law* was *treason* ... that the English should *not permit the Irish to pasture* or graze upon their lands, nor admit them to any ecclesiastical benefices or religious houses... (*Where were the Irish of the Pale to pasture their stock? At that time it was their main occupation!*)".   p. 224

232 *Bonaght*—a duty which the supreme and local kings, and also major clan chiefs in Ireland, levied on the smaller vassal chiefs for the maintenance of the troops. After the English conquest it was often paid to the English crown and its representatives in Ireland.   p. 225

233 *Coynye, livery*—taxes in kind the rank-and-file members of Irish clans paid to their chiefs in the form of food and equipment for the troops.   p. 225

234 A reference to the participation of Irish troops in the Hundred Years' War between England and France, which lasted, with interruptions, from 1337 to 1453. At the end of the 14th century only a few strongholds in France remained in English hands, but in 1415 King Henry V launched a new invasion, beating the French knights at Agincourt and capturing the entire north-western part of the country. In the course of a stubborn struggle, attended by a great upsurge of patriotic feeling (Joan of Arc), the French halted the advance of the English and gradually drove them from their land.   p. 225

235 At Wakefield, the army of Richard, Duke of York, claimant to the English crown, was beaten on December 27, 1460, by the supporters of the ruling house of Lancaster. The battle was one of the episodes in the Wars of the Roses (1455-85), caused by the struggle for the English throne between the houses of York and Lancaster. The war was so called after the white and red roses, that were the emblems of the Yorkist and Lancastrian parties respectively. The war was attended by the destruction of the feudal nobility and ended in the accession of the new, Tudor dynasty.   p. 226

236 *Degenerate English*—the name given to members of the Anglo-Irish families, who had long since settled in Ireland, become related to the clan élite, and assimilated many Irish customs.   p. 228

237 In the 15th century the power of the English colony in Ireland was at a low ebb. The English feudal lords were exhausted by the Hundred Years' War, and later owing to their feuds in the Wars of the Roses, the settlers in Ireland had great difficulty in withstanding the onslaught of the Irish clan chiefs. In order to get the latter to refrain from raids into the Pale they paid them an annual tribute, which became known as the "Black Rent".   p. 229

238 See Note 127.   p. 233

[239] See Note 180.   p. 235

[240] In view of the advance of the Reformation in England and the anti-Catholic policy of the government of Elizabeth I, Pope Pius V issued a special bull in February 1570, excommunicating Elizabeth and releasing her subjects from their oath of allegiance. Other acts of the Papal Curia against Elizabeth followed, and in 1576 she was deprived of her right to the Irish crown (see Note 112).   p. 236

[241] A reference to the restitution by James I of the Act of Uniformity passed in 1559 during the reign of Elizabeth I. The Act confirmed the principles of the Anglican Reformation (see Note 127) and decreed that worship was to be conducted according to a Book of Common Prayer sanctioned by the sovereign, as the head of the Church of England.   p. 241

[242] *Tanistry*—a system regulating the inheritance of chieftainship of the Celtic clans and septs (tribes) in Ireland. Like many other Irish customs, it was a relic of the tribal system. According to this custom, the successor of the clan chief, the tanist, was appointed during the lifetime of the chief from a definite family in the clan, whose members were considered the "eldest and worthiest".

Gavelkind—a term borrowed from the common law of the inhabitants of Kent and applied by English jurists to the Irish rules regulating the passing of the lands of a deceased member of the clan or sept into other hands. Ever since the time when tribal relations prevailed, land was regarded by the indigenous Irish not as private property but as a temporary tenure. Thus, after the death of its owner it did not pass to his descendants but was distributed among all free male kinsmen, including his sons out of wedlock. Although the lands of the chiefs and members of the clan élite were by that time no longer parcelled out after their death, they were not regarded as their private property and were not inherited by the family but passed to new ownership in accordance with the described tanistry principle.   p. 242

[243] Engels is referring to his excerpts from the book: G. Smith, *Irish History and Irish Character*, Oxford, London, 1861. The excerpts have never yet been published.   p. 243

[244] See Note 187.   p. 243

[245] The name given at that time to landowners among the colonists, and also to land speculators.   p. 244

[246] A reference to Engels's work, published in 1948 in Russian in the *Marx-Engels Archives*, Vol. X, under the heading "Excerpts on the History of Ireland in the 17th and 18th Centuries". These excerpts are based on material contained in the book: Matthew O'Conor, *The History of the Irish Catholics from the Settlement in 1691 with a View of the State of Ireland from the Invasion by Henry II to the Revolution*, Dublin, 1813. Engels supplemented this material with facts from many other works.

In particular, the reference is to the following passage (Engels's own remark is italicised): After the confiscation carried out in Ulster, the estates of the native Irish, in other parts of the kingdom, were invaded on the score of defective titles. "The confusion of the civil wars, and the uncertainty and fluctuation of Brehon tenures rendered them an easy prey to the rapacity of the administration; 66,000 acres between Dublin and Waterford, the properties of the Cavanaghs, Nolans, Byrnes, and O'Tooles were by inquisitions of office found to be the King's, and although a considerable portion of these escheated lands was regranted to the natives, yet the establishment of an English Protestant colony on 16,500 acres gave new vigour to old animosities, and inflamed the old proprietors with implacable hatred to the spoilers" (p. 22). *This happened apparently in 1612 or 1613.*

In 1614, "A commission issued to inquire into titles in the King's and Queen's counties, in Westmeath, Longford, and Leitrim, the counties of the O'Mulley's, O'Carroll's, M'Coughlan's, O'Doyne's, M'Geoghegan's, and O'Mallachlin's, 385,000 acres were in those districts found in the King, and planted as Ulster had been" (p. 24).

p. 244

[247] See Note 121.

[248] Engels is referring to the following place in his notes from O'Conor's book (the latter having borrowed the facts from Th. Carte, *A History of the Life of James, Duke of Ormonde,* vols. I-III, London, 1736):

"The incident with Phelim Bearn and his sons Brian and Turloug is illustrative. They owned the place of Ranelagh in County *Wicklow* according to a grant by Elizabeth (after the death of old Feag Bearn it had been regranted to Phelim) and James had issued orders on two occasions, one after another, that their rights should be accordingly respected. Nevertheless, Sir Richard Graham used counterfeited documents and invoked his connections in Dublin to seize part of the land belonging to Phelim, while Sir James Fitzpearce Fitzgerald tried to seize Brian's share for himself in like manner but did not succeed. At long last the case was submitted to a commission in England where Sir William Parsons, who had formerly in his capacity of judge in Dublin said that the contested land belonged to Phelim and not to any dummy freeholders of Graham, now asserted that the opposite was true. Since things still did not go smoothly enough, Graham and Parsons (who had by that time also become interested) declared that the land belonged to the *crown.* This put the matter in a new light. Lord Esmond gave evidence in their favour. A commission headed by Sir William Parsons was immediately appointed to investigate the matter. Although the King had ordered that the case should be heard in the last instance also by the English Council, Sir William Parsons succeeded in gaining possession of Phelim's land. He did not succeed, however, in seizing Brian's land. After all attempts had failed, Parsons, Esmond and others succeeded in having the two brothers, Brian and Turloug, gaoled in Dublin Prison on grounds of false evidence given by criminals and other persons who were forced to perjury by torture. The main accusation was that they had concealed several runaway

Irish rebels. From 1625 to 1628 there were unceasing attempts to have them convicted by resorting to false evidence and by reshuffling the composition of the jury, until, finally, Sir Francis Ennesli, later Lord of Mountnorris, and others came to their defence and a commission was appointed to investigate the charge. In December 1628, the commission found them not guilty and liberated them. However, the larger part of their possessions, notably Carrick Manor in Ranelagh, had by that time, by a grant of August 4, been handed over to *Sir William Parsons,* and they did not get it back!

"*All the above has been taken from Th. Carte (Life of Duke of Ormonde,* Vol. I, pp. 25-32)." p. 245

249 People who refused to conform to the established religion. In Ireland the name was applied to Roman Catholics, who were opposed to the Anglican Church. p. 245

250 In 1639, the war between England and Scotland ensued from an uprising of the Scots following the attempts by Charles I and his counsellors to extend absolutist ways to Scotland and introduce the Anglican prayer book to which the Scottish Calvinists, or Presbyterians, objected. After a series of Scottish victories Charles was obliged to conclude a peace treaty in the autumn of 1639 in Berwick. Meanwhile, however, he made secret preparations for revenge. In desperate need of money for his military schemes, Charles called a Parliament (the Short Parliament) in the spring of 1640, dissolving it almost immediately, upon its refusal to grant war subsidies. Thereupon he called a new Parliament—known as the Long Parliament. The King's conflict with that body eventually led to civil war—the English bourgeois revolution. Early in 1641, the Long Parliament declared the need for the establishment of lasting peace and closer union with Scotland. p. 248

251 A reference to the following passage Engels took from Matthew O'Conor's book: "1641. February. The deputies submitted to the King a remonstrance of grievances. There were complaints about 'fines, imprisonments and punishments in various shapes of torment and dishonour, for not joining in the established worship; the execution of martial law in the midst of profound peace; proclamations and acts of state made paramount to acts of the legislature; infringements of proclamations punished by imprisonment, by mutilation of members, and by confiscations, the constitution of Parliament subverted by the disfranchisement of cities and boroughs at the will of the court, the subversion of titles, and insecurity of all property by state inquisitions, by persecution of juries', etc." p. 249

252 Engels is referring to the following passage in his notes from Matthew O'Conor's book, which repudiates the slanderous inventions about "cruelty", "treachery", "conspiratorial tricks", etc., of the Irish rebels and their Ulster leader—Phelim O'Neill (Engels's own remarks are italicised):

"As regards the beating up of Protestants by Catholics, O'Conor maintains that the populous towns in the north remained in the hands of the English and thus served as refuges for the Protestant population of rural areas; many [Protestants] got safe to Derry, Enniskillen, Coleraine, and Carrickfergus, besides several thousands got safe to Dublin, 6,000 women and children were saved in Fermanagh, the Scots in Ulster did not come to harm, the capitulation of Bellyaghie was faithfully observed by the Catholics and generally at the commencement of the uprising no murders were committed (p. 33).

"Sir Phelim O'Neill was no coward; this can be seen from his 'constancy and fortitude in his last moments, his rejection of life and pardon, proffered to him on the terms of heaping dishonour and infamy on the grave of the late King'. (Carte, *Ormonde*, Vol. I, p. 181.)

"*The fact that at first (in October-December 1641) only the Irish who had been deprived of their possessions by James and ousted by the English settlers participated in the rebellion shows how badly it had been prepared.*" p. 250

253 A reference to the following passage in Engels's notes from Matthew O'Conor's book:

"To all intents and purposes the government drove the English of the Pale and the Anglo-Irish lords of Munster into the rebel army to have reasons for new confiscations. The Catholics of the Pale kept their faith to the *King* particularly zealously and with all their power resisted participation in the uprising but *had to* [join it]. The situation became particularly clear to them when the Irish Parliament which was to convene on November 9 and to confirm the 'graces' was suddenly, and *contrary to the King's orders*, postponed by the Lords Justices a few days before the date set for its convocation (p. 39). The session of Parliament was to have lasted only one day and it had been decided to submit to the King a remonstrance proposing that he should allow [the Irish Parliament] to suppress the uprising by its own forces. Lord Dillon, a Protestant who was sent to England, was arrested there by the [Long] Parliament and the remonstrance was destroyed. The impudence of the Lords Justices could be explained by the fact that the English Commons had voted £20,000, 4,000 foot, 2,000 horse for fully squashing the resistance in Ireland and the reinforcements were expected (Resolution of November 3, 1641)." p. 251

254 See Note 105. p. 255

255 Name of the agreement signed on September 25, 1643, between the Long Parliament and the Scottish Presbyterians; it reaffirmed the rights of the Scottish Calvinist Church and the freedoms and privileges of the Parliaments of both kingdoms; the terms of the agreement extended also to Scottish settlers in Ireland. p. 256

256 Owen Roe O'Neill's success at Benburb, which temporarily tipped the scale in the Irish Confederation in favour of radical elements who wanted to break not only with the Long Parliament but also with the King's party, was a major victory of the Irish rebels. However, as a result of the incessant quarrels and the clashes of interests in the Confederate camp, the moderate Anglo-Irish aristocrats soon gained the upper hand and signed a new agreement with

Ormond, the commander of the Royalist forces. This enabled Cromwell and his followers (who had by now defeated the Royalist forces in England, proclaimed a republic and beheaded Charles I) to organise a punitive expedition to Ireland on the pretext of destroying a Royalist stronghold. The true aim of the expedition was the colonial subjugation of the country. On August 15, 1649, Cromwell's army landed in Ireland and commenced the brutal suppression of the Irish rebellion, which was continued by Cromwell's successors—the Republican Generals Ireton and, later, Fleetwood. The last centres of resistance by the Irish, who had taken to guerilla warfare, were subdued in 1852. p. 258

257 This work of Engels's is a rough generalisation of the historical material compiled by him; Engels strove to reveal the main features of the English Government's policy in Ireland, which in the 16th and 17th centuries led to the subjugation of the entire island, mass evictions and the enslavement of the Irish population by new English landlords, making Ireland a mainstay of landlordism. Engels compiled his "Varia on the History of the Irish Confiscations" on the basis of numerous passages from a large number of books, and especially: J. Murphy, *Ireland, Industrial, Political and Social*, London, 1870. Of this book he made a special conspectus (as yet unpublished). In addition to Murphy he quoted or made references to passages from many other works. In the "Varia" Engels's own remarks are italicised. p. 258

258 See Note 180. p. 259

259 *Fee tail*—an estate the use of which is limited to a category of heirs stipulated in the grant; in practice it means life tenancy. p. 259

260 Engels is referring to the following passage he took from J. Davies, *Historical Tracts*, London, 1786. "Under Elizabeth only several Irish chiefs surrendered their estates and were regranted all their lands. However, the inferior chiefs and peasants as before held their several portions in course of tanistry and gavelkind, so that English law extended only to the *lords*. But James sent two special commissions [to Ireland]—'the one, for accepting surrenders and for regranting estates, ... the other, for strengthening of defective titles'. These commissions, in particular, took care to secure also the under-tenants [to the lord]. Before accepting each surrendered estate the commission had to enquire: 1) of the limits of the land; 2) how much the lord himself holds in demesne and how much is possessed by his tenants and followers; 3) what customs, duties and services he receives. After that the owner was returned the ownership of his demesne, his duties however were valued and reduced into certain sums of money, to be paid yearly in lieu thereof as rents, but the lands were left to them. In the case of defective titles like steps were taken before the title was confirmed." p. 261

261 See Note 242. p. 261

262 Engels is referring to the following passage in his excerpts from the first volume of Carte's book (Engels's own remark is italicised):

"*Plantation in Leinster*. Around the year 1608, the king's title had been found to 'all the lands between the river of Arckloe and that of Slane in the County of Wexford, and the former possessors thereof had to make surrenders of their lands into his hands. They amounted in all to 66,000 acres, 16,500 of which lying near the sea, the King determined to dispose of to an English colony, which was to be settled there, and to regrant the rest, in certain proportions, to the old proprietors under the like regulations and covenants as had been imposed on and submitted to by the planters of Ulster". (p. 22). After that came the turn of Longford and Leytrim, and also of the lands belonging to O'Carrols, O'Molloys, Mac-Coughlans, the Foxes, O'Doynes, Mac-Geoghegans, and O'Melaghlins in the Counties of the King, Queen and Westmeath. These regions became wild again and Irelandised; they caused a lot of trouble to [the English]—they were now safe receptacles of thieves and robbers. In 1614 it was decided 'to take a view of the counties and to enquire into the title which the Crown had to them or any part thereof', that is, *to take away these lands and to appropriate their incomes*. All this was done by a special commission.... 'It was an age of adventurers and projectors; the general taste of the world ran in favour of new discoveries and plantings of countries; and such as were not hardy enough to venture into the remote parts of the earth, fancied they might make a fortune nearer home by settling and planting in Ireland. The improvement of the King's revenues was the cover made use of by such projectors to obtain Commissions of enquiry into defective titles, and grants of concealed lands and rents belonging to the Crown, the great benefit of which was generally to accrue to the projector or discoverer, whilst the King was contented with an inconsiderable proportion of the concealment, or a small advance of the reserved rent'." p. 262

263 Engels is referring to the passage in his excerpts from M. O'Conor's *The History of the Irish Catholics*, already referred to in his "Chronology of Ireland" (see Note 246). In addition to the quotation given in that note, the relevant passage contains data on confiscations made in 1614 in County Longford, neighbouring on Connaught Province. These confiscations victimised the Irish aristocratic family of the O'Ferells and 25 clans, who lost their property which was parcelled out to English colonists; the other clans of the county were banished to mountainous and unfertile lands. Of the attempts to confiscate land in one of the counties of Connaught Province itself (Leitrim) the following is said: "In Leitrim immense possessions of Bryan na Murtha O'Rourke [see above (in O'Conor), p. 21] had been granted to his son Teige by patent in the first year of King James' reign *by the King himself*, and to the male heirs of his body. Teige died leaving several sons, their titles were clear, no plots or conspiracies could be urged to invalidate them. Then the commission declared them *all to be bastards* and confiscated their lands." p. 262

[264] *The Court of High Commission* was founded in England in 1559 by Elizabeth I to deal with cases of breaches of royal edicts and Acts of Parliament, instrumental in furthering the Reformation, and with offences against the Church of England. It was directed not only against the Catholics but also against the radical Protestant sects—the Puritans.   p. 264

[265] *The Star Chamber* was founded in England in 1487 by Henry VII as a special court for judging local barons. Under Elizabeth I it became one of the supreme judicial bodies investigating political crimes, a weapon in the ruthless struggle conducted against the opponents of feudal reaction and absolutism. Like the Court of High Commission, it was abolished by the Long Parliament in 1641.

In Ireland, the introduction by Strafford of similar institutions (one of them was called the Castle Chamber because it convened in Dublin Castle, the residence of the Lord Deputy) mainly served the purpose of expropriation and colonisation.   p. 264

[266] Ed. Spencer, "A View of the State of Ireland", in *Ancient Irish Histories,* Dublin, 1809. In Engels's excerpts from Spencer's book the following passage refers to the Irish clergy:

"...ye may find there ... gross simony, greedy covetousness, fleshly incontinency, careless sloth, and generally all disordered life in the common clergyman. And besides ... they do go and live like laymen, follow all kinds of husbandry and other worldly affairs as other Irishmen do. They neither read Scriptures, nor preach to the people, nor administer the Communion, but baptism they do, ... they take the tithes and offerings, and gather what fruit else they may of their livings, ... and some of them ... pay, as due, tributes and shares of their livings to their bishops...."

Engels added the following remark: *"All the above, apparently, refers to the Protestant priests of that time."*   p. 265

[267] A reference to the order given in 1641 by Lords Justices Parsons and Borlase to the English Commander, which contained instructions on the treatment of Irish rebels. The order instructed "to wound, kill, slay, and destroy all the rebels and their adherents and relievers, and burn, spoil, waste, consume, destroy, and demolish all the places, towns, and houses where the rebels were or have been relieved or harboured, and all the corn and hay there, and *to kill and destroy all the men there inhabiting able to bear arms".*   p. 265

[268] Drogheda, an ancient fortress in Eastern Ireland, was besieged on September 3, 1649, by Oliver Cromwell and taken by storm on September 12. In accordance with the order of the Commander-in-Chief to show no mercy to anyone caught with arms the three-thousand-strong Irish garrison was annihilated and many peaceful citizens were killed. Ruthless bloodshed by Cromwell's troops also attended the capture of Wexford on October 11, 1649.   p. 265

[269] Titles to plots of Irish land of definite size. They were given to soldiers of the Parliamentary army in lieu of wages. In many cases officers and speculators bought them from the soldiers for a song. p. 266

[270] On the Act of Settlement (1652) and the Act of Satisfaction (1653) see Note 98. p. 267

[271] The name given in the 16th-17th centuries to merchants and bankers, including speculators from the City of London. During the English bourgeois revolution in the 17th century "adventurers" loaned Parliament considerable sums of money for the war against the Royalists on the security of lands in Ireland. They engaged in the looting of these lands and also in the buying up of soldiers' debentures. Among the "adventurers" were many statesmen, members of the gentry and civil servants. p. 267

[272] Engels is referring to his notes from Matthew O'Conor's book, *The History of the Irish Catholics,* supplemented by excerpts from other sources. In this particular case the reference is to the passage dealing with the declaration made in 1660 by the government of Charles II at the outset of the Stuart Restoration (on the Irish policy of the post-Restoration Stuarts see Note 128). According to that declaration the "adventurers", the officers and men of the Parliamentary army retained their possessions in Ireland, while officers of Ormonde's Royalist army, who had served under him up to 1649 (hence the term "forty-nine officers"; in that year the majority of the defeated English Royalists left Ireland and the resistance to Cromwell's troops was continued mainly by the Irish rebels), received compensation out of the same fund of confiscated Irish lands. Indigenous Irishmen, who had fought under the King's banner during the Civil War and been deprived of their possessions because of it, received practically no compensation. p. 268

[273] The *Act of Settlement* was passed by the restored Stuart monarchy in 1662. The Act instituted a complicated procedure of enquiry into complaints and petitions for the return of lands to the Irish Catholics who had fought in civil war on the Royalist side. The satisfaction of complaints was encumbered by a whole system of casuistic objections and provisos. As a result, only a small part was considered and a still smaller satisfied (those who in fact received compensation for their forfeited lands were designated in the documents as "provisors"). The Act of Explanation passed in 1665 under pressure from the Protestant colonists cancelled all complaints not hitherto considered. It was called the "Black Act" in Ireland. p. 268

[274] Given below are data on the confiscations of Irish lands carried out by William III after the suppression of the 1689-91 Irish uprising, in violation of the surrender terms signed with the insurgents at Limerick (see notes 111 and 116). p. 269

[275] This article was written by Engels at the request of Marx's eldest daughter Jenny. It was intended as a preface to *Erins-Harfe,* a collection of songs on the words of Thomas Moore's *Irish Melodies,*

which was being prepared for publication in Hannover. Jenny Marx sent the article to Ludwig Kugelmann, Marx's friend in Hannover, asking him to hand it to Joseph Risse, the compiler of the collection. However, it did not appear in the collection which was printed in 1870, and was first published only in 1955.
p. 270

276 *Commission of Inquiry*—Engels's way of referring to the Special Commission of the Commons, appointed to study the effects of the Act on the Bank Restriction of 1797. The Act fixed a compulsory rate for banknotes issued by the Bank of England and abrogated their exchange for gold. Exchange was reintroduced by a new Act in 1819.
p. 272

277 An inaccuracy seems to have crept into Marx's statement. W. Blake's *Observations on the Principles which regulate the Course of Exchange; and on the Present Depreciated State of Currency,* investigating the difference between the nominal and real rate of bills of exchange, appeared in London in 1810. It was Henry Thornton's *An Enquiry into the Nature and Effects of the Paper Credit of Great Britain* that appeared in London, in 1802. In the sections on the real and nominal rates of bills of exchange Blake makes frequent references to Thornton's work.

The works of William Petty, relating to the difference between the nominal and real rate of bills of exchange are mentioned by Marx in his *Theories of Surplus-Value.*
p. 272

278 Marx compares Gladstone's cabinet (1868-74) with the "*ministry of all talents*" (see Note 36).
p. 272

279 A reference to A. Knox and J. Pollock, "Report of the Commissioners on the Treatment of the Treason-Felony Convicts in the English Convict Prisons", London, 1867.
p. 273

280 See Note 75.
p. 273

281 Engels mentions the issue of *Bee-Hive* of October 30, 1869. Its editorial "Ministers and the Fenian Prisoners" justified Gladstone's policy of repressions against the Irish Fenians. On the *Bee-Hive* see Note 176.
p. 274

282 See Note 144.
p. 274

283 In the first volume of *Capital* Marx quotes the work of George Ensor, *An Inquiry concerning the Population of Nations, containing a Refutation of Mr. Malthus's Essay on Population,* London, 1818.
p. 275

284 See Note 111.
p. 275

285 See Note 165.
p. 276

286 The amnesty ukase (edict) of May 25 (June 6), 1868, applied to some categories of people convicted for political crimes before January 1, 1866. It also affected some prisoners of foreign descent who, according to the Imperial Ukase, had been exiled from Russia for life. The amnesty entitled some Poles, who had been sentenced to

terms of imprisonment of less than twenty years, to return home.

The *Guelf conspiracy*—a reference to events after the Austro-Prussian war of 1866 in Hannover, which lost its independence and was annexed by Prussia. In the spring of 1867, in France, Georg V, the former King of Hannover, formed the Guelf legion, consisting of Hannoverian emigrants, in an attempt to restore the Guelf dynasty. On April 8, 1868, the Prussian judicial organs sentenced several officers who had had a part in forming the legion to ten years' imprisonment. Wishing to strengthen its positions in Hannover, however, early in May of the same year the Prussian Government granted an amnesty to the rank-and-file members of the Guelf legion.   p. 278

287 *Chetham Library*—one of England's oldest libraries, founded in 1653. Marx used it during his first stay in England in July and August 1845.   p. 279

288 *Der Volksstaat*—the central organ of the German Social-Democratic Workers' Party (the Eisenachers)—was published in Leipzig from October 2, 1869, to September 29, 1876. The newspaper, which expressed the views of the revolutionary wing in Germany's labour movement, was constantly persecuted by the government and the police. Although its editors changed overnight due to frequent arrests, general guidance was constantly given by Wilhelm Liebknecht. August Bebel had great influence in the newspaper. Marx and Engels contributed to it from the day it was founded, and constantly helped its editors in defining the paper's trend.   p. 280

289 See Note 63.   p. 281

290 On the instance of Marx, in November 1869, the General Council decided to arrange a debate on two questions: the attitude of the English Government to the Irish question and the attitude to this question of the English working class. Although the second question was never actually debated, Marx's point of view is known from his letters and the documents of the International. It is outlined in his letter to Engels of December 10, 1869, his letter to the Lafargues of March 5, 1870, and in the "Confidential Communication" of the General Council. Marx's main thesis was that the English working class could not achieve its own emancipation from capitalist oppression until it put an end to the colonial oppression of Ireland. p. 282

291 A reference to The United Irishmen, a patriotic society which prepared the 1798 uprising. See Note 111.   p. 285

292 Marx suggests that Engels should dedicate a separate chapter to this period in his book on the history of Ireland (see Note 178). Probably wishing to help Engels in his work on the book, Marx noted down passages from various sources on the history of Ireland between 1776 and 1801. The views he expresses in this letter are based on these excerpts. Engels intended to write a section "Rebellion and Union. 1780-1801" to be included in the chapter "English Rule".
                                                                p. 285

293 Gladstone addressed the House of Commons on February 15, 1870; his speech was published on the following day in *The Times*.   p. 287

[294] In February 1870, three candidates stood for election to Parliament from Southwark: Beresford for the Conservative Party, Odger for the workers, and Sydney Waterlow for the Liberal Party; Beresford polled 4,686 votes, Odger—4,382 and Waterlow—2,966.   p. 288

[295] See Note 222.   p. 289

[296] *Head Centre*—the name given to the leader of the secret organisation of the Fenian Brotherhood.   p. 290

[297] In the issue of *The Irishman* of March 5, 1870, Engels read a review by a Paris correspondent who highly praised the article Marx's daughter Jenny had written for *Marseillaise*. Jenny Marx came out in defence of the arrested Fenians, notably of O'Donovan Rossa, who were subjected to brutal treatment in "the prisons of humane England". The article was signed J. Williams. She borrowed this pen-name from Marx, who often signed his articles "Williams" but used the initial A. Therefore Engels writes that he "couldn't account for the first name".   p. 291

[298] *Shamrock*—Ireland's national emblem, generally depicted as a clover leaf and symbolising the Holy Trinity of the Christian faith.   p. 291

[299] At Marx's suggestion John Patrick MacDonnell, an Irish worker and former participant in the Fenian movement, was co-opted to the General Council on November 1, 1871. Preliminary enquiries had been made because some of the Irish nationalists spread malicious rumours about him. On July 4, 1871, Marx informed the General Council that nothing detrimental to MacDonnell's character had been discovered. Some functionaries of the Irish national movement, notably Murphy (a businessman and the owner of *The Irishman*), who claimed leadership of the movement and co-option to the General Council, continued the smear campaign.   p. 298

[300] O'Donovan Rossa, one of the Fenian leaders in whose defence Jenny Marx came out (see Note 297), spoke in the U.S.A., whereto he had emigrated after being amnestied, against the Paris communards, accusing them of murders.   p. 299

[301] On September 22, 1871, Marx addressed the London Conference of the International on the position of the Association in England. He particularly emphasised the need to organise independent Irish sections in the International, having in mind the specific relations between English and Irish workers, notably the antagonism between them which had for many years been fomented by the ruling classes.
                                                                     p. 301

[302] On May 14, 1872, a meeting of the General Council discussed the question of the relations between the Irish sections emerging in England and Ireland and the British Federal Council. Engels censured the chauvinistic positions held by Hales and several other English members of the General Council and the British Council, who obstructed the formation in the International of an independent Irish organisation and its struggle for Ireland's independence. In the

debate that ensued the majority of Council members   supported
Engels.

Engels's speech has been preserved in the form of notes he made
with a view to having the speech published in the press, and
also (in part) in the minutes of the General Council. The   speech
was not published because at the next meeting of the General Council
it was decided not to include the debate on the Irish question in the
report intended for the press: it was thought that some of the
speeches, notably that made by Hales, might be harmful to the
International.   p. 302

303 A reference to a clash between the Chartists and the Irish Repealers
in Manchester on March 8, 1842, provoked by the bourgeois nationalist leaders of the Irish National Repeal Association (champions
of the repeal of the Act of the Union of 1801), who were hostile
to the labour movement in England, notably to Chartism. O'Connor
and a group of Chartists were driven by the Repealers from the
Hall of Science, where O'Connor was to deliver a lecture.   p. 303

304 The *Hague Congress*, held from September 2 to 7, 1872, was called
to reaffirm by its decisions the resolutions of the London Conference
of 1871, notably, those on the political action of the working class
and the struggle against sectarianism. The preparations for the
congress involved a violent struggle between the Marxists and the
anarchists and their allies who rejected the basic principles of the
theory of scientific communism.

The Hague Congress was more representative than all previous
congresses. Sixty-five delegates from 15 national organisations attended. The inclusion by the Hague Congress in the General Rules
(Article 7) of the fundamentally important Marxist principle on
the need to found mass working-class parties and to establish a
proletarian dictatorship, and its decisions on organisational questions
were a major victory for the Marxists. The congress crowned with
success the persistent struggle Marx, Engels and their supporters
had waged against all sorts of petty-bourgeois sectarianism in the
working-class movement; the anarchist leaders (Bakunin and Guil-
laume) were expelled from the International. The decisions of the
Hague Congress laid the foundation for the setting up in each
country of an independent political party of the working class.
                                                             p. 305

305 Engels's *Letters from London* appeared in *La Plebe*, the newspaper
of the International's sections in Italy, early in April 1872, and
continued throughout the year. Early in 1873, Engels's co-operation with *La Plebe* was temporarily interrupted due to government
reprisals against the paper's editors. *La Plebe* was published under
the editorship of E. Bignami in Lodi between 1868 and 1875, and
in Milan between 1875 and 1883. Up to the early seventies the
newspaper followed a bourgeois-democratic line, later it became
socialist. In 1872-73 *La Plebe* played an important role in the
struggle against the anarchist influence in the Italian working-class
movement. Engels's contributions greatly promoted the paper's suc-

cess. In 1882, the first independent party of the Italian proletariat—
the Workers' Party—formed around *La Plebe*.   p. 306

[306] See Note 172.   p. 306

[307] By the "last" General Council Engels means the London Council
that existed before the Hague Congress of the International at which
a decision was adopted to transfer the seat of the General Council
to New York.   p. 308

[308] In the fourth article of the *Letters from London* series: "Meeting
in Hyde Park.—The Position in Spain", written on December 11,
1872, Engels reported that the Justice of the Peace could do no
more than impose the smallest possible fine, and since his decision
anyway ran contrary to the rules governing behaviour in Hyde
Park the accused demanded that the case be brought before a court
of appeal.   p. 308

[309] In December 1872, a split occurred in the British Federal Council.
By refusing to recognise the decisions of the 1872 Hague Congress
the Council's Right wing, headed by J. Hales, was, according to
the Rules of the International, making itself liable to expulsion from
the Association. This was confirmed by a decision of the General
Council of May 30, 1873. The Left wing of the British Federal
Council established itself as the British Federal Council and was
recognised by the majority of sections of the British Federation as
their leading body. In January 1873, the self-styled Federal Council
attempted to organise a congress of the Federation but only 12
delegates, representing a small portion of the British sections, arrived.
Soon after the failure of the congress this British Council disinte-
grated.   p. 309

[310] Engels had in mind the debate on the Mines Regulation Act
of 1872.   p. 310

[311] See Note 148.   p. 311

[312] *Dialectics of Nature*—Engels's outstanding philosophical work which
gives a detailed dialectical materialist interpretation of the most
important theoretical problems of natural science. Engels did not
succeed in completing the book. *Dialectics of Nature* was first published in the Soviet Union in 1925 in German and Russian.   p. 313

[313] In this work Engels subjected to devastating criticism the views of
the German petty-bourgeois philosopher and economist Eugen Dühring,
who criticised Marxism and claimed to have evolved a new system
of philosophy, political economy and socialism. In his polemics with
Dühring, Engels expounded all the three component parts of the
Marxist teaching: dialectical and historical materialism, political economy and the theory of scientific communism.   p. 314

[314] G. L. Maurer's works (12 volumes) analyse the agrarian, urban and
state systems of medieval Germany. They are: *Einleitung zur Geschichte der Mark-, Hof-, Dorf- und Städt-Verfassung und der
öffentlichen Gewalt*, München, 1854; *Geschichte der Markenverfas-*

sung in Deutschland, Erlangen, 1856; *Geschichte der Fronhöfe, der Bauernhöfe und der Hofverfassung in Deutschland,* Bd. I-IV, Erlangen, 1862-1863; *Geschichte der Dorfverfassung in Deutschland,* Bd. I-II, Erlangen, 1865-1866; *Geschichte der Städtverfassung in Deutschland,* Bd. I-IV, Erlangen, 1869-1871. The first, second and fourth of these works make a special study of the system of the German *Mark.*   p. 314

315 Engels's article "American Food and the Land Question" was printed in *The Labour Standard,* the organ of the London Trades Union Council, which appeared weekly between 1881 and 1885 under the editorship of J. Shipton. Between March and August 1881 it carried eleven articles by Engels directed against the narrow aims pursued by the trade union movement, which tended to reduce the working-class struggle to everyday economic demands. In these articles Engels expounded the principles of Marxist political economy.   p. 316

316 A reference to the *Anti-Socialist Law* introduced by Bismarck's government with the support of a majority in the Reichstag on October 21, 1878, for the purpose of fighting the socialist and working-class movement. The law deprived the Social-Democratic Party of Germany of its legal status; it prohibited all its organisations, workers' mass organisations and the socialist and workers' press, decreed confiscation of socialist literature, and subjected Social-Democrats to reprisals. The law was extended every 2-3 years. Despite this policy of reprisals the Social-Democratic Party increased its influence among the masses. Under pressure of the mass working-class movement the law was repealed on October 1, 1890.   p. 318

317 *Irish National Land League*—a peasant organisation founded by Irish revolutionary democrats in 1879. It was headed by Michael Davitt, a former Fenian. The Land League was supported by the urban poor. The most progressive section of the Irish national bourgeoisie—the Home Rulers, headed by Parnell—also associated themselves with the League. The Left wing of the Land League's leadership (Davitt, Dillon, Devoy and others) demanded full independence for Ireland, the abolition of landlordism and the transfer of the land to the peasants. The Rights confined themselves to the demand for Home Rule—the granting to Ireland of self-government within the framework of the British Empire and the normalisation of relations between the landlords and tenants. The Land League was very active and resorted to diverse methods of struggle: boycott of the supporters of the English Government, refusal to pay rent, etc. In 1881, the English Government prohibited the League and many of its leaders were arrested, but the League continued its activity almost up to the end of the decade.   p. 318

318 Marx made this synopsis of J. R. Green's book in the last years of his life, while working on a chronology of world history.
    He began to make excerpts not from the beginning of the book but from the second half of the first volume.   p. 320

319 See Note 231.   p. 321

320 Engels's article on Jenny Longuet was published in the newspaper *Sozialdemokrat*.

Der Sozialdemokrat—a German weekly, the central organ of the Socialist Workers' Party of Germany, was published during the time the Anti-Socialist Law was in force, from September 1879 to September 1888 in Zürich, and from October 1888 to September 27, 1890, in London. p. 323

321 Isaac Butt's letter from Dublin was read at the meeting of the General Council of the First International on January 4, 1870. Butt offered his offices in bringing about a union between English and Irish workers. p. 325

322 The fact that A. Regnard, a French petty-bourgeois journalist and historian, approached Marx's daughter, Jenny Longuet, about his articles on Irish history, is explained by the popularity she had won by writing articles censuring Gladstone's policy towards the Fenians for the French newspaper *La Marseillaise*. p. 326

323 *The Scottish Covenanters*—supporters of the National Covenant, the agreement signed in 1638 in Scotland after the successful uprising in 1637 against the absolutist government of Charles I. Under the banner of protection of the Presbyterian (Calvinist) religion against bishopry, the participants in the Covenant fought for Scotland's national autonomy, against all attempts to implant absolutist ways in the country. The war accelerated the outbreak of the bourgeois revolution in England. See also Note 250. p. 326

324 *Vendée*—a department in the west of France, where a counter-revolutionary uprising flared up in March 1793, during the French bourgeois revolution. The rebels were mostly backward peasants, incited and led by counter-revolutionary priests and noblemen. The uprising was put down in 1795, but attempts to renew it were made in 1799 and in later years.

Vendée has become a synonym for reactionary uprisings and counter-revolutionary hotbeds. p. 327

325 See Note 119. p. 328

326 In 1795, Pitt's government helped to found the Irish Catholic college in the town of Maynooth and granted heavy subsidies to it. This policy was intended to draw the élite of Irish landowners, bourgeoisie and clergymen over to the English side and thereby split the Irish national liberation movement. p. 328

327 *"Rotten boroughs"*—the name given to the electoral districts in rural areas in England where there were very few voters (mainly because the rural population was moving to the towns), and where the local landowner arbitrarily disposed of the votes of people dependent on him. p. 328

328 A reference to the school system introduced in Ireland in 1831 by Stanley (Earl of Derby), the then Chief Secretary for Ireland. Joint schools were set up for Catholics and Protestants and only religious subjects were taught separately. p. 328

329 Apparently a reference to the resolution adopted by the Commons
at Gladstone's proposal on February 3, 1881, to introduce a new
procedure in the British Parliament. Since the obstruction tactics
resorted to by the Irish opposition in the House of Commons
prevented the passing by Parliament of a Bill introducing coercion
laws in Ireland, Gladstone proposed according the Speaker the
right to interrupt speeches of orators and in case of insubordination
to evict them from the premises.   p. 329

330 The spread of peasant action against English landlords moved
Parliament to adopt, early in 1881, two bills on the introduction
of coercion laws in Ireland. These laws suspended constitutional
guarantees and introduced a state of siege in the country; troops
were sent to help the landlords evict tenants refusing to leave.
     The Land Bill for Ireland, proposed by Gladstone's Liberal
government at the end of 1880, was an attempt to divert the Irish
peasants from the revolutionary struggle by somewhat restricting
the arbitrary rule of the English landlords over the peasant tenants.
It was finally passed on August 22, 1881. According to the Land
Act of 1881 a landlord was not allowed to evict a tenant from
the land if he paid rent in time, the size of the rent being stipulated for 15 years in advance. Although the Land Act gave
the landlords the opportunity to sell their land profitably to the
state and the size of the rent fixed by it continued to be extremely
high, the English landlords obstructed its implementation because
they wanted to preserve their unlimited power in Ireland.   p. 329

331 The meeting of English landlords was held in Dublin on January 3,
1882, with the Duke of Abercorn in the chair. It was called to
discuss the activities of the assistant commissioners, officials appointed to implement measures connected with the 1881 Land Act
for Ireland (see Note 330). Referring to the lack of proper qualifi-
cations and the inexperience of these officials and also to the
absence of Parliamentary decisions defining their competency, the
landlords accused the assistant commissioners of adopting biassed
decisions on lowering the rents collected by the landlords. In an
attempt to sabotage the Land Act, the landlords demanded that
the government consider their appeals without delay and pass a
law on compensation for losses they might incur if the government
sanctioned a reduction of rents.   p. 331

332 See Note 110.   p. 333

333 The *Democratic Federation*—an association of various British radical
societies of a semi-bourgeois, semi-proletarian trend, set up on
June 8, 1881, under the guidance of H. M. Hyndman. The Federation adopted a bourgeois-democratic programme containing 9 points:
universal suffrage, a three-year Parliament, a system of equal
electoral districts, the abolition of the House of Lords as a legislative body, independence for Ireland in the field of legislation, nationalisation of the land, etc.
     At the inaugural conference of the Democratic Federation
Hyndman's pamphlet *England for All* was distributed among the

participants. In its two chapters (Chapter II—"Labour", and Chapter III—"Capital") Hyndman included whole sections from the first volume of *Capital* as programme principles of the Federation. He made no reference to either the author or the book, and in many cases distorted Marx's propositions.

In 1884 the Democratic Federation was reorganised as the Social-Democratic Federation.   p. 333

334 The mass action of the Irish peasants led by the Land League and various secret societies forced Gladstone to repeal the emergency measures introduced in 1881. On May 2, 1882, the Irish M.P.s, the leaders of the Land League (see Note 165) Parnell, Davitt, Dillon and O'Kelly, were released from gaol. At the same time the champions of the emergency measures—F. T. Cowper, the Viceroy for Ireland, and W. Forster, Chief Secretary for Ireland—had to resign. Lord Cavendish was appointed Chief Secretary for Ireland.
                                                                p. 333

335 Gladstone's repressions in Ireland intensified the activities of various secret societies which resorted to terror against the landlords and their managers, and against government officials. As a result many estate owners left Ireland.   p. 333

336 Engels wrote this letter after reading "Die Situation in Irland", an article by Eduard Bernstein signed "Leo", in the May 18, 1882, issue of *Sozialdemokrat*. Bernstein gave Engels's letter to W. Liebknecht, who published a large portion of it in the same newspaper on July 13, 1882, in the form of an article entitled "Zur irischen Frage", in which he inserted his editorial comments. He also appended Engels's text with an introduction and a conclusion by the editorial board. In his letter to Bernstein of August 9, 1882, Engels expresses his indignation with Liebknecht's misrepresentation of his views on the Irish question (see p. 337).   p. 333

337 The *Alabama affair*—a conflict between the U.S.A. and England due to the military help rendered by the latter to the Southern States during the Civil War of 1861-65. The English Government built and equipped cruisers for the Southern States, including the *Alabama*, which did considerable damage to the Northern States. After the war the U.S. Government demanded of the English Government full compensation for the losses inflicted by the *Alabama* and other vessels. The tribunal of arbitration in Geneva adjudged on September 14, 1872, that England should pay the United States $15,500,000 damages. England submitted to the tribunal's decision because she wanted the U.S.A. to keep out of Irish affairs and to stop supporting the Irish revolutionaries.   p. 335

338 Lord Cavendish, the newly appointed Chief Secretary for Ireland, and Thomas Henry Burke, the former Under-Secretary, were assassinated on May 6, 1882, in Phoenix Park in Dublin by members of the terrorist organisation "The Invincibles", which incorporated some former Fenians. Marx and Engels did not approve of the terrorist tactics of these epigoni of Fenianism; in their view, such

anarchistic acts could not in the least affect England's colonial policy towards Ireland but only involved unnecessary sacrifices on the part of the Irish revolutionaries and disorganised the national liberation movement. p. 336

339 In 1878, attempts to assassinate Kaiser Wilhelm I were made by Max Hödel, an apprentice from Leipzig, and by Karl Nobiling, an anarchist. These attempts became the pretext for the institution of the Anti-Socialist Law. See Note 316. p. 336

340 The conquest of Wales by the English was completed in 1283. However, Wales retained its autonomy after that, and was finally united with England in the mid-16th century. p. 338

341 Engels is referring to his work on the history of Ireland which remained uncompleted (see Note 178). In studying the history of the Celts he also looked into the ancient Welsh laws. p. 338

342 In September 1891, Engels made a trip to Scotland and Ireland. p. 341

343 See Note 218. p. 341

344 L. H. Morgan, *Ancient Society,* London, 1877, pp. 357, 358. p. 341

345 Beda Venerabilis, *Historia ecclesiastica gentis Anglorum,* Book I, Chapter I. p. 341

346 Engels gave this interview to a reporter of the *New Yorker Volkszeitung* on September 19, 1888, after a trip round the U.S.A. Engels travelled incognito and wanted to avoid all contacts with the press. Jonas, the editor of the *New Yorker Volkszeitung,* however, got to know of Engels's stay in New York and sent T. Cuno, a former functionary of the First International, to him on behalf of the paper. The interview was published in the paper without preliminary discussion of its text with Engels. On October 13 the interview was reprinted in *Sozialdemokrat,* apparently without any objections being voiced by Engels. p. 343

347 This preface was written by Engels for the English edition of his book *The Condition of the Working-Class in England,* published in London in 1892. The first edition of the authorised English translation appeared in New York in 1887. Most of the preface—with a few editorial changes and a few deletions—consists of the appendix to the American edition written by Engels in 1886 and his article "England in 1845 and 1885", which it included. The concluding part of the preface was written by Engels specially for the 1892 English edition. p. 344

348 "*Little Ireland*"—a workers' district in a southern suburb of Manchester inhabited mainly by Irishmen. It is described in Engels's work *The Condition of the Working-Class in England* (see p. 38). "*Seven Dials*"—a workers' district in central London. p. 344

349 This article was written by Engels for the journal *Die Neue Zeit,* a theoretical organ of the German Social-Democrats, published in Stuttgart from 1883. Speeches made by the German Right-wing So-

cial-Democrat G. Vollmar on the agrarian question prompted Engels to write it. He felt that it was necessary to explain the fundamentals of the revolutionary proletarian attitude towards the peasant question in a special article and to criticise Vollmar's opportunist views and deviations from the Marxist theory in the agrarian programme of the French socialists, adopted at the Marseilles Congress (September 1892) and supplemented at the Nantes Congress (September 1894).
p. 346

350 The general election in England was held between November 23 and December 19, 1885. As a result of this first election after the 1884 Parliamentary Reform, the Liberals obtained 331 seats, losing 20, the Conservatives—249 and supporters of Home Rule for Ireland—86.
p. 348

351 *Centre*—a political party of the German Catholics founded in 1870-71. It generally held intermediate positions, manoeuvring between the parties supporting the government and the Left opposition factions in the Reichstag. Under the banner of Catholicism it united various sections of the Catholic clergy, landowners, bourgeoisie, some of the peasants, predominantly in the small and medium-sized states in West and South-West Germany—that is, people of very different social status—and supported their separatist trends. The Centre was in opposition to Bismarck's government but voted for his measures directed against the labour and socialist movement.
p. 348

352 *"New Manchester School"*—in the late seventies and early eighties, when England encountered growing competition from the U.S.A. and Germany on the world market, the English bourgeoisie who had hitherto supported the "Manchester School" (see Note 40) began to change their attitude and press for the introduction of protective tariffs.
p. 348

353 The debates on the Irish Arms Bill mentioned by Engels were held during its second reading in the House of Commons on May 20, 1886. The Bill was to prolong the ban established by the 1881 law on the sale, import and carrying of arms in some districts of Ireland. John Morley, the Secretary for Ireland, in bringing the Bill before Parliament, said that it was particularly important for Northern Ireland (Ulster), where open agitation was being conducted among the Protestant population for the organisation of armed resistance against the introduction of self-government in Ireland on a Home Rule basis. Randolph Churchill said in his speech that these actions were legitimate and referred to Althorp and Robert Peel, who in 1833 had said that civil war could be morally justified in the face of a threat to the integrity of the British Empire. In his reply Gladstone reproached Churchill for supporting resistance to government measures. The Bill was passed by the House of Commons by 353 votes to 89.
p. 349

354 During the first half of April 1887, the House of Commons discussed the draft Crimes Bill for Ireland, which provided for the introduc-

tion there of a simplified judicial procedure with a view to quelling the growing peasant disturbances. The executive organs were to be granted the right to outlaw various societies, and sentences on charges of conspiracy, illegal meetings, insubordination, etc., could be passed by the judiciary without a jury. Mass meetings in protest against the Bill, held on April 11, 1887, in Hyde Park, were attended by 100,000-150,000 people. The meetings called by various organisations were addressed by speakers from the Liberal Party (Gladstone and others), the Social-Democratic Federation (Bateman, Williams, Burns and others), the Socialist League (Eleanor Marx-Aveling, Edward Aveling and others) and from other organisations.

In its report on the meeting entitled "Irish Crimes Bill, Great Demonstration in Hyde Park, Processions and Speeches" the *Daily Telegraph* said on April 12, 1887, that Eleanor Marx-Aveling's speech had evoked lively interest and had been greeted enthusiastically.

p. 350

355 Engels is referring to differences within the Liberal Party. In 1886, the wing opposed to the granting of self-government to Ireland split away to form the Liberal Unionist Party under J. Chamberlain. On most issues the Liberal Unionists supported the Conservatives.

p. 350

356 See Note 72.

p. 351

357 A reference to the stand of the Progressist Party in the Reichstag elections in February 1887. During the second ballot the supporters of the Progressist Party voted for the candidates of the "cartel"— the bloc of both conservative parties and the National-Liberals— against the Social-Democrats, thereby helping that bloc, which supported Bismarck's government, to victory.

p. 351

358 In April 1886, hoping to win the support of the Irish M.P.s, Gladstone tabled the Home Rule Bill providing for self-government for Ireland within the framework of the British Empire. This Bill led to a split in the Liberal Party and the break away of the Liberal Unionists (see Note 355). The Bill was defeated.

p. 352

359 The *National Union of Gasworkers and General Labourers of Great Britain and Ireland,* founded in April 1889, had over 100,000 members. It was the first trade union in the English and Irish labour movements to organise unskilled workers. Its chief demand was the introduction of an eight-hour working day. Eleanor Marx-Aveling played a major role in its organisation and leadership.

The active dissemination of socialist ideas among the trade union members by Eleanor Marx and her comrades helped the Union exert a major influence on Ireland's working-class movement. Its example promoted the formation of the dockers', agricultural workers' and other trade unions.

p. 353

360 The *Second Congress of the National Union of Gasworkers and General Labourers of Great Britain and Ireland* was held on May

17, 1891, in Dublin. The Congress adopted a decision on the participation of the Union in the forthcoming International Socialist Workers' Congress in Brussels: and Eleanor Marx-Aveling and William Thorne were elected delegates.   p. 353

361 The gasworks owners in Leeds demanded that workers should be hired for a term of four months and not be entitled to strike during that period. They also demanded that the volume of work done during an 8-hour shift be 25 per cent greater than it was when the working day was longer. These conditions were tantamount to the destruction of the gasworkers' trade union in Leeds and the abolition of the 8-hour working day. They caused a storm of indignation among the workers and were rejected by them. Early in July 1890 clashes occurred between the strikers and strikebreakers, who were supported by troops. The staunch resistance of the strikers forced the strikebreakers and the troops to retreat, and the bosses were compelled to waive their conditions.   p. 354

362 Engels is referring to the success of the workers and socialists in the Parliamentary elections in England in the summer of 1892. The English workers' and socialist organisations nominated a large number of candidates, three of whom—Keir Hardie, John Burns and J. H. Wilson—were elected to Parliament. The elections were won by the Liberals.   p. 355

363 Engels is referring to the persecution by English and Irish reactionaries of C. S. Parnell, the leader of the Irish national movement. At the end of 1889, the Liberal Unionists (former members of the Liberal Party, who left it in 1886 because they opposed Home Rule) had Parnell brought to court on a charge of adultery. The court (November 1890) found Parnell guilty and this let loose a smear campaign against him. Both Liberal and Conservative M.P.s demanded that he be removed from the post of leader of the Irish Parliamentary faction. The attacks against Parnell, which played on bourgeois hypocrisy in questions of morals, pursued the aim of removing him from the political scene and weakening the Irish national movement. The smear campaign against Parnell was supported by the Right wing of the Irish faction and the Irish Catholic clergy, who feared his influence and did not share his aspirations for Home Rule. All this led to a split of the Irish Parliamentary faction and weakened the Irish national movement. The campaign was largely responsible for Parnell's early death in 1891.   p. 355

364 The *Fabian Society* was founded in 1884. The name was derived from Quintus Fabius Maximus, a Roman general of the 3rd century B.C., nicknamed the "Cunctator" (or Delayer) because he achieved success in the second Punic war against Hannibal by avoiding direct battle and using dilatory tactics. Most of the Fabians were bourgeois intellectuals, chief among whom were Sidney and Beatrice Webb. They rejected Marx's teaching on the class struggle of the proletariat and the socialist revolution and maintained that a transition from capitalism to socialism could be effected by petty reforms and the gradual transformation of society, through so-

called municipal socialism. The Fabian Society diffused bourgeois influence among the working class and propagated reformist ideas in the English labour movement. Lenin defined Fabianism as "the most consummate expression of opportunism and of liberal-labour policy". In 1900 the Fabian Society was incorporated in the Labour Party. "Fabian socialism" is still one of the sources of the ideology of class conciliation.   p. 355

365 The *Social-Democratic Federation*—an English socialist organisation founded in August 1884, on the basis of the Democratic Federation. It united heterogeneous socialist elements, mainly intellectuals. The Federation was for a long time led by reformists, with Hyndman at the head, who followed an opportunist and sectarian policy. The group of revolutionary Marxists in the Federation (Eleanor Marx-Aveling, Edward Aveling, Tom Mann and others) opposed Hyndman's line and fought for the establishment of close links with the mass working-class movement. After the split in the autumn of 1884 and the formation in December 1884 by the Left-wingers of an independent organisation—the Socialist League—the opportunists became more influential in the Federation. Under the influence of the revolutionary-minded masses, however, revolutionary elements kept forming in the Federation and dissatisfaction with the opportunistic leadership grew.

The *Socialist Labour Party of America* was founded in 1876. Most of its members were immigrants (chiefly Germans) who had little contacts with the native American workers. As its programme the party proclaimed the struggle for socialism, but, owing to the sectarian policy of its leadership, which ignored work in the American proletariat's mass organisations, it did not become a genuinely revolutionary Marxist party.   p. 356

366 *The Labour Leader*—an English monthly founded in 1887 as *Miner*. From 1889, under its new name, it appeared as the organ of the Scottish Labour Party, and in 1893 it became the organ of the Independent Labour Party. James Keir Hardie was its editor up to 1904.   p. 356

367 General Parliamentary elections were held in England from July 12 to 29, 1895, and were won by the Conservatives with a majority of more than 150 seats. Many candidates of the Independent Labour Party, including Keir Hardie, were blackballed.   p. 356

368 This document was drawn up by Peter Fox, a member of the General Council, following the debate of the question of Irish political prisoners at the Council meetings on February 20 and March 6, 1866. On the decision of the General Council it was published under Odger's name in the newspaper *Commonwealth* No. 157, March 10, 1866.   p. 361

369 *Van Diemen's Land* (Tasmania)—a penal colony to which English courts exiled political convicts sentenced to hard labour for life.
                          p. 363

370 Hyde Park was the scene of mass meetings organised by the

Reform League, which led the struggle for the election reform in 1865-67. The tens of thousands of workers attending them wanted decisive action and the leaders of the League were unable to keep them within the "bounds of the law". The workers clashed with the police, broke into the territory of the Park despite the ban on entry and smashed windows in houses belonging to M.P.s opposing the reform. In May 1867, a new wave of mass meetings began in Hyde Park. This made the ruling circles rush to carry out the reform. p. 368

371 This address, which is in fact the manifesto of the Land and Labour League (see Note 165), founded in October 1869, was drawn up by Eccarius around November 14, 1869. It was edited by Marx. p. 373

372 In agitating for the repeal of the Corn Laws, the speakers of the Anti-Corn Law League endeavoured to prove to the workers attending the meetings that with the introduction of free trade their real wages would rise and their loaf of bread would be twice as large. p. 373

373 This category of taxpayers includes people deriving their income from trade, and people of the free professions. p. 374

374 The Poor Law adopted by Parliament in 1834 abolished all relief to the poor, which had until then existed in parishes; all the needy, including children under age, were now sent to special workhouses. Because of the prison regime in them, the people called these houses Bastilles for the poor. p. 375

375 These articles were written by Marx's daughter Jenny for the French Republican newspaper *Marseillaise* and dealt with the questions raised in Marx's article "The English Government and the Fenian Prisoners". The third article was written together with Marx. All except the second article were signed J. Williams. See also Note 297. p. 379

376 Gladstone's speech appeared in *The Times* on March 4, 1870. p. 381

377 The author paraphrases Voltaire's words: "All genres are good except the boring one." p. 386

378 The demonstration demanding an amnesty for the Fenians detained in English prisons was held in Hyde Park on October 24, 1869. See Note 144. p. 388

379 An anonymous article in *The Times* of March 16, 1870, written by Henry Bruce, Home Secretary in the Liberal Government, attempted to disprove the facts adduced by O'Donovan Rossa. p. 388

380 George Moore's speech in the House of Commons and Gladstone's reply on March 17, 1870, were published in *The Times* on March 18, 1870. p. 389

381 See Note 141. p. 389

382 See Note 148. p. 390

383 See Notes 177 and 173.   p. 392

384 A reference to the book: F. T. H. Blackwood, *Mr. Mill's Plan
    for the Pacification of Ireland Examined,* London, 1868.   p. 392

385 A quotation from *Reynolds's Newspaper* of March 20, 1870. The
    article was signed "Gracchus".   p. 392

386 The author paraphrases Shakespeare. See *King Henry VI,* Part I,
    Act I, Scene 2.   p. 398

387 Lawyer Laurier made this speech on March 25, 1870, at the trial
    of Prince Pierre Bonaparte, who was accused of the murder
    of the journalist Victor Noir. The speech was published in the
    French newspaper *Marseillaise* No. 97, March 27, 1870.   p. 400

388 At the meeting of the General Council of the International Working
    Men's Association on April 2, 1872, MacDonnell, the Correspond-
    ing Secretary for Ireland, reported on the persecution to which
    the Irish sections in Dublin, Cork and other places were being
    subjected by the police. A commission made up of Marx, Mac-
    Donnell and Milner was charged with drawing up a special declara-
    tion in this connection. On April 9, MacDonnell submitted to the
    General Council a declaration on police terror in Ireland. The
    text was approved and it was decided to print 1,000 copies in the
    form of a leaflet for distribution in Ireland.
        The text of the General Council's declaration was also printed
    in the Spanish newspaper *Emancipacion* with a foreword by the
    editors quoting MacDonnell's report of April 2, 1872.   p. 404

389 See Note 73.   p. 405

390 The *Universal Federalist Council* was formed early in 1872 of repre-
    sentatives of the 1871 French section which had not been accepted
    into the International, of various bourgeois and petty-bourgeois
    organisations, Lassalleans, who had been expelled from the London
    German Workers' Educational Association, and other elements. This
    body pretended to the leadership of the international working-class
    movement, including the International. The organisers of the
    Universal Council claimed, in particular, that the General Council
    of the International Working Men's Association was not a legiti-
    mate body.   p. 407

391 Eleanor Marx-Aveling's and William Thorne's letter was addressed
    to the Chairman of the association of the trade unions of American
    workers—the American Federation of Labour (A.F.L.). The authors,
    who expressed the sentiments of the revolutionary forces acting
    under Engels's leadership, advocated the unity of the international
    labour movement, and did all they could to bring it about. p. 414

NAME INDEX

A

Abercorn, James Hamilton, Duke of (1811-1885)—Lord Lieutenant of Ireland (1866-68 and 1874-76)—123, 135, 138, 143, 144, 275, 331

Aberdeen, George Hamilton-Gordon, Earl of (1784-1860) —British statesman, Tory; from 1850 leader of Peelites, Foreign Secretary (1828-30, 1841-46) and Prime Minister (1852-55)—69, 362

Acland, Sir Thomas Dyke—English bourgeois radical—144

Adam of Bremen (d. c. 1085)— chronicler, author of Gesta Hammaburgensis Ecclesiae Pontificum—209

Adrian IV (Nicholas Breakspear) (d. 1159)—Pope (1154-59), of English origin—127, 216, 320

Albinus (4th cent.)—Irish Christian missionary—201

Albinus (latter half of 8th cent.) —Irish scholar—202

Alen, John—Lord Chancellor of Ireland (1538-46 and 1548-50)—233

Alexander the Great. See Alexander of Macedon

Alexander of Macedon (356-323 B.C.)—soldier and statesman of Ancient Greece—192

Alexander II (1818-1881)—Russian Emperor (1855-81)—336

Alexander III (1845-1894)—Russian Emperor (1881-94)—351

Alfred the Great (849-899)— King of West Saxons (871-899)—201

Alison, Sir Archibald (1792-1867)—English historian and economist, Tory—40

Allen, William Philip (1848-1867)—Irish Fenian, sentenced to death by an English court and executed—118, 145, 146, 372

Althorp, John Charles Spencer, Viscount (1782-1845)—British Whig statesman, Chancellor of Exchequer (1830-34)—349

Ammianus Marcellinus (c. 332-c. 400)—Roman historian—201

Anne (1665-1714)—Queen of Great Britain and Ireland (1702-14)—123, 129, 140, 148

Anselm of Canterbury (1033-1109)—theologian, representative of early scholasticism —203

Applegarth, Robert (1833-1925) —British trade union leader, General Secretary of Amalgamated Society of Carpenters and Joiners (1862-71) member of London Council of Trade Unions, member of General Council of International (1865, 1868-72)—278, 279, 405

Argyll, George Douglas Campbell, Duke of (1823-1900)— British statesman, Liberal, Lord Privy Seal (1880-81) in Gladstone's government— 330

Arnaud, Antoine (1831-1885)— French revolutionary, Blanquist, member of Paris Commune, member of General Council of International (1871-72)—407

Aveling, Edward (1851-1898)— English socialist, writer and publicist, one of translators of first volume of Capital into English; member of Social-Democratic Federation from 1884, then a founder of Socialist League and organiser of general labourers union, husband of Marx's daughter Eleanor—353, 355

B

Bagenal, Henry (1556-1598)—Marshal of Ireland—239

Baker, Robert—British factory inspector in 1850s and 1860s—112

Baltinglass, Eustachius (d. 1585)—a leader of Irish uprising against English rule—237

Barrett, Michael (d. 1868)—Irish Fenian, sentenced to death by English authorities—396

Barry, David de (mid-13th cent.)—Lord Justice of Ireland—219

Barry, Maltman (1842-1909)—English journalist, socialist; member of General Council (1871-72) and British Federal Council (1872-74) of International; contributed to the conservative newspaper Standard; in 1890s supported so-called socialist wing of Conservative Party—325, 405, 407, 408

Beaufort, Daniel Augustus (1739-1821)—Irish geographer and clergyman of French origin, author of Memoir of a Map of Ireland—180

Bebel, August (1840-1913)—outstanding figure in German and international working-class movement, one of founders and leaders of German Social-Democratic Party, associate and friend of Marx and Engels—318, 319, 351, 355, 414

Becker, Johann Philipp (1809-1886)—prominent figure in German and international working-class movement, organiser of German sections of First International in Switzerland, associate and friend of Marx and Engels—349

Beda the Venerable (c. 673-735)—Anglo-Saxon ecclesiastic, scholar and historian—201, 341

Bellingham, Edward (d. 1549)—Lord Deputy of Ireland from 1548—259

Benignus (d. 468)—Irish priest; according to tradition, one of compilers of Senchus Mor, a collection of ancient laws—194

Bernard of Clairvaux (c. 1091-1153)—French theologian, fanatical champion of Catholicism—196

Bernstein, Eduard (1850-1932)—German Social-Democrat, publicist, editor of the newspaper Sozialdemokrat (1881-90); in latter half of 90s, after Engels's death, openly advocated revision of Marxism—329, 330, 333, 337, 349

Besson, Alexandre—French émigré, lived in London; member of General Council of International (1866-68), Corresponding Secretary for Belgium—119

Bismarck, Otto, Prince (1815-1898)—statesman of Prussia and Germany, representative of Prussian Junkers; Prime Minister of Prussia (1862-71) and Chancellor of German Empire (1871-90)—318, 319, 332, 405

Blackburne, Francis (1782-1867)—Irish lawyer and statesman, held high posts in English judiciary in Ireland—132, 144

Burgh, Richard de (c. 1269-1326) —ruler of Ulster and Connaught—220

Burgh, Ulick de, Marquis of Clanricarde (1604-1657)— representative of Anglo-Irish aristocracy, Commander-in-Chief of King's army in Ireland in 1650-52—251

Burke, Richard (d. 1870)—Irish Fenian, served in American army; one of organisers of 1867 uprising in Ireland; died in prison—165, 381, 385, 386, 388, 389, 394, 395, 396

Burke, Thomas F. (b. 1840)— Irish Fenian, general of Southern army in American Civil War; one of organisers of 1867 uprising in Ireland; sentenced to life imprisonment in April 1867—165

Burke, Thomas Henry (1829-1882)—from 1868 permanent Irish Under-Secretary; assassinated by members of Irish terrorist organisation Invincibles on May 6, 1882 —336

Burke (Burgh) Ulick, Lord Clanricarde (d. 1504)—representative of Anglo-Irish aristocracy in Ireland—229

Burnet, Gilbert (1643-1715)— English bishop, author of a number of historical works— 264

Burns, John (1858-1943)—active participant in British working-class movement; one of leaders of new trade unions in 80s but in 90s adopted positions of Liberal trade unionism—414

Burns, Lydia (Lizzy) (1827-1878) —Irish working woman, participant in Irish national liberation movement; Engels's wife—273, 278, 291

Burt, Thomas (1837-1922)— English trade unionist, Secretary of Miners' Union of Northumberland, Liberal M.P. (1874-1918)—311

Butler, Benjamin Franklin (1818-1893)—American politician and general; commander of Northern army during American Civil War—98

Butler, Edmund (d. 1337)—Lord Justice of Ireland (1315-16) —222

Butler, Edmond (d. 1551)—son of Piers Butler, Earl of Ormonde; Archbishop of Cashel—234

Butler, Sir Piers, Earl of Ormonde (d. 1539)—Lord Justice of Ireland (1528)—230, 232

Butler, Thomas—Lord Justice of Ireland (1408-09)—225

Butt, Isaac (1813-1879)—Irish lawyer and politician, Liberal M.P., defended Fenian prisoners in state trials in 60s; one of organisers of Home Rule movement—72, 153, 155, 156, 276, 283, 299, 311, 325

Buttery, G. H.—member of General Council of First International—405

Byrne, Hugh—one of leaders of insurgent army of Irish Confederates in 1640s—249, 252

C

Caesar (Gaius Julius Caesar) (c. 100-44 B.C.)—Roman general, statesman and writer, author of *De bello Gallico*— 209, 340

Caird, James (1816-1892)—Scottish agriculturist, Liberal M.P., author of a number of works on agrarian question in England and Ireland—181

Cairnech (5th cent.)—Christian missionary in Ireland; according to tradition, one of compilers of *Senchus Mor*, a collection of ancient laws— 194

Camden, William (1551-1623)— English historian—197, 260

Campion, Edmund (1540-1581)— Catholic missionary in England, author of Historie of Ireland—197

Canning, George (1770-1827)— British statesman, a leader of Tories, Foreign Secretary (1807-09, 1822-27), and Prime Minister (1827)—70

Carew, George (1555-1629)— English statesman, Lord President of Munster—240

Carey, Martin Henley—Irish journalist, Fenian; in 1865 was sentenced to five years penal servitude—154, 165, 381

Carlyle, Thomas (1795-1881)— Scottish writer and historian, idealist philosopher, aligned with Tories—41

Carnarvon, Henry Howard Molyneux Herbert, Earl of (1831-1890)—British statesman, Conservative—98

Carolan (O'Carolan), Torlogh (1670-1738)—Irish folk singer, author of many folk songs—270

Carte, Thomas (1686-1754)— English historian—245, 261-63, 265, 267

Casey, John—Irish Fenian, sentenced to five years penal servitude in 1866—399, 400

Castlehaven, James Touchet, Earl of (1617-1684)—English Royalist, supporter of Charles I; Commander-in-Chief of insurgent army of Irish Confederation—256, 265

Castlereagh, Robert Stewart, Viscount (1769-1822)—British statesman, Tory, one of organisers of suppression of Irish uprising in 1798; Chief Secretary for Ireland

(1799-1801), Secretary for War and Colonies (1805-06, 1807-09) and Foreign Secretary (1812-22)—88, 98, 125, 157, 275, 393

Cavendish, Frederick Charles, Lord (1836-1882)—British statesman, Liberal, Chief Secretary for Ireland; assassinated on May 6, 1882 by members of Irish terrorist organisation Invincibles—336

Cavour, Camillo Benso, di, Count (1810-1861)—Italian statesman, leader of liberal-monarchist bourgeoisie and bourgeoisified nobility; head of Sardinian government (1852-59 and 1860-61) and of all-Italian government (1861)—332

Celestius (mid-4th-early 5th cent.)—Icelandic monk and missionary—201

Chamberlain, Joseph (1836-1914) —British statesman, Liberal, then Unionist Liberal, member of British cabinet over a number of years, supporter of active colonial policy, opposed Home Rule for Ireland—348

Champion, Henry Hyde (1859-1928)—British socialist, member of Social-Democratic Federation up to 1887, editor and publisher of the newspaper Labour Elector; emigrated to Australia in 90s—356

Charlemont, Toby Caulfeild, Baron (d. 1642)—English aristocrat, killed during Irish uprising—250

Charles the Bald (823-877)—King of France (840-77)—203

Charles the Great (c. 742-814)— King of Franks (768-800), Emperor (800-14)—202

Charles I (1600-1649)—King of England (1625-49), executed

during English bourgeois revolution of 17th century—127, 128, 245, 248, 252-57, 262, 263, 327

Charles II (1630-1685)—King of England (1660-85)—128, 267

Charles V (1500-1558)—Emperor of Holy Roman Empire (1519-56), King of Spain (1516-56) under name of Charles I—232

Chichester, Arthur, Lord of Belfast (1563-1625)—Lord Deputy of Ireland (1604-14)—241, 262

Churchill, George Charles Spencer, Duke of Marlborough (1844-1892)—British aristocrat, elder brother of Randolph Churchill—349

Churchill, Randolph Henry Spencer, Lord (1849-1895)—British statesman, a Conservative leader, Secretary of State for India (1885-86), advocate of colonial expansion, opponent of Home Rule for Ireland—349

Clanricarde. See *Burgh, Ulick de*

Clanricarde, Ulick John de Burgh, Marquis of (1802-1874)—British diplomat and statesman, Whig, Ambassador to St. Petersburg (1838-41)—67

Clare, Richard (Strongbow), Earl of Pembroke (d. 1176)—Anglo-Norman feudalist, owned lands in South Wales; one of chief organisers of conquest of Ireland and of English colonies in southwestern part of island—217, 218, 320

Clare, Thomas (d. 1287)—Earl of Gloucester—220

Clarence, George, Duke of (1449-1478)—brother of King Edward IV—227

Clarence, Lionel of Antwerp, Duke of (1338-1368)—son of

Edward III, Lord Lieutenant of Ireland—224

Clarendon, George William Frederick Villiers, Earl of (1800-1870)—British statesman, Whig and later Liberal; Lord Lieutenant of Ireland (1847-52); organised suppression of Irish national liberation movement in 1848; Foreign Secretary (1853-58, 1865-66 and 1868-70)—45, 82

Claudianus, Claudius (4th cent.)—Roman poet of Greek origin—201

Clinton, Henry Pelham Fiennes Pelham, Duke of Newcastle (1811-1864)—British statesman, Peelite, Chief Secretary for Ireland (1846), Secretary for War and Colonies (1852-54)—69

Cobbett, William (1763-1835)—English radical politician and publicist; published *Cobbett's Weekly Political Register* from 1802—157, 275

Cobden, Richard (1804-1865)—English manufacturer, bourgeois politician, one of leaders of Free Traders, M.P.—54

Cogan, Milo de (d. 1182)—Anglo-Norman feudalist, participated in conquest of Ireland, ruler of Dublin—217

Colcraft—English executioner, who hanged Irish Fenians Allen, Larkin and O'Brien, sentenced by an English court, in Manchester, October 23, 1867—145

Columba (c. 521-597)—Irish Christian missionary in Scotland—202

Conary I (2nd cent.)—King of Ireland—213

Concobar—King of Ireland (818-33)—213

Coote, Charles (d. 1642)—Governor of Dublin, participated in suppressing Irish uprising —252

Coote, Charles, the Younger (d. 1661)—commander in Parliamentary Army in Ireland and organiser of suppression of Irish uprising in 1641-52; supported Restoration—257-58

Corc (5th cent.)—King of Munster, according to a legend from Irish chronicles, participated in compiling *Senchus Mor*, a collection of ancient laws—194

Cormac Mac-Culinan (836-908)—bishop and King of Cashel (901-08)—213

Cormac Ulfadha (3rd cent.)—King of Ireland—200, 213

Costello, Augustin—Irish Fenian, officer in American army; went to Ireland in 1867 to take part in uprising but was arrested and sentenced to 12 years penal servitude—272, 386

Courcy, John de (d. 1219)—Anglo-Norman feudalist, Lord Deputy of Ireland (1185-89)—217, 218

Cournet, Frédérick-Etienne (1839-1885)—French revolutionary, Blanquist, member of Paris Commune and of General Council of International (1871-72)—407

Cowen, Joseph (1831-1900)—British politician and journalist, radical, adhered to Chartists, became M.P. in 1874—330

Cowper, Francis Thomas de Grey, Lord (1834-1905)—British statesman, Liberal, Lord Lieutenant of Ireland (1880-82)—333

Cowper-Temple, William Francis (1811-1888)—British states-

man, M.P., held portfolio in a number of Liberal cabinets; Palmerston's stepson—333

Cremer, William Randall (1828-1908)—prominent figure in British trade union movement, leader of Amalgamated Society of Carpenters and Joiners, member of General Council of International and its General Secretary (1864-66), subsequently bourgeois pacifist, Liberal M.P.—144, 351, 367

Cromwell, Henry (1628-1674)—son of Oliver Cromwell, general in English Parliamentary Army; in 1650 participated in punitive expedition to Ireland, in 1654 commander of army in Ireland; Lord Deputy of Ireland (1658-59)—128, 267

Cromwell, Oliver (1599-1658)—leader of bourgeoisie and bourgeoisified nobility during English bourgeois revolution of 17th century; became Commander-in-Chief and Lord Lieutenant in Ireland in 1649, named Lord Protector of England, Scotland and Ireland in 1653—123, 126-28, 140, 147, 250, 252, 262, 265, 270, 274, 275, 281, 285, 287, 299, 326

Cromwell, Thomas (1485-1540)—Vicar-General of King Henry VIII, one of leaders of Anglican Reformation—232, 233

Cunninghame Graham, Robert Bontine (1852-1936)—Scottish writer of aristocratic origin; in 80s-90s participated in socialist movement, M.P.—414

Curran, John Philpot (1750-1817)—Irish lawyer, bourgeois radical, Member of Irish

Fitzgerald, John Oge (d. 1569)— member of Anglo-Irish clan of Geraldines (branch of Desmonds), known as the White Knight—234

Fitzgerald, Maurice (c. 1194-1257)—Lord Justice of Ireland (1232-45)—219

Fitzgerald, Raymond (d. c. 1182) —constable of Leinster—217

Fitzgerald, Thomas, Earl of Kildare (d. 1478)—Lord Deputy (1455-59, 1461-62) and Lord Lieutenant (1468-75) of Ireland—227, 228

Fitzgerald, Thomas, Earl of Kildare (1513-1536)—son of Gerald Fitzgerald, Earl of Kildare; raised revolt against English—231, 232

Fitzgibbon. See *Fitzgerald, John Oge*

Fitzgibbon, Gerald (1793-1882)— Irish lawyer and bourgeois publicist—328

Fitzmaurice, James Fitzgerald (d. 1579)—member of Anglo-Irish feudal clan of Geraldines (branch of Desmonds), took part in anti-English rebellion—236, 237

Fitzsymons, Walter (d. 1511)— Archbishop of Dublin, Lord Deputy of Ireland (1492-1503)—228

Fitzthomas, James Fitzgerald (d. 1608)—member of Anglo-Irish feudal clan of Desmonds in Southern Ireland, had title of Earl of Desmond —240

Fitzthomas, Maurice, Earl of Desmond (d. 1356)—Lord Justice of Ireland (1355-56) —222, 223

Fleetwood, Charles (d. 1692)— general in Parliamentary Army during English bourgeois revolution of 17th century, Commander-in-Chief of English army in Ireland (from 1652), Lord Deputy of Ireland (1654-57) —128

Flourens, Gustave (1838-1871)— French revolutionary and naturalist, Blanquist, member of Paris Commune, brutally assassinated by Versaillais in April 1871—324

Forster, William Edward (1818-1886)—British factory owner and Liberal politician, Chief Secretary for Ireland (1880-82); pursued a policy of ruthlessly suppressing national liberation movement— 318, 319, 333

Fortescue-Parkinson, Chichester Samuel (1823-1898)—British Liberal statesman, Chief Secretary for Ireland (1865-66 and 1868-70)—153

Foster, John Leslie (c. 1780-1842) —Irish lawyer, Tory—272

Fourier, Charles (1772-1837)— French utopian socialist—199

Fox, Peter (Peter Fox André) (d. 1869)—participant in British democratic and working-class movement, journalist, member (1864-69) and General Secretary (September-November 1866) of General Council of International, Corresponding Secretary for America (1866-67)—144, 147, 366

Francis I (1494-1547)—King of France (1515-47)—230

Frankel, Leo (1844-1896)—prominent figure in Hungarian and international working-class movement, jeweller by profession, member of Paris Commune and of General Council of International (1871-72), a founder of General Workers' Party of Hungary, associate and friend of Marx and Engels —405, 407

bourgeois historian and statesman; from 1840 to February 1848 revolution was virtually head of conservative government—52

H

Hales, John (b. 1839)—participant in British working-class movement, member of General Council of International (1866-72) and its Secretary (from May 1871); headed reformist wing of British Federal Council from beginning of 1872—277, 302, 406

Hales, William—member of General Council of First International—405, 407, 408, 409, 410, 411, 412, 413

Hall, Edward (c. 1498-1547)—English historian and lawyer, supporter of absolute monarchy—228

Hallam, Henry (1777-1859)—English historian—257

Halliday, Thomas (b. 1835)—a British trade union leader, Secretary of Joint Miners' Association—310

Hancock, U. Nelson—Irish lawyer, with O'Mahony published two volumes of *Senchus Mor,* a collection of ancient laws—193

Hanmer, Meredith (1543-1604)—English clergyman and historian, author of *The Chronicle of Ireland*—197

Harald I Hårfagr (c. 850-c. 933)—King of Norway (872-930)—204

Hardie, James Keir (1856-1915)—prominent figure in British working-class movement, reformer, organiser and leader of Scottish Labour Party (from 1888) and Independent Labour Party (from 1893); M.P. (1892-95)—*355, 356*

Hardy. See *Gathorne-Hardy*

Harney, George Julian (1817-1897)—prominent figure in British working-class movement, a leader of revolutionary wing of Chartism; friendly with Marx and Engels—278

Harris, George—participant in British working-class movement, member of Chartist National Reform League, member (1869-72) and Financial Secretary (1870-71) of General Council of International—277, 282, 405

Hegel, Georg Wilhelm Friedrich (1770-1831)—German classical philosopher, objective idealist—202

Hennessy, John Pope (1834-1891)—Irish politician, Conservative M.P., proposed several reforms in Ireland in early 60s—139, 148, 361

Henry I (1068-1135)—King of England (1100-35)—196

Henry II (1133-1189)—King of England (1154-89)—127, 216, 218, 247, 320

Henry III (1207-1272)—King of England (1216-72)—218, 219

Henry IV (1367-1413)—King of England (1399-1413)—225

Henry V (1387-1422)—King of England (1413-22)—225, 321

Henry VI (1421-1471)—King of England (1422-71)—225, 227

Henry VII (1457-1509)—King of England (1485-1509)—227, 228, 229, 244

Henry VIII (1491-1547)—King of England (1509-47)—229, 233, 235, 246, 258, 264

Herman, Alfred—participant in Belgian working-class movement, member of General Council of International and Corresponding Secretary for Belgium (1871-72)—405

Jones, Ernest Charles (1819-1869)—a leader of revolutionary Chartism, proletarian poet and publicist, friend of Marx and Engels—143

Jukes, Joseph Beete (1811-1869) —English geologist—172, 173, 174

Jung, Hermann (1830-1901)—prominent figure in Swiss and international working-class movement, member of General Council of International, Corresponding Secretary for Switzerland (November 1864-72) and Treasurer of General Council (1871-72); following Hague Congress, joined reformist wing of International—119, 147, 151, 368, 406, 407

Juvenal (Decimus Junius Juvenalis) (b. in 60s-d. after 127) —Roman satirical poet—289

K

Kane, Robert John (1809-1890) —Irish scholar, professor of chemistry and physics, also dealt with problems of Irish economy—174, 187

Kautsky, Karl (1854-1938)—a leader and theoretician of German Social-Democratic Party and Second International; subsequently a Centrist; in 1914 betrayed Marxism and revolutionary working-class movement —332

Kay-Schuttleworth, James Philips (1804-1877)—English physician, bourgeois public figure —39

Keen, Charles—participant in British working-class movement, member of General and British Councils of International (1872)—405

Kelly, Thomas (b. c. 1831)—an Irish Fenian leader—145

Kelly-Wischnewetzky, Florence (1859-1932)—American socialist, translated into English Engels's book *The Condition of the Working-Class in England*—350

Kennedy (10th cent.)—King of Munster—213

Kenneth MacAlpin (d. 860)—founder of Scottish royal dynasty who united Scots and Picts under his rule in middle of 9th century —201

Keogh, William Nicholas (1817-1878)—Irish lawyer and politician, a leader of Irish group in Parliament; repeatedly held high posts in English administration in Ireland—75, 76

Kératry, Emile, Comte de (1832-1905)—French reactionary politician, Prefect of Upper Garonne Department in 1871—324

Kickham, Charles Joseph (1826-1882)—Irish Fenian, participant in national liberation movement of 40s, an editor of the newspaper *Irish People* (1865); sentenced to fourteen years penal servitude in 1865, released in 1869—164, 386, 402

Kildare. See Fitzgerald

Kilian (d. 697)—Irish Christian missionary in Eastern Franconia, first bishop of Würzburg—202

Kimbaoth (3rd cent. B.C.)—ruler of Ulster mentioned in chronicles—192, 213

Knox, Alexander Andrew (1818-1891)—English journalist and magistrate, member of commission which reported to Parliament in 1867 on treatment of political prison-

in prison the same year—383, 399, 400, 401

Lyons, Robert Spencer Dyer (1826-1886)—Irish physician, Liberal, member of commission of inquiry (1870) into condition of Irish political prisoners—381

M

Macaulay, Thomas Babington (1800-1859)—English bourgeois historian and politician, Whig, M.P.—211, 264

MacCarthy, Florence (c. 1562-c. 1640)—Irish feudalist, was persecuted by English authorities—240

M'Carthy, Justin (1830-1912)—Irish writer and politician, Liberal M.P. (1879-1900), Vice-Chairman of Irish Home Rule Party in House of Commons; opposed Parnell's leadership in 1890—353

MacDonald, Alexander (1821-1881)—a British trade union leader, Secretary of National Association of Miners, M.P. from 1874; adhered to Liberal Party—310

MacDonnell, J. Patrick (c. 1845-1906)—active in Irish working-class movement, member of General Council of International and Corresponding Secretary for Ireland (1871-72); emigrated in 1872 to U.S.A. where he participated in American working-class movement—298, 299, 306, 405, 407, 408, 409, 410

M'Donnell—prison doctor in Dublin dismissed because of his protest against cruel treatment of Fenian prisoners—154, 167-69, 390, 391

Mac-Dowal, Duncan—ruler of Galloway, end of 13th and beginning of 14th century—221

Mac-Geoghegan, Jacques (James) (1702-1762)—French abbot of Irish origin, author of *History of Ireland*—261

Machiavelli, Niccolo (1469-1527) —Italian politician, historian and writer, an ideologist of Italian bourgeoisie of initial period of capitalist relations—383

MacMahon, Hugh (c. 1606-1644) —Irish feudalist, participated in 1641 uprising—249, 250, 252

Mac-Morrough, Art (d. 1417)—head of Irish clan from County Cavan; for 50 years led resistance movement of natives of Leinster and Southern Ulster against English—225

Mac-Morrough, Donald (14th cent.)—head of Irish clan from County Cavan, descendant of ancient kings of Leinster; led uprising of Irish clans against English in 1328—222

Mac-Murchad, Dermot (c. 1110-1171)—King of Leinster from 1126 to 1171—216

Macpherson, James (1736-1796) —Scottish poet—200, 270

Maelseachlainn (Mael Sechnaill II) (949-1022)—King of Ireland (980-1002 and 1014-22) —207, 214, 215

Magon (d. 976)—King of Munster from 964 to 976—213, 214

Maguire—Irish feudalist, participant in 1641 uprising—249, 250

Maguire, Thomas—Irish sailor, arrested in 1867 in Manchester on a charge of attempting to organise escape of Fenian prisoners and sentenced to death; sentence was rescinded for lack of evidence—118

Malachias (c. 1094-1148)—Irish archbishop—196

Malachy. See *Maelseachlainn*

Malmesbury, James Howard Harris, Earl of (1807-1889) —British statesman, Tory but subsequently a prominent leader of Conservative Party—69

Malthus, Thomas Robert (1766-1834)—English clergyman, economist, ideologist of landed nobility who had adopted bourgeois ways and methods, author of reactionary theory of overpopulation—106

Maolmordha (d. 1014)—King of Leinster (999-1014)—205, 207

Maolmua (c. 930-978)—King of Desmond—214

Martin, Constant—French revolutionary, Blanquist, participant in Paris Commune, member of General Council of International (1871-72)—301, 405, 407

Martin, John (1812-1875)—Irish politician, participant in national liberation movement, Honorary Secretary of Home Rule League, M.P. (1871-75)—144, 401

Martin, William (b. c. 1832)—Irish Fenian; sentenced by a Manchester court in 1867—144

Marx, Eleanor (Tussy) (1855-1898)—youngest daughter of Karl Marx, prominent figure in British and international working-class movement, wife of Edward Aveling—273, 275, 290, 323, 353, 414, 415

Marx, Jenny, née von Westphalen (1814-1881)—wife of Karl Marx—331

Marx, Jenny (1844-1883)—Marx's eldest daughter, journalist, active in international working-class movement, married Charles Longuet in 1872—289-92, 295, 323-24

Mary Tudor (1516-1558)—Queen of England (1553-58)—233, 235, 249, 259

Maurer, Georg Ludwig (1790-1872)—German bourgeois historian, studied social system of ancient and medieval Germany—314

Maurice, Zévy—member of General Council of International (1866-72), Corresponding Secretary for Hungary (1870-71)—405

Mayo, Henry—active in British working-class movement, member of General Council (1871-72) and British Federal Council (1871-72) of International, whose reformist wing he joined—405, 407, 409

Mayo. See *Naas, Richard Southwell Bourke,* Earl of Mayo

Mazzini, Giuseppe (1805-1872)—Italian revolutionary, bourgeois democrat, a leader of national liberation movement in Italy—54

Meagher, Thomas Francis (1823-1867)—participant in Irish national liberation movement, one of founders of Irish Confederation (1847); arrested in 1848 for taking part in preparing uprising and sentenced to hard labour for life; escaped to America in 1852; led Irish volunteer brigade on side of Northerners during Civil War (1861-65)—132, 139, 141, 148

Measor—British official, Deputy Governor of Chatham Prison—364

chief of Breffny in East Connaught—216

O'Shea, Henry—Irish public figure, defender of imprisoned Fenians in 1869—156, 277

O'Sullivan, Daniel (1560-1618)—Irish feudalist, took part in anti-English uprising led by Tyrone and Tyrconnel—241

Outlaw, Roger (d. 1340)—Lord Justice of Ireland (1328-32 and 1340)—222

P

Pakington, John Somerset (1799-1880)—British Tory statesman, later joined Conservative Party—363-64

Palmerston, Henry John Temple, Viscount (1784-1865)—British statesman, Tory at beginning of his activity and a Whig leader from 1830 on; Foreign Secretary (1830-34, 1835-41 and 1846-51), Home Secretary (1852-55) and Prime Minister (1855-58 and 1859-65)—53, 70-71, 75, 76, 78, 98, 117, 125, 280, 333, 398

Paparo, John—papal legate at Holy Synod in Ireland in 1152—216

Parks, William—371

Parnell, Charles Stewart (1846-1891)—Irish bourgeois politician, participant in national liberation movement, elected M.P. in 1875, Home Rule Party leader from 1877, joined a bloc with Irish radicals, supported Land League (1879)—325, 330, 333, 348, 351, 353, 355

Parsons, William (c. 1570-1650)—Lord Justice of Ireland (1640-48), inspired policy of Ireland's colonial subjugation—250, 263, 265, 326

Parsons, William, Earl (1800-1867)—English astronomer, in 1867 published a pamphlet on relations between Irish landowners and tenants—276

Patrick or *Patricius* (c. 373-c. 463)—preached Christianity in Ireland, founded and became first bishop of Catholic Church in Ireland—194, 200, 202, 213, 396

Patterson, William—Irish physician, author of a book about climate of Ireland—187, 188

Peel, Robert (1788-1850)—British statesman, leader of moderate Tories (Peelites), Home Secretary (1822-27 and 1828-30) and Prime Minister (1834-35 and 1841-46); with support of Liberals repealed Corn Laws in 1846—34, 35, 72, 80, 81, 82, 90, 181, 328, 349

Pelagius the Heretic (c. 360-c. 420)—British theologian, condemned as a heretic for his teaching on man's free will —200, 201

Pembroke, William Marshal, Earl of (d. 1219)—Regent of England during Henry III's infancy—218

Percy, Henry, Earl of Northumberland (1342-1408)—big English feudalist—321

Percy, Henry (called *Hotspur*) (1364-1403)—English feudalist, son of Henry Percy—Earl of Northumberland; participant in barons' revolts against English Crown—321

Perrot, Benjamin-Pierre (1791-1865)—French general, took part in suppressing June 1848 uprising—52

Perrot, John (c. 1527-1592)—Lord President of Munster (1570-73), Lord Deputy of Ireland (1584-88)—236, 237, 238, 239, 260

Petrie, George (1789-1866)—Irish

lish economist, outstanding representative of classical bourgeois political economy —55, 57, 63, 272

Richard, Duke of York (1411-1466)—Lord Lieutenant of Ireland (1449-60)—226

Richard II (1367-1400)—King of England (1377-99)—224, 225, 321

Richard III (1452-1485)—King of England (1483-85)—227

Rinuccini, Giovanni Battista (1592-1653)—papal nuncio in Ireland—256

Roach, John—participant in British working-class movement, member of General Council of International (1871-72), Corresponding Secretary of British Federal Council (1872), adhered to its reformist wing—405, 407, 409

Roberts, William R.—a leader of Fenian movement in U.S.A., American officer, one of organisers of Fenian invasion of Canada, end of May 1866 —147

Robespierre, Maximilien (1758-1794)—prominent figure in French bourgeois revolution at end of 18th century, Jacobin leader—371

Rochat, Charles (b. 1844)—active in French working-class movement, Corresponding Secretary of General Council (of International) for Holland (1871-72)—405

Rochefort, Victor-Henri (1831-1913)—French publicist and politician, Left Republican, monarchist from end of 80s —289, 291, 323, 379, 398

Roden, Robert Jocelyn, Earl of (1788-1870)—English aristocrat, Conservative—68

Roscoe—English lawyer, legal

adviser of big trade unions —408

Rossa (5th cent.)—according to tradition, one of compilers of *Senchus Mor,* a collection of ancient laws—194

Rossa. See *O'Donovan Rossa*

Rosse. See *Parsons, William*

Rühl, J.—German worker, member of General Council of International (1870-72)—405, 407

Russell, John (1792-1878)—British statesman, Whig leader, Prime Minister (1846-52 and 1865-66), Foreign Secretary (1852-53 and 1859-65)—79, 80, 81, 82, 98, 361, 362, 365

Rutty, John (1698-1775)—Irish physician and meteorologist —186, 187

S

Sadleir, John (1814-1856)—Irish banker and politician, a leader of Irish faction in Parliament, member of government in 1853—78

Sadler, Michael Thomas (1780-1835)—English economist and politician, bourgeois philanthropist, opponent of Malthusianism, close to Tory Party—106

Sadler, Thomas—member of General Council of International (1871-72)—405

Saintleger, Anthony—Lord Deputy of Ireland (1540-48, 1550-51 and 1553-56)—233, 234, 235

Saintleger, William (d. 1642)—Lord President of Munster—251

St. Leonards. See *Sugden, Edward Burtenshaw*

Saxo Grammaticus (mid-12th-beginning of 13th cent.)—Danish chronicler, author of *Gesta Danorum (Historia Danica)*—209

American working-class movement, socialist, member of Communist League and of International; emigrated to U.S.A. in 1867, where was an organiser of sections of International Working Men's Association; supporter of Marx and Engels—292

W

Waddington, H.—British Home Office official—367

Wakefield, Edward (1774-1854)— English bourgeois statistician and agronomist, author of the book *An Account of Ireland, Statistical and Political*—179, 180, 181, 183, 186, 188, 189, 190, 199, 200, 276, 279

Wakley, Thomas (1795-1862)— English physician and politician, bourgeois Radical—45

Walsh, John Benn (1798-1881)— British Tory politician, M.P. —98

Warbeck, Perkin (1474-1499)— pretender to English throne, posed as son of Edward IV —228, 229

Waterlow, Sydney Hudley (1822-1906)—British Liberal politician—288

Wellington, Arthur Wellesley, Duke of (1769-1852)—British general and statesman, Tory, Prime Minister (1828-30)— 70

Wentworth, Thomas, Earl of Strafford (1593-1641)— British statesman, inspirer of policy of absolutism, Lord Deputy of Ireland (1632-40) —246, 247, 248, 249, 251, 262, 263, 264

Weston, John—participant in British working-class movement, follower of Owen,

member of General Council of International (1864-72), an organiser of Land and Labour League—119, 144, 146, 277, 278, 368, 371, 378, 405, 407

Whiteside, James (1804-1876)— Irish lawyer, Tory M.P., Attorney-General for Ireland (1858-59 and 1866)—87

Wilkinson—owner of St. George's Hall, London—408

William III, Stadtholder of Holland. See *William III of Orange*

William III of Orange (1650-1702)—Stadtholder of Netherlands (1672-1702), King of England (1689-1702) —126, 128, 129, 140, 269, 270

William IV (1765-1837)—King of Great Britain and Ireland (1830-37)—72

Wilson, James (1805-1860)—British economist and politician, Free Trader, founder and editor of *Economist*, Financial Secretary to Treasury in 1853-58—63

Wilton, Arthur Grey (1536-1593) —Lord Deputy of Ireland (1580-82)—237

Wogan, John (d. 1321)—Lord Justice of Ireland (1295-1312)—220

Wolsey, Thomas (c. 1475-1530)— English prelate and statesman—230

Worcester. See *Tiptoft*

Wróblewski, Walery (1836-1908) —Polish revolutionary democrat, a leader of Polish national uprising of 1863-64, general of Paris Commune, member of General Council of International, Corresponding Secretary for Poland (1871-72)—406

Y

Z

SUBJECT INDEX

A

Agrarian history of Ireland—327

Agrarian revolution in Ireland—
76, 97, 99-107, 110-11, 114-
16, 121-23, 133-36, 138-39,
141-42, 147-48, 190

Agricultural competition
—effect for small peasants—
317

Agricultural labourers in Ireland
—39-40, 56, 63, 108-12, 285,
296-97, 325, 355

Agriculture in Ireland—39, 99-
105, 141-42

Ancient Order of Foresters—155

Anglo-Irish Parliament—129-31,
140, 244, 246-47, 248, 258

B

Battle of Clontarf—205-08, 215

Boycott—334

Brehon Laws—243, 279

C

Castle Chamber—264

Catholic clergy in Ireland—35,
76, 124, 126, 140-41, 146, 253,
401-02

Catholic Emancipation—70, 74-
75, 124, 131, 334

Chartists
—and Irish question—35, 49-
50, 284

Christianity in Ireland—200-01

Clans in Ireland—245, 333, 340-
41, 352

Clearing of estates in Ireland—
53, 56, 71, 84, 88, 123, 135,
142, 143-44, 147-48, 271, 297,
330, 335-36

Communal ownership of land—
53, 279, 284, 340-41

Community in Ireland—314-15,
352

Commutation Bill of 1838—41

Corn-acre system—122

Corn Laws—81, 133, 141
—repeal of—95, 123, 134

Court of High Commission—263-
64

Court of Wards—263

Crimes in Ireland—92-94

Criminal Justice Act of 1855—
124

Crisis of 1847—44

Cromwellian colonisation of Ireland—127-28, 140, 250, 265-
67, 274, 281, 285, 326-27

D

Domestic industry—108, 355

E

Economy of Ireland—105-06

Emigration—44, 54-58, 63, 66,
76, 84, 95, 99, 106-07, 112-
16, 121, 131, 134-35, 136, 137,
162, 271, 313, 335-36, 375

Encumbered Estates Act—67-69,
76, 77, 134, 147-48, 297

Encumbered Estates Commission
—62